Pigments of Flowering Plants

Pigments of Flowering Plants

Contributors

Reena Kushwaha, Pankaj Srivastava et al.

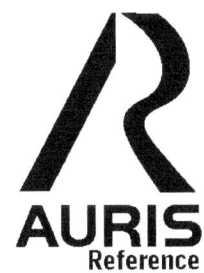

www.aurisreference.com

Pigments of Flowering Plants

Contributors: Reena Kushwaha, Pankaj Srivastava et al.

Published by Auris Reference Limited
www.aurisreference.com

United Kingdom

Pigments of Flowering Plants

ISBN: 978-1-78154-868-4

British Library Cataloguing in Publication Data
A CIP record for this book is available from the British Library

Printed in the United Kingdom

Exclusively distributed by CBS Publishers & Distributors Pvt. Ltd.

Sales & Distribution Rights only for India, Pakistan, Bangladesh, Sri Lanka, Nepal and Bhutan. This book is not to be sold outside these territories.

Contents

List of Abbreviations

3D	Three-Dimensional
AOMT	Anthocyanin O-methyltransferase
Ch7Neo	Chrysoeriol 7-O-neohesperidose
CHS	Chalcone Synthase
Cy	Cyanidin
Cy3G	Cyanidin 3-O-glucoside
Cy3G5G	Cyanidin 3,5-di-O-glucoside
DAD	Diode-Array Detection
DEG	Differentially Expressed Gene
Del	Delphinidin
DFR	Dihydroflavonol 4-reductase
DHK	Dihydrokaempferol
DHQ	Dihydroquercetin
Dp3G	Delphinidin 3-O-glucoside
DSSCs	Dye-sensitized solar cells
DW	Dry Weight
ESI	Electrospray Ionization
F3′5′H	Flavonoid 3′5′-hydroxylase
F3′H	Flavanone 3′-hydroxylase
FDR	False Discovery Rate
FLS	Flavonol Synthase
HPLC	high-performance liquid chromatography
IPCE	Incident Photon-To-Current Conversion Efficiencies
Is3G7G	Isorhamnetin 3,7-di-O-glucoside
KEG	Kyoto Encyclopedia of Genes and Genomes
Km3G	Kaempferol 3-O-glucoside
LAR	Leucoanthocyanidin reductase
Lu7G	Luteolin 7-O-glucoside
Lu7Neo	Luteolin 7-O-neohesperidose
MS	Mass Spectrometry
NCBI	National Center for Biotechnology Information
OMT	O-methyltransferase
ORF	Open Reading Frame
PEC	Photoelectrochemical
Pg3G	Pelargonidin 3-O-glucoside
Pn3G	Peonidin 3-O-glucoside
Pn3G5G	Peonidin 3,5-di-O-glucoside
qPCR	Quantitative PCR
Qu3G7G	Quercetin 3,7-di-O-glucoside
Qu3R	Quercetin 3-O-rutinoside

SAM   S-adenosylmethionine.
SOAP   Short Oligonucleotide Analysis Package

List of Contributors

Reena Kushwaha
Department of Chemistry, Faculty of Science, Banaras Hindu University, Varanasi 221005, India

Pankaj Srivastava
Department of Chemistry, Faculty of Science, Banaras Hindu University, Varanasi 221005, India

Lal Bahadur
Department of Chemistry, Faculty of Science, Banaras Hindu University, Varanasi 221005, India

Qianqian Shi
State Key Laboratory of Tree Genetics and Breeding, Key Laboratory of Tree Breeding and Cultivation of State Forestry Administration, Research Institute of Forestry, Chinese Academy of Forestry, Beijing 100091, China

Lin Zhou
State Key Laboratory of Tree Genetics and Breeding, Key Laboratory of Tree Breeding and Cultivation of State Forestry Administration, Research Institute of Forestry, Chinese Academy of Forestry, Beijing 100091, China

Yan Wang
State Key Laboratory of Tree Genetics and Breeding, Key Laboratory of Tree Breeding and Cultivation of State Forestry Administration, Research Institute of Forestry, Chinese Academy of Forestry, Beijing 100091, China

Kui Li
State Key Laboratory of Tree Genetics and Breeding, Key Laboratory of Tree Breeding and Cultivation of State Forestry Administration, Research Institute of Forestry, Chinese Academy of Forestry, Beijing 100091, China

Baoqiang Zheng
State Key Laboratory of Tree Genetics and Breeding, Key Laboratory of Tree Breeding and Cultivation of State Forestry Administration, Research Institute of Forestry, Chinese Academy of Forestry, Beijing 100091, China

Kun Miao
State Key Laboratory of Tree Genetics and Breeding, Key Laboratory of Tree Breeding and Cultivation of State Forestry Administration, Research Institute of Forestry, Chinese Academy of Forestry, Beijing 100091, China

Jian Zheng
College of Landscape Architecture, Beijing University of Agriculture, Beijing, 102206, China
Beijing Engineering Research Center of rural landscape planning and design, Beijing, 102206, China

Zenghui Hu
College of Landscape Architecture, Beijing University of Agriculture, Beijing, 102206, China
2 Beijing Engineering Research Center of rural landscape planning and design, Beijing, 102206, China

Xuelian Guan
College of Landscape Architecture, Beijing University of Agriculture, Beijing, 102206, China

Dequan Dou
College of Landscape Architecture, Beijing University of Agriculture, Beijing, 102206, China

Guo Bai
College of Landscape Architecture, Beijing University of Agriculture, Beijing, 102206, China

Yu Wang
College of Landscape Architecture, Beijing University of Agriculture, Beijing, 102206, China

Yingtian Guo
College of Landscape Architecture, Beijing University of Agriculture, Beijing, 102206, China

Wei Li
College of Landscape Architecture and Forestry, Qingdao Agricultural University, Qingdao, 266100, China

Pingsheng Leng
College of Landscape Architecture, Beijing University of Agriculture, Beijing, 102206, China

Takashi Nakatsuka
Iwate Biotechnology Research Center, 22-174-4 Narita, Kitakami, Iwate 024-0003, Japan

Eri Yamada
Iwate Biotechnology Research Center, 22-174-4 Narita, Kitakami, Iwate 024-0003, Japan

Hideyuki Takahashi
Iwate Biotechnology Research Center, 22-174-4 Narita, Kitakami, Iwate 024-0003, Japan

Tomohiro Imamura
Iwate Biotechnology Research Center, 22-174-4 Narita, Kitakami, Iwate 024-0003, Japan

Mariko Suzuki
Department of Biotechnology and Life Science, Faculty of Engineering, Tokyo University of Agriculture and Technology, 2-24-16 Naka-cho, Koganei, Tokyo 184-8588, Japan

Yoshihiro Ozeki
Department of Biotechnology and Life Science, Faculty of Engineering, Tokyo University of Agriculture and Technology, 2-24-16 Naka-cho, Koganei, Tokyo 184-8588, Japan

Ikuko Tsujimura
Iwate Biotechnology Research Center, 22-174-4 Narita, Kitakami, Iwate 024-0003, Japan

Misa Saito
Iwate Biotechnology Research Center, 22-174-4 Narita, Kitakami, Iwate 024-0003, Japan

Yuichi Sakamoto
Iwate Biotechnology Research Center, 22-174-4 Narita, Kitakami, Iwate 024-0003, Japan

Nobuhiro Sasaki
Department of Biotechnology and Life Science, Faculty of Engineering, Tokyo University of Agriculture and Technology, 2-24-16 Naka-cho, Koganei, Tokyo 184-8588, Japan

Masahiro Nishihara
Iwate Biotechnology Research Center, 22-174-4 Narita, Kitakami, Iwate 024-0003, Japan

Chiyomi Uematsu
Botanical Gardens, Graduate School of Science, Osaka City University, Osaka 576-0004, Japan

Hironori Katayama
Food Resources Research and Education Center, Kobe University, Hyogo 675-2103, Japan

Izumi Makino
Botanical Gardens, Graduate School of Science, Osaka City University, Osaka 576-0004, Japan

Azusa Inagaki
Botanical Gardens, Graduate School of Science, Osaka City University, Osaka 576-0004, Japan

Osamu Arakawa
Faculty of Agriculture and Life Science, Hirosaki University, Aomori 036-8561, Japan

Cathie Martin
Department of Metabolic Biology, John Innes Centre, Norwich NR4 7UH, UK

Hui Du
Key Laboratory of Plant Resources/ Beijing Botanical Garden, Institute of Botany, Chinese Academy of Sciences, Beijing 100093, PR China
University of Chinese Academy of Sciences, Beijing 100049, PR China

Jie Wu
Key Laboratory of Plant Resources/ Beijing Botanical Garden, Institute of Botany, Chinese Academy of Sciences, Beijing 100093, PR China
University of Chinese Academy of Sciences, Beijing 100049, PR China

Kui-Xian Ji
Key Laboratory of Plant Resources/ Beijing Botanical Garden, Institute of Botany, Chinese Academy of Sciences, Beijing 100093, PR China

Qing-Yin Zeng
State Key Laboratory of Systematic and Evolutionary Botany, Institute of Botany, Chinese Academy of Sciences, Beijing 100093, PR China

Mohammad-Wadud Bhuiya
MOgene Green Chemicals, Saint Louis, MO 63132, USA

Shang Su
Key Laboratory of Plant Resources/ Beijing Botanical Garden, Institute of Botany, Chinese Academy of Sciences, Beijing 100093, PR China
University of Chinese Academy of Sciences, Beijing 100049, PR China

Qing-Yan Shu
Key Laboratory of Plant Resources/ Beijing Botanical Garden, Institute of Botany, Chinese Academy of Sciences, Beijing 100093, PR China

Hong-Xu Ren
Key Laboratory of Plant Resources/ Beijing Botanical Garden, Institute of Botany, Chinese Academy of Sciences, Beijing 100093, PR China

Zheng-An Liu
Key Laboratory of Plant Resources/ Beijing Botanical Garden, Institute of Botany, Chinese Academy of Sciences, Beijing 100093, PR China

Liang-Sheng Wang
Key Laboratory of Plant Resources/ Beijing Botanical Garden, Institute of Botany, Chinese Academy of Sciences, Beijing 100093, PR China

Takashi Nakatsuka
Iwate Biotechnology Research Center, 22-174-4 Narita, Kitakami, Iwate 024-0003, Japan

Misa Saito
Iwate Biotechnology Research Center, 22-174-4 Narita, Kitakami, Iwate 024-0003, Japan

Eri Yamada
Iwate Biotechnology Research Center, 22-174-4 Narita, Kitakami, Iwate 024-0003, Japan

Kohei Fujita
Iwate Biotechnology Research Center, 22-174-4 Narita, Kitakami, Iwate 024-0003, Japan

Yuko Kakizaki
Iwate Biotechnology Research Center, 22-174-4 Narita, Kitakami, Iwate 024-0003, Japan

Masahiro Nishihara
Iwate Biotechnology Research Center, 22-174-4 Narita, Kitakami, Iwate 024-0003, Japan

Mamatha Hanumappa
Kumho Life and Environmental Science Laboratory, Gwangju 500-712, Korea

Goh Choi
Kumho Life and Environmental Science Laboratory, Gwangju 500-712, Korea

Sunhyo Ryu
Kumho Life and Environmental Science Laboratory, Gwangju 500-712, Korea

Giltsu Choi
Department of Biological Sciences, KAIST, Daejeon 305-701, Korea

Qian Lou
College of Horticulture, Northwest A & F University, Yangling 712100, Shaanxi, PR China

Key Laboratory of Biology and Genetic Improvement of Horticultural Crops (Northwest Region), Ministry of Agriculture, Yangling, Shaanxi 712100, PR China
State Key Laboratory of Crop Stress Biology in Arid Areas, Northwest A&F University, Yangling

Yali Liu
Key Laboratory of Biology and Genetic Improvement of Horticultural Crops (Northwest Region), Ministry of Agriculture, Yangling, Shaanxi 712100, PR China
State Key Laboratory of Crop Stress Biology in Arid Areas, Northwest A&F University, Yangling 712100, Shaanxi, PR China
College of Forestry, Northwest A & F University, Yangling 712100, Shaanxi, PR China

Yinyan Qi
College of Horticulture, Northwest A & F University, Yangling 712100, Shaanxi, PR China
Key Laboratory of Biology and Genetic Improvement of Horticultural Crops (Northwest Region), Ministry of Agriculture, Yangling, Shaanxi 712100, PR China
State Key Laboratory of Crop Stress Biology in Arid Areas, Northwest A&F University, Yangling

Shuzhen Jiao
Key Laboratory of Biology and Genetic Improvement of Horticultural Crops (Northwest Region), Ministry of Agriculture, Yangling, Shaanxi 712100, PR China
State Key Laboratory of Crop Stress Biology in Arid Areas, Northwest A&F University, Yangling 712100, Shaanxi, PR China
College of Forestry, Northwest A & F University, Yangling 712100, Shaanxi, PR China

Feifei Tian
Key Laboratory of Biology and Genetic Improvement of Horticultural Crops (Northwest Region), Ministry of Agriculture, Yangling, Shaanxi 712100, PR China
State Key Laboratory of Crop Stress Biology in Arid Areas, Northwest A&F University, Yangling 712100, Shaanxi, PR China
College of Forestry, Northwest A & F University, Yangling 712100, Shaanxi, PR China

Ling Jiang
Key Laboratory of Biology and Genetic Improvement of Horticultural Crops (Northwest Region), Ministry of Agriculture, Yangling, Shaanxi 712100, PR China
State Key Laboratory of Crop Stress Biology in Arid Areas, Northwest A&F University, Yangling 712100, Shaanxi, PR China
College of Forestry, Northwest A & F University, Yangling 712100, Shaanxi, PR China

Yuejin Wang
College of Horticulture, Northwest A & F University, Yangling 712100, Shaanxi, PR China
Key Laboratory of Biology and Genetic Improvement of Horticultural Crops (Northwest Region), Ministry of Agriculture, Yangling, Shaanxi 712100, PR China
State Key Laboratory of Crop Stress Biology in Arid Areas, Northwest A&F University, Yangling 712100, Shaanxi, PR China

Preface

Biological pigments, also known simply as pigments or biochromes are substances produced by living organisms that have a color resulting from selective color absorption. Biological pigments include plant pigments and flower pigments. Many biological structures, such as skin, eyes, feathers, fur and hair contain pigments such as melanin in specialized cells called chromatophores. There are two major classes of flower pigments: carotenoids and flavonoids. Carotenoids include carotene pigments (which produce yellow, orange and red colors). Flavonoids include anthocyanin pigments (which produce red, purple, magenta and blue colors). Usually, the color a flower appears depends on the color of the pigments in the flower, but this can be affected by other factors. The book, Pigments of Flowering Plants, deals with the pigments associated to flower colors. In first chapter, we report the performance of four natural dyes extracted from the leaves of teak (Tectona grandis), tamarind (Tamarindus indica), eucalyptus (Eucalyptus globulus), and the flower of crimson bottle brush (Callistemon citrinus). Transcriptomic analysis of *Paeonia Delavayi* wild population flowers to identify differentially expressed genes involved in purple-red and yellow petal pigmentation has been presented in second chapter. Third chapter provides fundamental information on the genes and pathways involved in flower secondary metabolism and development in *S. oblata*, providing a useful database for further research on *S. oblata* and other plants of genus *Syringa*. Fourth chapter attempts to reconstruct the betalain biosynthetic pathway as a self-contained system in an anthocyanin-producing plant species. Fifth chapter focuses on peace, a myb-like transcription factor, regulates petal pigmentation in flowering peach 'genpei' bearing variegated and fully pigmented flowers. Sixth chapter characterizes the subclass of type II OMTs by integrating biochemical, molecular, and phytochemical analysis, which will support an understanding of the anthocyanin methylating mechanism and shed light on its influence on flower coloration. In seventh chapter, two P1 orthologues, Gt MYBP3 and Gt MYBP4, were isolated and characterized in Japanese gentian. Eighth chapter confirm the importance of the methionine residue and demonstrate the utility of the dominant-negative CHS in modulating flower colour intensity even in a distantly related species. In ninth chapter, the first RNA-Seq project for *M. armeniacum* and its white variant was performed using the Illumina sequencing technique. Through a combination of chemical analysis with bioinformatics, the major metabolic pathways related to *Muscari* flower pigmentation were deduced and the candidate genes targeting the loss of pigmentation in the plants were examined.

Chapter 1

NATURAL PIGMENTS FROM PLANTS USED AS SENSITIZERS FOR TIO$_2$ BASED DYE-SENSITIZED SOLAR CELLS

Reena Kushwaha, Pankaj Srivastava, and Lal Bahadur

Department of Chemistry, Faculty of Science, Banaras Hindu University, Varanasi 221005, India

ABSTRACT

Four natural pigments, extracted from the leaves of teak (Tectona grandis), tamarind (Tamarindus indica), eucalyptus (Eucalyptus globulus), and the flower of crimson bottle brush (Callistemon citrinus), were used as sensitizers for TiO$_2$ based dye-sensitized solar cells (DSSCs). The dyes have shown absorption in broad range of the visible region (400–700 nm) of the solar spectrum and appreciable adsorption onto the semiconductor (TiO$_2$) surface. The DSSCs made using the extracted dyes have shown that the open circuit voltages (V_{oc}) varied from 0.430 to 0.610 V and the short circuit photocurrent densities (J_{sc}) ranged from 0.11 to 0.29 mA cm^{-2}. The incident photon-to-current conversion efficiencies (IPCE) varied from 12–37%. Among the four dyes studied, the extract obtained from teak has shown the best photosensitization effects in terms of the cell output.

INTRODUCTION

Harvesting energy from sunlight using photovoltaic technology is one of the most important research areas because of an ever increasing global energy need. The conventional solid-state silicon based solar cells, though highly efficient, are yet to become popular for mass applications as they are highly expensive. The necessity for developing low cost devices for harvesting solar energy was, therefore, very much desirable. A new hope was generated in this direction when O'Regan and Gräetzel reported to have achieved an unprecedented high energy conversion efficiency (η) of 7.1% through a dye-sensitized solar cell (DSSC) developed by using nanocrystalline TiO$_2$ thin film electrode sensitized by a highly efficient Ru(II) polypyridyl complex [1]. This has proven that significantly high light-to-electricity conversion efficiency can be achieved through DSSCs as well. Once this was established, such cells attracted greater attention of the scientists particularly because of two reasons; first, their production cost was expected to be quite low due to ease of their fabrication, and second, they are more environment friendly as compared to conventional solid-state silicon based photovoltaic devices [2]. Being optimistic that DSSCs have the potential to become a commercially viable alternative to expensive silicon solar cells, extensive studies have been conducted on such devices during last two decades.

A dye-sensitized solar cell is usually composed of a dye-capped nanocrystalline porous semiconductor electrode, a metal counter electrode, and a redox electrolyte mediating electron transfer processes occurring in the cell. The performance of the cell is primarily dependent on the material and quality of the semiconductor electrode and the sensitizer dye used for the fabrication of the cell. For their application in DSSCs, many wide band-gap metal oxide semiconductors have been studied but most extensively employed semiconductors are TiO_2 and ZnO [3–8]. Titanium dioxide (TiO_2) has several advantages, including long-term thermal and photostability. The essential properties of semiconductor can be changed significantly by using different techniques for their deposition on the substrate [9]. The sensitizer (dye) plays a key role in absorbing light, and in this respect the highest efficiency obtained so for is with Ru (II) polypyridyl complexes [10, 11]. However, the ruthenium complexes are expensive due to the paucity of the Ru metal and the complexity of preparation procedure limiting the production of low cost DSSC. This has stimulated the search for potential alternative metal complex sensitizers. Simultaneously, organic dyes [12, 13] and natural dyes [14–20] extracted from plants were also studied to explore the possibility of their application as photosensitizer. Organic dyes have been reported to meet the efficiency as high as 9.8% [12]. However, these dyes have been fraught with problems, such as complicated synthetic routes and low yields. On the other hand, the natural dyes found in flowers, leaves, and fruits of plants can be extracted by simple procedures and then employed in DSSCs. The advantages of natural dyes, resembling in functionalities to organic dyes, are their easy availability, nontoxicity, complete biodegradability, and temperature compatibility. Several of natural dyes such as tannin [21], carotene [22], anthocyanin [23], betalain [24], and chlorophyll [25, 26] have been extensively investigated as sensitizers in dye-sensitized solar cells [27].

In this paper, we report the performance of four natural dyes extracted from the leaves of teak (Tectona grandis), tamarind (Tamarindus indica), eucalyptus

(Eucalyptus globulus), and the flower of crimson bottle brush (Callistemon citrinus). The basic structures of the coloring components found in these extracts are given in Figure 1. Tannin, that is, gallic acid [3,4,5–trihydroxybenzoic acid] and ellagic acid [2,3,7,8-tetrahydroxybenzopyrano(5,4,3-cde)benzopyran-5,10-dione] are the main constituents of these natural dyes along with some minor components [28–30]. Teak extract mainly contains tectoleafquinone, 1,4,5,8-tetrahydroxy-2 isopentadienyl anthraquinone and tannin [28]. To the best of our knowledge, the use of these plant extracts is being reported for the first time as sensitizers for TiO₂ based dye-sensitized solar cells (DSSCs).

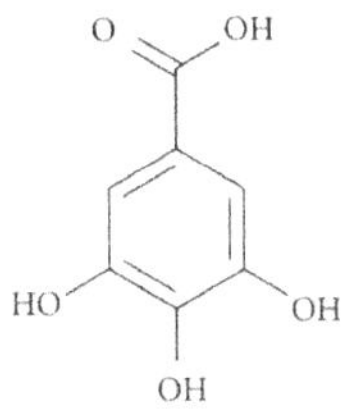

3,4,5-Trihydroxybenzoic acid
(gallic acid)

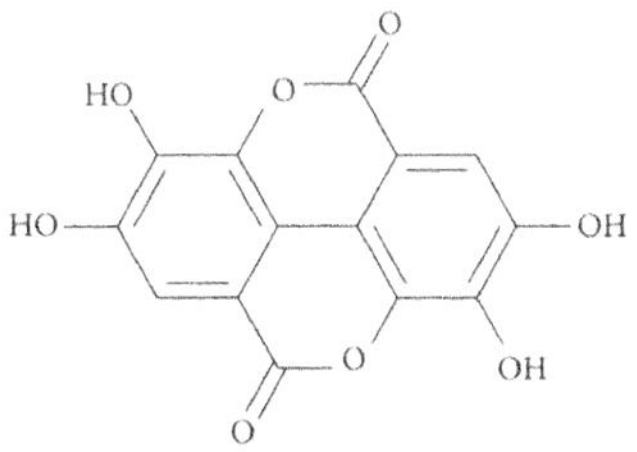

[2,3,7,8-Tetrahydroxy(1)benzopyrano-
-(5,4,3-cde)(1)benzopyran-5, 10-dione]
(ellagic acid)
(b)

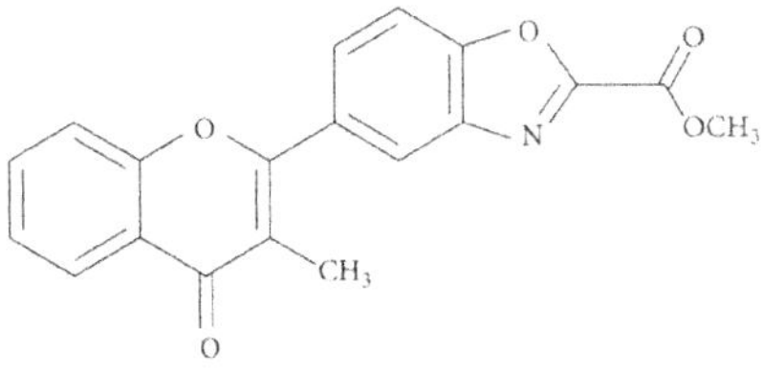

1,4,5,8-Tetrahydroxy-2 isopentadienyl-
-anthraquinone
(c)

Figure 1: Basic molecular structure for the main components of the extracts.

EXPERIMENTAL

Materials

Ethanol (A.R. grade, 99.9%, Merck) was used for extracting natural dyes from plants. Titanium paste (HT), platinum catalyst (T/SP), and the sealing tape (SX1170–60, 50 μm thick) were obtained from Solaronix. Propylene carbonate (>99%, Merck) was taken as the medium of cell electrolyte. Anhydrous lithium iodide (99.9%, Aldrich) and iodine (G. R. grade, 99.8%, BDH) were used as redox couple in photoelectrochemical (PEC) experiments without any further purification. FTO-coated (Fluorine-doped tin oxide) conductive glass slides (surface resistivity 15 $\Omega/\square$, thickness 2.2 mm) obtained from Pilkington, USA, were used as substrates for preparing TiO_2 thin film electrode and Platinum counter electrode.

Apparatus and Instruments

A bipotentiostat (model number AFRDE 4E, Pine Instrument Company, USA) and e-corder (model 201, eDAQ, Australia) were used for current-potential measurements. For photoelectrochemical (PEC) measurements, a 150 W Xenon arc lamp with lamp housing (model number 66057) and power supply (model number 68752), all from Oriel Corporation, USA, was used as the light source. The semiconductor electrode was illuminated after passing the collimated light beam through a 6-inch long water column (to filter IR part of the light) and condensing it with the help of fused silica lenses (Oriel Corporation, USA). The UV part of this IR-filtered light (referred to as "white light") was cut off by using a long pass filter (model number 51280, Oriel Corporation, USA) and the light obtained this way is mentioned as "visible light." The light was monochromatised, when required, by using a grating monochromator (Oriel model 77250 equipped with model 7798 grating).

The width of the exit slit of the monochromator was kept at 0.5 mm. To obtain the action spectrum (j_{photo}) of the dye-sensitized TiO_2 electrode, monochromatic light-induced photocurrent was measured with the help of a digital multimeter (Philips Model number 2525) in combination with the potentiostat. The intensities of light were measured with a digital photometer (Tektronix model J16 with model J 6502 sensor) in combination with neutral density filters (model number 50490-50570, Oriel, USA). The absorption spectrums were recorded on Shimadzu UV-1700 spectrophotometer. The FT-IR spectra were recorded by Varian 3100 FT-IR spectrometer.

Preparation of Natural Dye Solutions (Extracts)

The natural dyes were extracted with ethanol employing the following procedure: fresh leaves of teak (Tectona grandis), tamarind (Tamarindus indica), eucalyptus (Eucalyptus globulus), and the flower of crimson bottle brush (Callistemon citrinus) were washed with water and dried. After crushing them into small pieces in a mortar, these were kept in glass bottles and filled with ethanol; these solutions were kept for one week in the dark at room temperature. Then, the residual (solid) parts were filtered out and the resulting filtrates were used as dye solutions.

Preparation of TiO$_2$ Electrode (Photo Anode) and Counter Electrode

TiO$_2$ thin film electrodes (photoanodes) were prepared by spreading highly transparent paste of TiO$_2$(Titanium-HT) on FTO-coated conductive glass plate by the doctor's blade method. On the conducting side of glass substrate, a U-shaped frame of adhesive tape was applied to control the thickness of the film and to provide noncoated area for electrical contact.

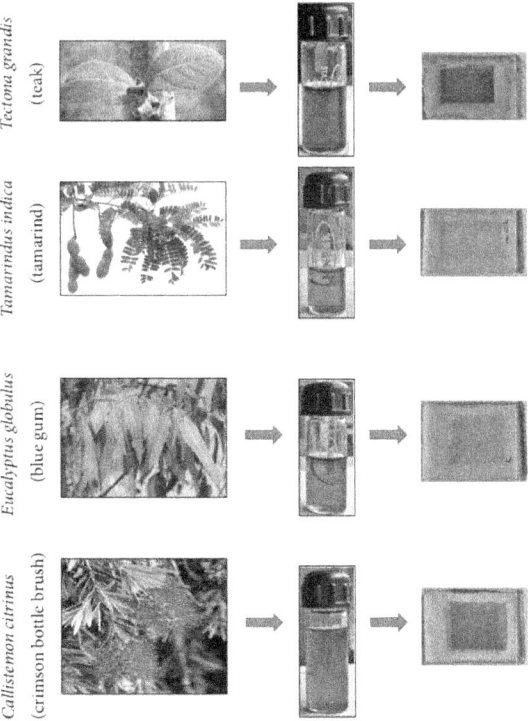

Figure 2: Plants, extracted dyes, and the dye-loaded TiO$_2$ electrode.

After spreading TiO_2 paste, the adhesive tapes were carefully removed and films were annealed at 450°C in air for half an hour in a tubular furnace. This resulted in TiO_2 film of ~6 μm thickness. The dyes were anchored onto the surface of the TiO_2 thin film electrode by immersing it into ethanol solution of natural dye for overnight. The nonadsorbed dye was washed up with anhydrous ethanol. The dye-coated films were air dried and used as photoelectrode in the cell (Figure 2). The platinum counter electrode was prepared on another FTO-coated glass substrate by depositing platinum catalyst (T/SP, Solaronix) using screen printing method and annealing at 400°C for half an hour in air. The electrolyte consisted of 0.2 M lithium iodide and 0.02 M iodine in propylene carbonate.

Fabrication of Sandwich Type DSSCs

The photo-electrode (dye-coated TiO_2 film) was put over platinum counter electrode in such a way that the conductive side of both the electrodes faced each other, and the cell was sealed from three sides using spacer/sealing tape (heating it at ~80°C); one side was left open for the injection of electrolyte. The cell electrolyte was injected through open side and was drawn into the space between the electrodes by capillary action. Thereafter, the open side of the cell assembly was sealed properly with Araldite and the contacts were made by copper wires using silver paste (Figure 3).

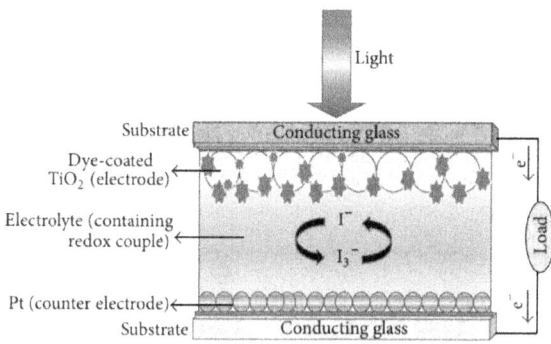

Figure 3: Schematic diagram of dye-sensitized solar cell (DSSC) assembly.

RESULTS AND DISCUSSION

Absorption Spectra of Natural Dyes

Figure 4 shows the absorption spectra of the ethanol extracts of Tectona grandis, Tamarindus indica, Eucalyptus globulus, and Callistemon citrinus.

From this figure, it is evident that these natural extracts absorb in the visible region of light spectrum and hence fulfill the primary criterion for their use as sensitizers in DSSCs. To be more specific, Tectona grandis exhibited broad absorption band in the range 425–550 nm besides showing a sharp absorption peak at 662 nm. Tamarindus indica and Eucalyptus globulus have absorption peaks at 410 nm and 472 nm, respectively. Each of them has a common peak at 663 nm which is consistent with the characteristic absorption band of chlorophyll [25, 26]. Callistemon citrinus absorbs in the wide range of 410–600 nm with an absorption peak at 450 nm. The differences and variations in the absorption characteristics of dyes can be attributed to the different colors of the extracts due to respective pigments present in them.

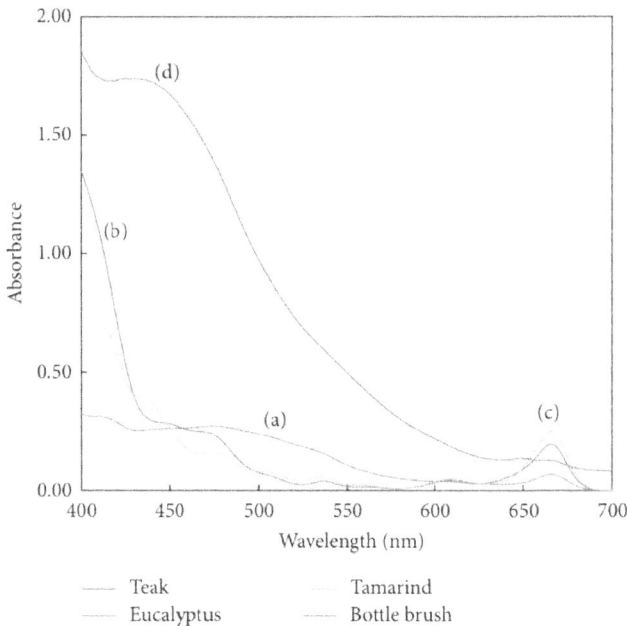

Figure 4: Absorption spectra of ethanol solution of natural dyes extracted from (a) teak, (b) eucalyptus, (c) tamarind, and (d) bottle brush, respectively.

FTIR Spectra

The infrared spectra of these four natural extracts were obtained by pressing them in pellets with KBr. The respective FTIR spectra were recorded in the range from 4000 to 400 cm⁻¹ and shown in Figure 5. An examination of the spectra reveals that they exhibit broad absorption in the range 3000–3700 cm⁻¹ with a wide and strong band at 3407 cm⁻¹ which is attributed to the –OH stretching and due to the wide variety of hydrogen bonding between OH. In these spectrums, a sharp peak at around 2927 and a small shoulder at 2855 cm⁻¹ associated with the symmetric and antisymmetric –C–H– stretching vibrations of CH_2 and CH_3 groups, respectively, is observed. Also, the signal characteristics bands of C=O (carbonyl) stretching vibration at 1730–1705 cm⁻¹ and C–O at 1100–1300 cm⁻¹ can be observed due to presence of some aromatic esters. The bands observed in the range 1669–1400 cm⁻¹ are due to aromatic ring vibrations, while the ones at 1190 and 1052 cm⁻¹ are due to ester linkage. The band at around 751 cm⁻¹ is assigned to aromatic C–H bending vibration. Hence, the IR spectra of extracts contain bands that can be assigned to the coloring components found in these extracts as given in Figure 1 Tannin, that is, gallic acid, ellagic acid, and tectoleafquinone, 1,4,5,8-tetrahydroxy-2 isopentadienyl anthraquinone.

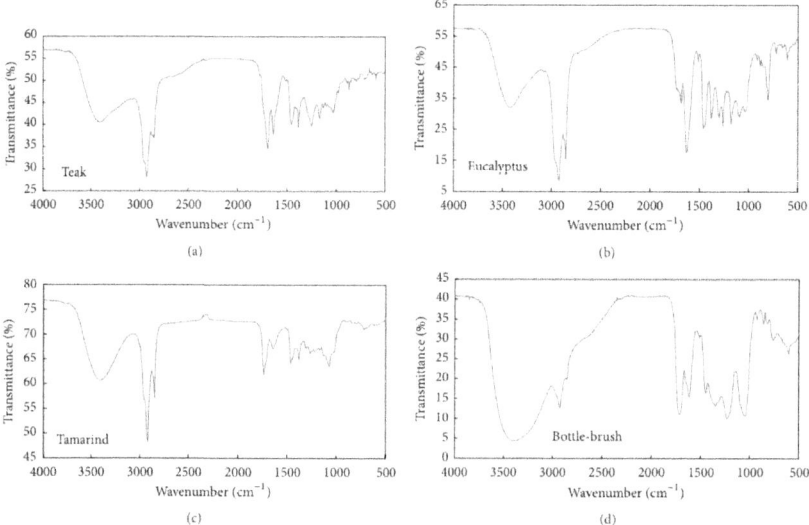

Figure 5: Infra-red spectra of extracts obtained from (a) teaks (b) tamarinds (c) eucalyptuss and (d) bottle brush.

Photoelectrochemical Studies

Current-Potential (J-V) Curves

The photovoltaic performances of DSSCs using natural dyes as photosensitizer (TiO₂-dye/electrolyte containing I, /I₃⁻ counter electrode) were determined by recording the current-potential (J-V) curves under visible light illumination and displayed in Figure 6. The similar curve for the cell using bare TiO_2 electrode determined under identical experimental conditions is also shown in the figure (curve (e)). Almost insignificant current is observed in this case as expected, since visible light is incapable of exciting wide band-gap TiO_2. The values of photovoltaic parameters derived from these curves are given in Table 1.

Table 1: The cell output of DSSCs sensitized by four kinds of natural dyes: (a) teak, (b) tamarind, (c) eucalyptus, and (d) bottle brush under visible light ($256\,mW/cm^2$) illumination.

Natural extract	Peak wavelength λ (nm)	I_{sc} (mA/cm²)	V_{oc} (mV)	IPCE (%)	P_{max} (mW/cm²)	FF
Teak (*Tectona grandis*)	470, 662	0.29	460	37	0.105	79
Tamarind (*Tamarindus indica*)	410, 663	0.18	610	33	0.061	56
Eucalyptus (*Eucalyptus globulus*)	472, 663	0.15	500	12	0.070	93
Bottle brush (*Callistemon citrinus*)	450	0.11	430	34	0.030	63

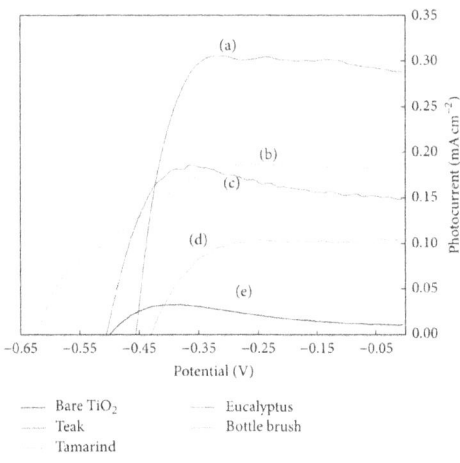

Figure 6: Photocurrent-voltage (J-V) curves for the DSSCs sensitized by four kinds of natural dyes: (a) teak, (b) tamarind, (c) eucalyptus, and (d) bottle brush under visible light illumination of intensity $256\,mW/cm^2$ (electrolyte composition: $0.2\,M\,LiI$, $0.02\,M\,I_2$ in propylene carbonate). Curve (e) is the same for bare TiO_2 electrode.

With DSSCs using these dyes, open circuit voltage (V_{oc}) from 0.430 to 0.610 V and the short circuit photocurrent densities (J_{sc}) in the range of 0.11–0.29 mA/cm² could be achieved. The highest V_{oc} (0.610 V) was obtained with tamarind extract-sensitized DSSC, whereas maximum J_{sc} (0.29 mA/cm²) was obtained with the DSSC sensitized by teak extract.

Transient Photocurrent-Time (J_{photo}-t) Profile

The transient current-time profiles were recorded to know the sustainability of the photocurrent observed initially on illumination of the DSSCs with desired intensity of light. For such an assessment, initially the dark current was monitored for a few seconds; then the semiconductor electrode was illuminated and the short circuit photocurrent was monitored as a function of time. The photocurrent-time (J_{photo}-t) profile obtained under visible light ($256\,mW/cm^2$) illumination of natural dye sensitized DSSCs are shown in Figure 7.

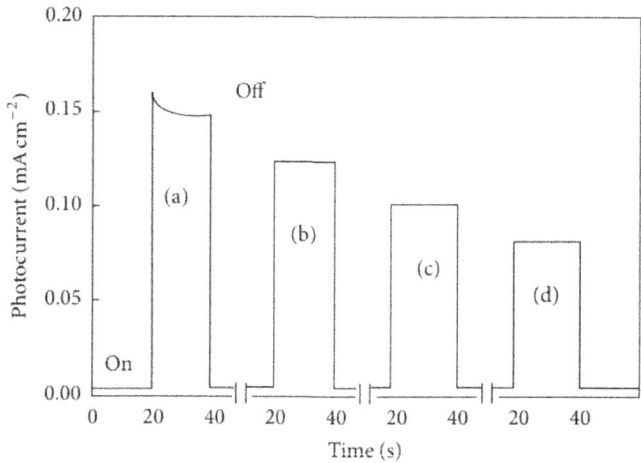

Figure 7: Transient current-time (J_{photo}-t) profiles obtained under visible light illumination (intensity $256\,mW/cm^2$) for the four DSSCs sensitized by (a) teak, (b) tamarind, (c) eucalyptus, and (d) bottle brush, respectively. Electrolyte composition and intensity are the same as in Figure 6.

Except for the curve (a), in all the other cases, ideal behavior (no decay in photocurrent) was observed. In case of curve (a), initially the photocurrent reached maximum, but the same was not sustained and it decayed to ~93% of its initial value before getting stabilized. This may be the result of slowness of dye regeneration process as compared to rate of charge carriers' injection by the excited dye molecule.

Photocurrent Action Spectrum (IPCE)

In order to conclusively ascertain the sensitization of photocurrent by the dyes under investigation, the short-circuit photocurrent (J_{photo}) spectra of dye modified TiO_2 electrodes were determined. From the values of J_{photo} and the intensity of the corresponding monochromatic light (I_{inc}), the incident photon-to-current conversion efficiency (IPCE) was calculated at each excitation wavelength (λ) using the following relation:

$$\text{IPCE (\%)} = \frac{1240 J_{photo}\left(A/cm^2\right)}{\lambda\,(nm) \cdot I_{inc}\left(W/cm^2\right)} \times 100.$$

(1)

The IPCE versus wavelength (λ) curves for different cases (the natural dyes) are shown in Figure 8. It is clearly seen from this figure that there is close resemblance of the nature of IPCE curve with the absorption spectrum of the respective dye providing clear evidence of the sensitization of photocurrent by dye. The IPCE values observed at the characteristic wavelengths of the dyes ranged from 12% to 37%, decreasing in the order Tectona grandis > Callistemon citrinus > Tamarindus indica > Eucalyptus globules. The variation in IPCE values for different natural dyes could be due to the varied amount of dye loaded onto the TiO$_2$ thin film, different degree of charge carrier's recombination, different energy levels of excited dye molecule, and the quenching of excited state.

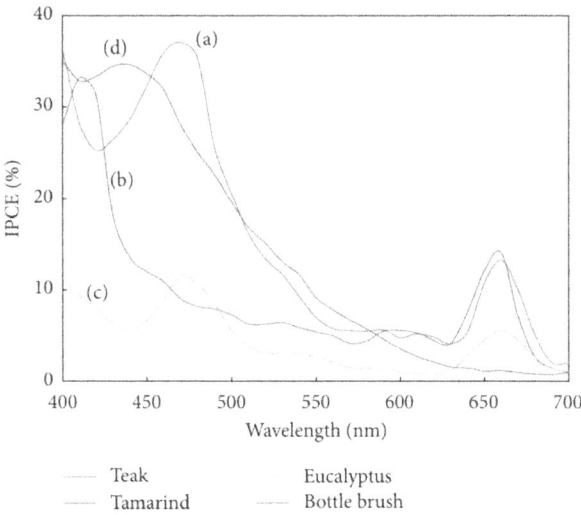

Figure 8: Action spectra of solar cell sensitized by the extracts (a) teak, (b) tamarind, (c) eucalyptus, and (d) bottle brush.

Power Conversion Efficiency (η) and Fill Factor (FF)

The power conversion efficiency and the fill factor of dye-sensitized solar cells were determined from the (J-V) curve of the respective cell under illumination by visible light. From the experimentally determined J-V curves (Figure 6), the values of fill factor (FF) and power conversion efficiency (η) were evaluated using the following relations:

$$FF = \frac{P_{max}}{P_{ideal}} = \frac{J_{max}\left(A/cm^2\right) \times V_{max}\,(V)}{J_{sc}\,(A/cm^2) \times V_{oc}\,(V)},$$

$$\eta\,(\%) = \frac{J_{max}\left(A/cm^2\right) \times V_{max}\,(V)}{I_{inc}\,(W/cm^2)} \times 100.$$

$$(2)$$

Here, J_{sc}, V_{oc}, and I_{inc} are short-circuit photocurrent, open-circuit potential, and intensity of incident light, respectively. With the use of these dyes power conversion efficiency follows the order (Tectona grandis >Eucalyptus globulus > Tamarindus indica > Callistemon citrinus), while fill factor is obtained as (Eucalyptus globulus > Tectona grandis > Callistemon citrinus > Tamarindus indica).

The maximum output power (P_{max}) is obtained by choosing a point on experimentally determined (J-V) curve corresponding to which the product of current (J_{max}) and potential (V_{max}) gives the maximum value. Figure 9shows the (power versus potential) curves for the natural dye(s)-sensitized solar cells, and the corresponding powers (P_{max}) obtained from various extracts are revealed in Table 1. The maximum photopower was obtained in the case of teak leaf extract, however low conversion responses may be due to poor interaction of sensitizers with the semiconductor electrode that restricts the transport of electrons from the excited dye molecules to the TiO_2 film.

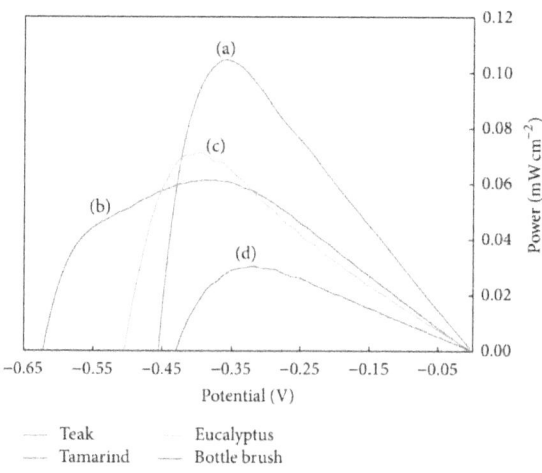

Figure 9: Power versus voltage curves of the DSSCs using the natural dyes extracted from the (a) teak, (b) tamarind, (c) eucalyptus, and (d) bottle brush.

CONCLUSIONS

Four natural dyes extracted from the leaves or flowers of the plants were used as sensitizer and their photovoltaic characteristics were studied. The extracted dyes contain tannins as the major coloring component along with some other minor components. Chlorophyll is the common component present in all the dyes extracted from the leaves. Tectoleafquinone is the key component present in the teak leaf extract. The chemical adsorption of these dyes becomes possible because of the condensation of hydroxyl and methoxy protons with the hydroxyl groups on the surface of nanostructured TiO$_2$. The DSSCs made using the extracted dyes showed the open circuit voltages () varying between 0.430 and 0.610 V, and the short circuit photocurrent densities () ranged from 0.11 to 0.29 mA cm^{-2}. The incident photo-to-current conversion efficiencies (IPCEs) varied from 12 to 37%. Among the four dyes studied, the extract obtained from teak has shown the best photosensitization effects in terms of the cell output as against the expectation arising from the apparent matching profile of the bottle brush extract with the solar spectrum. The natural dye extracts are, generally, a mixture of several pigments. Therefore, the possible reason for the observed differences in sensitization actions of dyes is their varied abilities towards adsorption onto the semiconductor surface. The impact of the different rates of electron transfer from the dye molecule to the conduction band of semiconductor electrode (energy levels alignments) is also reflected. Sometimes, a complication such as dye aggregation on semiconductor film produces absorptivity that results in either the nonelectron injection or the steric hindrance preventing the dye molecules from effectively arraying on the semiconductor film. This leads to the weaker binding and greater resistance, resulting in the low output of cells. Addition of appropriate additives for improving without causing dye degradation might result in further enhancement of the cell performances. Hence, though photocurrent densities, photovoltages, and IPCE obtained with these dyes are somewhat low, they are quite useful for their nontoxicity, greater availability, and very low cost of production opening up a perspective of feasibility for inexpensive and environmentally friendly dye cells.

Acknowledgments

Reena kushwaha acknowledges the financial support received from the University Grant Commission, New Delhi, and the Ministry of New and Renewable Energy (MNRE), New Delhi, for this work.

REFERENCES

1.   B. O'Regan and M. Gräetzel, "A low-cost, high-efficiency solar cell based on dye-sensitized colloidal TiO$_2$ films," Nature, vol. 353, pp. 737–740, 1991.

2. M. Grätzel, "Dye-sensitized solar cell," Journal of Photochemistry and Photobiology C, vol. 4, pp. 145–153, 2003.

3. M. Grätzel, "Sol-gel processed TiO_2 films for photovoltaic applications," Journal of Sol-Gel Science and Technology, vol. 22, no. 1-2, pp. 7–13, 2001. · ·

4. J. Jiu, S. Isoda, M. Adachi, and F. Wang, "Preparation of TiO_2 nanocrystalline with 3–5 nm and application for dye-sensitized solar cell," Journal of Photochemistry and Photobiology A, vol. 189, no. 2-3, pp. 314–321, 2007. · ·

5. C.-S. Chou, F.-C. Chou, and J.-Y. Kang, "Preparation of ZnO-coated TiO_2 electrodes using dip coating and their applications in dye-sensitized solar cells," Powder Technology, vol. 215-216, pp. 38–45, 2012. · ·

6. T. S. Senthil, N. Muthukumarasamy, D. Velauthapillai, S. Agilan, M. Thambidurai, and R. Balasundaraprabhu, "Natural dye (cyanidin 3-O-glucoside) sensitized nanocrystalline TiO_2 solar cell fabricated using liquid electrolyte/quasi-solid-state polymer electrolyte," Renewable Energy, vol. 36, no. 9, pp. 2484–2488, 2011. · ·

7. S. Kushwaha and L. Bahadur, "Characterization of synthetic Ni(II)-xylenol complex as a photosensitizer for wide-band gap ZnO semiconductor electrodes," International Journal of Photoenergy, vol. 2011, Article ID 980560, 9 pages, 2011. · ·

8. S. S. Kanmani and K. Ramachandran, "Synthesis and characterization of TiO_2/ZnO core/shell nanomaterials for solar cell applications," Renewable Energy, vol. 43, pp. 149–156, 2012.

9. F. C. Krebs, "Fabrication and processing of polymer solar cells: a review of printing and coating techniques," Solar Energy Materials and Solar Cells, vol. 93, no. 4, pp. 394–412, 2009.

10. Y. Chiba, A. Islam, Y. Watanabe, R. Komiya, N. Koide, and L. Han, "Dye-sensitized solar cells with conversion efficiency of 11.1%," Japanese Journal of Applied Physics, vol. 45, no. 24–28, pp. L638–L640, 2006.

11. R. Buscaino, C. Baiocchi, C. Barolo et al., "A mass spectrometric analysis of sensitizer solution used for dye-sensitized solar cell," Inorganica Chimica Acta, vol. 361, no. 3, pp. 798–805, 2008.

12. G. Zhang, H. Bala, Y. Cheng et al., "High efficiency and stable dye-sensitized solar cells with an organic chromophore featuring a binary π-conjugated spacer," Chemical Communications, no. 16, pp. 2198–2200, 2009.

13. P. Srivastava and L. Bahadur, "Dye-sensitized solar cell based

on nanocrystalline ZnO thin film electrodes combined with a novel light absorbing dye Coomassie Brilliant Blue in acetonitrile solution,"International Journal of Hydrogen Energy, vol. 37, no. 6, pp. 4863–4870, 2012.

14. S. Hao, J. Wu, Y. Huang, and J. Lin, "Natural dyes as photosensitizers for dye-sensitized solar cell," Solar Energy, vol. 80, no. 2, pp. 209–216, 2006.

15. K. Wongcharee, V. Meeyoo, and S. Chavadej, "Dye-sensitized solar cell using natural dyes extracted from rosella and blue pea flowers," Solar Energy Materials and Solar Cells, vol. 91, no. 7, pp. 566–571, 2007. · ·

16. G. Calogero and G. D. Marco, "Red Sicilian orange and purple eggplant fruits as natural sensitizers for dye-sensitized solar cells," Solar Energy Materials and Solar Cells, vol. 92, no. 11, pp. 1341–1346, 2008. · ·

17. K. E. Jasim, S. Al-Dallal, and A. M. Hassan, "Natural dye-sensitised photovoltaic cell based on nanoporous TiO$_2$," International Journal of Nanoparticles, vol. 4, no. 4, pp. 359–368, 2011.

18. C. Sandquist and J. L. McHale, "Improved efficiency of betanin-based dye-sensitized solar cells," Journal of Photochemistry and Photobiology A, vol. 221, no. 1, pp. 90–97, 2011. · ·

19. S. Sönmezoğlu, C. Akyürek, and S. Akin, "High-efficiency dye-sensitized solar cells using ferrocene-based electrolytes and natural photosensitizers," Journal of Physics D, vol. 45, Article ID 425101, 2012.

20. L. U. Okoli, J. O. Ozuomba, A. J. Ekpunobi, and P. I. Ekwo, "Anthocyanin-dyed TiO$_2$ electrode and its performance on dye-sensitized solar cell," Research Journal of Recent Sciences, vol. 1, pp. 22–27, 2012.

21. R. Espinosa, I. Zumeta, J. L. Santana et al., "Nanocrystalline TiO$_2$ photosensitized with natural polymers with enhanced efficiency from 400 to 600 nm," Solar Energy Materials and Solar Cells, vol. 85, no. 3, pp. 359–369, 2005.

22. E. Yamazaki, M. Murayama, N. Nishikawa, N. Hashimoto, M. Shoyama, and O. Kurita, "Utilization of natural carotenoids as photosensitizers for dye-sensitized solar cells," Solar Energy, vol. 81, no. 4, pp. 512–516, 2007.

23. H. Zhu, H. Zeng, V. Subramanian, C. Masarapu, K.-H. Hung, and B. Wei, "Anthocyanin-sensitized solar cells using carbon nanotube films as counter electrodes," Nanotechnology, vol. 19, no. 46, Article ID 465204, 2008. · ·

24. D. Zhang, S. M. Lanier, J. A. Downing, J. L. Avent, J. Lum, and J. L.

McHale, "Betalain pigments for dye-sensitized solar cells," Journal of Photochemistry and Photobiology A, vol. 195, no. 1, pp. 72–80, 2008.

25.   W. H. Lai, Y. H. Su, L. G. Teoh, and M. H. Hon, "Commercial and natural dyes as photosensitizers for a water-based dye-sensitized solar cell loaded with gold nanoparticles," Journal of Photochemistry and Photobiology A, vol. 195, no. 2-3, pp. 307–313, 2008.

26.   A. R. Hernández-Martínez, M. Estevez, S. Vargas, F. Quintanilla, and R. Rodríguez, "Natural pigment based dye sensitized solar cells," Journal of Applied Research and Technology, vol. 10, pp. 38–47, 2012.

27.   M. R. Narayan, "Review: dye sensitized solar cells based on natural photosensitizers," Renewable and Sustainable Energy Reviews, vol. 16, no. 1, pp. 208–215, 2012. ··

28.   R. Aradhana, K. N. V. Rao, D. Banji, and R. K. Chaithanya, "A review on Tectona grandis.linn: chemistry and medicinal uses," Herbal Tech Industry, vol. 6, no. 11, 2010.

29.   R. Mongkholrattanasit, J. Kryštůfek, J. Wiener, and J. Studničková, "Natural dye from Eucalyptus leaves and application for wool fabric dyeing by using padding techniques," in Natural Dyes, E. A. Kumbasar, Ed., chapter 4, 2011.

30.   E. D. Caluwé, K. Halamová, and P. V. Damme, "Tamarindus indica L.: a review of traditional uses, phytochemistry and pharmacology," Afrika Focus, vol. 23, pp. 53–83, 2010.

Chapter 2

TRANSCRIPTOMIC ANALYSIS OF PAEONIA DELAVAYI WILD POPULATION FLOWERS TO IDENTIFY DIFFERENTIALLY EXPRESSED GENES INVOLVED IN PURPLE-RED AND YELLOW PETAL PIGMENTATION

Qianqian Shi, Lin Zhou, Yan Wang, Kui Li, Baoqiang Zheng, Kun Miao

State Key Laboratory of Tree Genetics and Breeding, Key Laboratory of Tree Breeding and Cultivation of State Forestry Administration, Research Institute of Forestry, Chinese Academy of Forestry, Beijing 100091, China

ABSTRACT

Tree peony (*Paeonia suffruticosa* Andrews) is a very famous traditional ornamental plant in China. *P. delavayi* is a species endemic to Southwest China that has aroused great interest from researchers as a precious genetic resource for flower color breeding. However, the current understanding of the molecular mechanisms of flower pigmentation in this plant is limited, hindering the genetic engineering of novel flower color in tree peonies. In this study, we conducted a large-scale transcriptome analysis based on Illumina HiSeq sequencing of cDNA libraries generated from yellow and purple-red *P. delavayi* petals. A total of 90,202 unigenes were obtained by *de novo* assembly, with an average length of 721 nt. Using Blastx, 44,811 unigenes (49.68%) were found to have significant similarity to accessions in the NR, NT, and Swiss-Prot databases. We also examined COG, GO and KEGG annotations to better understand the functions of these unigenes. Further analysis of the two digital transcriptomes revealed that 6,855 unigenes were differentially expressed between yellow and purple-red flower petals, with 3,430 up-regulated and 3,425 down-regulated. According to the RNA-Seq data and qRT-PCR analysis, we proposed that four up-regulated key structural genes, including *F3H*, *DFR*, *ANS* and *3GT*, might play an important role in purple-red petal pigmentation, while high co-expression of *THC2'GT*, *CHI* and *FNS*

II ensures the accumulation of pigments contributing to the yellow color. We also found 50 differentially expressed transcription factors that might be involved in flavonoid biosynthesis. This study is the first to report genetic information for *P. delavayi*. The large number of gene sequences produced by transcriptome sequencing and the candidate genes identified using pathway mapping and expression profiles will provide a valuable resource for future association studies aimed at better understanding the molecular mechanisms underlying flower pigmentation in tree peonies.

BACKGROUND

Tree peony belongs to section *Moutan* of the genus *Paeonia*, family Paeoniaceae, is an important traditional ornamental and medicinal plant in China, and has been named 'the king of flowers' for its large, showy and colorful flowers [1]. There are nine wild species of tree peony, *P. suffruticosa*, *P. cathayana*, *P. jishanensis*, *P. qiui*, *P. ostii*, *P. rockii*, *P. decomposita*, *P.delavayi* and *P. ludlowii* [2–4], and about 1,500 cultivars with a wide range of flower colors that have been produced across the world by conventional breeding. Among these species, *P.delavayi* is very special, with extreme variability both within and between populations in the number, length, and width of leaf segments and the number, size, and color of all its floral parts [5]. These plants are distributed mainly in the northwest of Yunnan Province, southwest of Sichuan Province and southeast of Tibet, China [5,6]. Some papers and our previous study have reported various petal colors in the same population of *P. delavayi*, including yellow, orange, red, dark red, or purple-red [5,7]. Plants with yellow flowers are considered the most precious resource for cultivar breeding because traditional Chinese cultivar flowers are usually purple, pink, red or white, but lacking in pure yellow. Moreover, Li et al. [8] reported that the antioxidant activity of *P. delavayi* was higher than that of other species with yellow flowers, which have been considered potential resources for the development of new drugs or functional foods, demonstrating the great economic significance of research on this species.

Among the three major groups of pigments, flavonoids, particularly anthocyanidins, are the most widely distributed in higher plants and contribute to a wide range of colors, from pale yellow to red, purple and blue [9,10]. To date, more than 30 different flavonoids, including anthocyanins and multiform glycosides of flavones and flavonols, have been identified and quantified from different groups and several wild species of tree peony [11–15]. Wang et al. [12] analyzed the composition and content of flower pigments in seven wild tree peonies and found no pelargonidin (Pg)-based anthocyanins in any accessions in subsection Delavayanae; *P.delavayi* with purple flowers mainly contained peonidin-3,5-glucosides (Pn3G5G). In a recent study, Li et al. [8] examined the composition and content of flavonoids from yellow flowers of *P.delavayi*, and identified the main compound as chalcone 2'-glucoside (isosalipurposide, ISP). In our previous research [16], we investigated the pigment composition of yellow petals from a wild *P. delavayi* population in Yunnan Province, China, and found that chalcones, flavones and flavonols were the main components, including ISP, kaempferol, quercetin, isorhamnetin, chrysoeriol and apigenin-glycopyranoside [16]. These studies provide a physiological and biochemical basis for future research on the molecular mechanism of *P. delavayi* flower pigmentation.

Flavonoids are synthesized through the phenylpropanoid biosynthesis pathway, which can be divided into three stages. The first stage is the conversion of phenylalanine to coumaroyl-CoA, which is shared in many secondary metabolism pathways. The second stage is the synthesis of dihydroflavonols such as dihydrokaempferol (DHK), dihydroquercetin (DHQ) and dihydromyricetin (DHM) from one molecule of coumaroyl-CoA and three molecules of malonyl-CoA catalyzed by a series of enzymes including chalcone synthase (CHS), chalcone isomerase (CHI), flavanone 3-hydroxylase (F3H), flavonoid 3'-hydroxylase (F3'H), flavonoid 3'5'-hydroxylase (F3'5'H), flavone synthase (FNS) and flavonol synthase (FLS). The third stage is the synthesis of various anthocyanidins from dihydroflavonols catalyzed by dihydroflavonol 4-reductase (DFR) and anthocyanidin synthase (ANS), followed by the formation of stable anthocyanins through a series of glycosylation and methylation reactions catalyzed by anthocyanidin 3-*O*-glucosyltransferase (3GT), anthocyanidin 5-*O*-glucosyltransferase (5GT) and anthocyanin *O*-methyltransferase (AOMT) [17,18]. In the flavonoid biosynthetic pathway, the transcription levels of flavonoid biosynthesis genes are regulated by various transcription factors, which can be classified into three families: R2R3-MYB, basic-Helix-Loop-Helix (bHLH) and WD40 [19]. To date, a large number of structural genes as well as some regulatory genes have been well characterized and isolated in mutants or crossed lines of snapdragon, maize, petunia and *Arabidopsis* as model plants [17,18]. Meanwhile, numerous studies on the molecular mechanism of pigmentation have been performed in different ornamental plants, including *P. suffruticosa*, in which five structural genes (*PsCHS1*, *PsCHI1*, *PsANS1*, *PsF3H1* and *PsDFR1*) were cloned and identified as the most important genes involved in tree peony flower pigmentation [20–22]. However, despite the economic and breeding importance of *P.delavayi*, flower pigmentation in this plant has previously been studied only at the physiological level [8,16]. Therefore, studying the molecular biology of *P. delavayi* color formation is of great importance to accelerate the use of the unique flower genes in this species.

For woody plants, especially those of high heterozygosity, such as tree peony, whole genome sequencing requires long-term and expensive investment. Transcriptome sequencing based on the Illumina/Solexa high-throughput sequencing platform is a rapid and convenient way to obtain information on the expressed fraction of genome [23,24]. RNA-Seq is not restricted by a reference sequence and is suitable for non-model organisms without genomic sequences [25–28]. It has been widely applied to model as well as non-model organisms in various studies, including transcript profiling, single nucleotide polymorphism discovery and the identification of genes that are differentially expressed between samples [29–34].

In the present work, individuals with purple-red and yellow flowers within a wild *P. delavayi*population in Yunnan Province, China were used as experimental materials, and two transcriptomes (of petals of each color mixed at different flower opening stages) were sequenced on the Illumina HiSeq 2000 platform. Additionally, by analyzing the data with various bioinformatics tools, we discovered some potential differentially expressed genes involved in purple-red and yellow petal pigmentation, and analyzed the expression profiles of seven candidate genes through real-time PCR. To the best of our knowledge, this is the first exploration of the petal transcriptome of the wild tree peony species *P. delavayi*. These transcriptome sequences may provide a theoretical basis to understand the molecular mechanisms behind the pigment composition of *P. delavayi* and a valuable resource for identifying genes expressed in this wild population.

RESULTS

Illumina Sequencing and *de novo* Transcriptome Assembly

To achieve a broad survey of the genes associated with *P. delavayi* flower pigmentation, two cDNA libraries (Pl and Pd) were generated for RNA-Seq. Each included a mixture of equal amounts of RNA from petals at five flower opening stages (Pl, yellow petals; Pd, purple-red). The mRNA was isolated, enriched, sheared into smaller fragments, and reverse-transcribed into cDNA, which was subjected to sequencing on an Illumina HiSeq 2000 sequencing platform. In total, we got 52.2 million raw reads with an average length of 83.42 nt and 54.3 million reads with an average length of 84.94 nt from the Pl and Pd sequencing libraries, respectively. After removal of adaptor sequences, ambiguous reads, and low-quality reads, 48,401,848 high-quality clean reads comprising 4,356,166,320 nucleotides (4.35 Gb) with a Q20 percentage of 98.31% and a GC percentage of 45.47% were generated from Pl transcriptome sequencing, while 51,300,426 high-quality clean reads comprising

4,617,038,340 nucleotides (4.61 Gb) with the same Q20 percentage as that of Pl and a GC percentage of 45.69% were generated from Pd transcriptome sequencing (Table 1). All high-quality reads were assembled *de novo* by the Trinity program [35]. From the Pl dataset, 149,470 contigs were produced with an N50 of 542 nt (i.e. 50% of the assembled bases were incorporated into contigs of 542 nt or longer), while 144,205 contigs with an N50 of 579 nt were generated from the Pd sample. After linking the contigs together, we obtained 88,330 unigenes with an average length of 603 nt from the Pl dataset, most of which were 200–3000 bp long, and 82,720 unigenes with an average length of 615 nt from the Pd sample. Finally, 90,202 non-redundant All-unigenes were obtained by splicing and reducing the redundancy of the unigene sequences, with average length and N50 values of 721 nt and 1093 nt, respectively (Table 1). Of the total unigenes, only 814 from Pl and 823 from Pd were longer than 3 kb. The length distributions of the unigenes are shown in S1A Fig. These results showed that the throughput and sequencing quality was high enough for the following analyses.

Table 1. Summary of the sequence assembly after Illumina sequencing.

	Samples	Total number	Total length (nt)	Mean length (nt)	N50	Q20 (%)	GC (%)
Read	Pl	48,401,848	########	90	-	98.31	45.47
	Pd	51,300,426	########	90	-	98.31	45.69
Contig	Pl	149,470	47,741,711	319	542	-	-
	Pd	144,205	46,839,327	325	579	-	-
Unigene	Pl	88,330	53,279,410	603	989	-	-
	Pd	82,720	50,846,368	615	1018	-	-
	All	90,202	65,006,636	721	1093	-	-

doi:10.1371/journal.pone.0135038.t001

The all-unigene sequences were aligned against protein databases using Blastx (e-value < 0.00001) in the following order: the non-redundant protein database (NR), the Swiss-Prot protein database (Swiss-Prot), the Kyoto Encyclopedia of Genes and Genomes database (KEGG), the Cluster of Orthologous Groups of proteins database (COG), the Gene Ontology database (GO) and the non-redundant nucleotide database (NT) [36]; a total of 44,811 (49.68%) were annotated (S1 Table). The size distribution of the CDSs and predicted proteins are shown in S1B Fig. Additionally, the CDSs of 4,083 all-unigenes that had no hits in Blastx were predicted by ESTScan and then translated into peptide sequences [37]; their size distribution is shown in S1C Fig.

Functional Annotation

Unigene annotation including COG clusters, GO classification and KEGG pathway annotation provides information on expression and function. In the annotation results obtained by Blastx with e-value < 0.00001, 42,699,

26,496, 24,262 and 15,073 All-unigenes were annotated to the NR, Swiss-Prot, KEGG and COG databases, accounting for 47.3%, 29.4%, 26.9% and 16.7%, respectively (S1 Table). Based on the NR annotations and the e-value distribution, 33.43% of the mapped All-unigenes showed strong homology (< 1 e-100), and 66.57% ranged from 1e-5 to 1e-100 (S2A Fig). Additionally, 25.40% of the All-unigenes had > 80% identity with the NR database and 44.01% had similarity ranging from 60–80% (S2B Fig). The majority of sequences (57.91%) shared the highest homology with *Vitis vinifera*, followed by *Ricinus communis* (11.04%), *Populus trichocarpa* (10.74%), *Glycine max* (4.98%), *Medicago truncatula*(2.35%), and *Arabidopsis thaliana* (1.08%); the remaining 4,747 All-unigenes, which consisted of 11.12% of our unique transcripts, showed less than 0.78% similarity to other species (S2 Table).

When searched against the COG database, 15,073 All-unigenes (17.1%) were annotated and classified into 25 different functional classes (S3 Fig). Among these categories, the cluster for 'General function prediction', which contained 4,963 All-unigenes (32.93%), represented the largest group, followed by 'Transcription' (2,648, 17.57%), 'Replication, recombination and repair' (2,496, 16.56%), 'Posttranslational modification, protein turnover, chaperones' (2,151, 14.27%) and 'Signal transduction mechanisms' (2,038, 13.52%), while only one All-unigene was assigned to 'Nuclear structure' as the smallest group (S3 Fig).

In GO analysis, 33,249 All-unigenes were categorized into the three main GO ontologies (molecular function, cellular component and biological process) including 55 functional groups. In the biological process category, the dominant groups were 'cellular process' (21,224, 63.83%) and 'metabolic process' (20,478, 61.58%). Under the cellular component category, 'cell' and 'cell part' were the most frequent terms, with the same percentage (26,088, 78.46%). In the molecular function category, 'binding' (15,950, 47.97%) and 'catalytic activity' (16,117, 48.47%) were predominant, whilst fewer than 10 genes each were categorized into the 'channel regulator activity' (7), 'metallochaperone activity' (5), 'protein tag' (4) and 'translation regulator activity' (4) groups (Fig 1).

The KEGG database can help study the complicated biological behaviors of genes and provide pathway annotations [38]. A total of 24,262 All-unigenes were mapped into 128 KEGG pathways, representing metabolism, genetic information processing, organism systems and cellular processes (S3 Table). The two largest pathways were 'metabolic pathways' and 'biosynthesis of secondary metabolites' with 5,288 and 2,552 All-unigenes, respectively. Two pathways contained only four All-unigenes each, 'caffeine metabolism' and 'betalain biosynthesis'; these were the least represented categories. Many All-unigenes corresponded to flower pigmentation related pathways involved in the biosynthesis of secondary metabolites, including 'flavonoid biosynthesis' (179, 0.74%), 'flavone and flavonol biosynthesis' (100, 0.41%), 'anthocyanin biosynthesis' (10, 0.04%) and 'isoflavonoid biosynthesis' (65, 0.27%). These annotations provide a valuable resource for the investigation of specific processes, functions and pathways in *P. delavayi*, and the genes in these flavonoid biosynthesis-related pathways might play an important role in flower coloration.

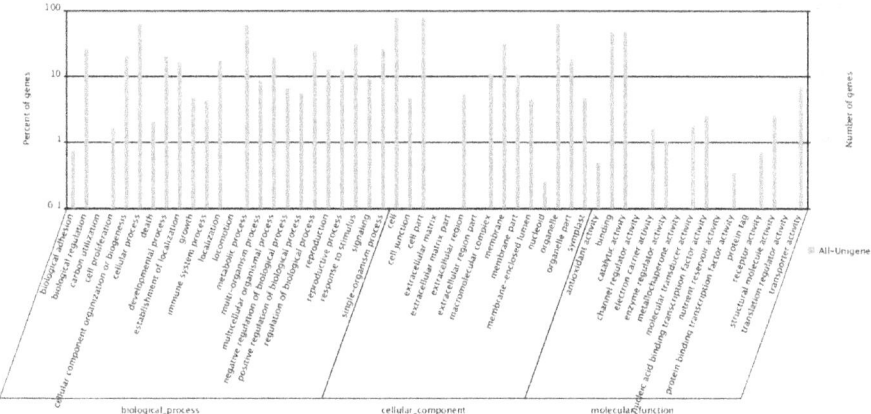

Figure 1. Gene ontology (GO) classification for the *P. delavayi* transcriptome. The transcripts (33,249) were categorized into 55 function groups. The right y-axis indicates the number of genes in a category, whereas the left y-axis indicates the percentage of a specific category of genes in the corresponding main category. doi:10.1371/journal.pone.0135038.g001

Transcripts that Encode Specific Genes Associated with Flower Pigmentation

To obtain unique sequences related to flower pigmentation, the non-redundant transcripts associated with anthocyanin biosynthesis were analyzed. As shown in Table 2, we found 66 All-unigenes encoding 11 enzymes involved in the 'flavonoid biosynthesis' pathway and potentially related to flower coloration, including 12, 8, 7, 13, 11 and 10 All-unigenes annotated as CHS, CHI, F3H, F3'H, DFR, and FLS, respectively, 2 All-unigenes annotated as flavone synthase II (FNS II), and only 1 All-unigene each aligned to 2'4'6'4-tetrahydroxychalcone 2'-glucosyltransferase (THC2'GT), F3'5'H, and ANS. Additionally, 10 All-unigenes encoding two enzymes involved in the 'anthocyanin biosynthesis' pathway, including 3GT and 5GT, were found.

Among these All-unigenes, one transcript of each multi-gene enzyme, CHS, CHI, DFR, F3H and ANS, had high sequence similarity to genes isolated from *P. suffruticosa*, *PsCHS1*(GenBank: GQ483511), *PsCHI1* (GenBank: GQ984161), *PsDFR1* (GenBank: HQ283448),*PsF3H1* (GenBank: HQ283449) and *PsANS1* (GenBank: HQ283446), respectively [20–22]. This is first time that any of these unigenes have been identified in *P. delavayi*.

Table 2. Putative unigenes related to flower pigmentation.

Pathway	Gene	Enzyme	Unigene number
Flavonoid biosynthesis	CHS	Chalcone synthase	12
	THC2'GT	2'4'6'4- tetrahydroxychalcone 2'-glucosyltransferase	1
		Chalcone isomerase	
	CHI	Flavanone 3-hydroxylase	8
	F3H	Flavanone 3'-hydroxylase	7
	F3'H	Flavonoid 3',5'-hydroxylase	13
	F3'5'H	Flavonol synthase	1
	FLS	Flavone synthase	10
	FNS II	Dihydroflavonol-4-reductase	2
	DFR	Anthocyanidin synthase	11
	ANS		1
Anthocyanin biosynthesis	3GT	Anthocyanin 3-O-glycosyltransferase	7
	5GT	Anthocyanin 5-O-glucosyltransferase	3

doi:10.1371/journal.pone.0135038.t002

Analysis of genes differentially expressed between the transcriptomes of Pl and Pd

To identify the candidate genes controlling *P. delavayi* yellow and purplered flower pigmentation, we performed differentially expressed gene (DEG) analysis by calculating fragments per kb per million fragments (FPKM) values [39].

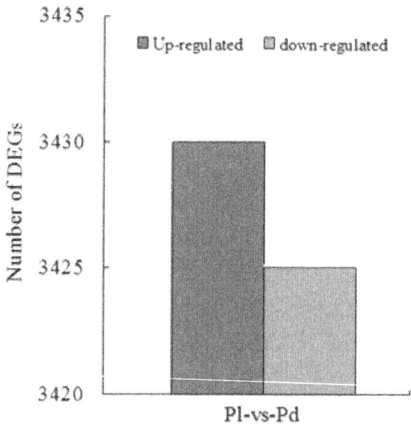

Figure 2. Summary of differentially expressed genes in pairwise comparisons between Pl and Pd.

doi:10.1371/journal.pone.0135038.g002

All-unigene differential expression analysis was carried out by comparing the two samples to predict genes with different expression levels. The thresholds false discovery rate (FDR) ≤ 0.001 and absolute value of $\log_2$ratio ≥ 1 were used to judge the significance of the DEGs (Pl vs. Pd). A total of 18,784 All-unigenes were identified as differentially expressed, among which 6,855 (36.5%) were aligned with the NR database, with 3,430 up-regulated and 3,425 down-regulated (Fig 2).

Functional analysis of DEGs

GO analysis provided functional classification annotation as well as functional enrichment analysis for the DEGs, and pathway analysis helped to further understand their main biological functions. As shown in Fig 3, 5,052 DEGs were categorized into 55 functional groups, including 24, 15 and 16 groups in the biological process, cellular component and molecular function categories, respectively. In biological process, the major classifications for these All-unigene products were 'metabolic process' (3,108, 61.5%) and 'cellular process' (3,052, 60.4%). The most frequent cellular component terms were 'cell' and 'cell part', each with the same number of All-unigenes (3,616, 71.6%). Most of the All-unigenes were classified into the molecular functions 'catalytic activity' (2,609, 51.6%) and 'binding' (2,303, 45.6%).

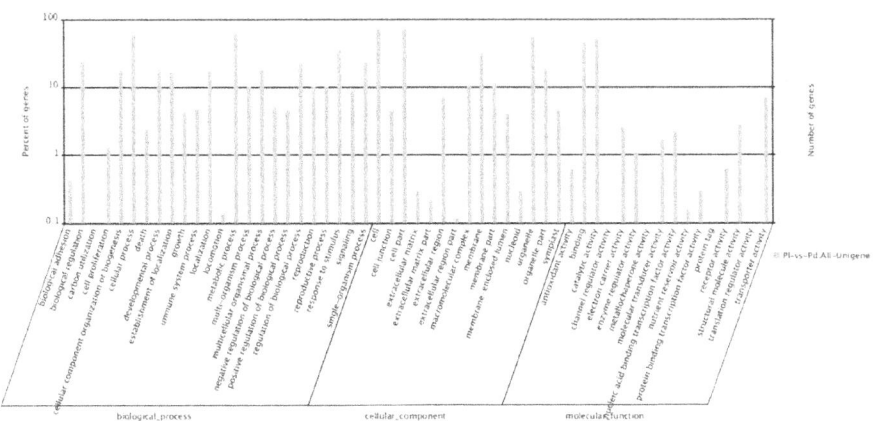

Figure 3. Gene ontology (GO) classification for the DEGs.

Many studies have shown that the coloration of yellow and purple-red *P. delavayi* petals is closely related to flavonoids [8,12,16]. Using the KEGG database, we mapped 3,969 of 18,784 DEGs to 127 pathways, and noted that 59, 39, and 5 DEGs were involved in 'flavonoid biosynthesis' (ko00941), 'flavone and flavonol biosynthesis' (ko00944) and 'anthocyanin biosynthesis' (ko00942), respectively, and all belonged to the flavonoid biosynthesis pathway. Based on NR, NT and Swiss-Prot annotations, GO functional analysis and KEGG pathway analysis, we subsequently predicted 15 up-regulated and 12 down-regulated All-unigenes involved in petal coloration (Table 3). The annotations for these unigenes included most of the structural genes in the anthocyanin biosynthetic pathway, and the proposed pathway showed that most steps in anthocyanin biosynthesis were up-regulated in purple-red petals (Fig 4).

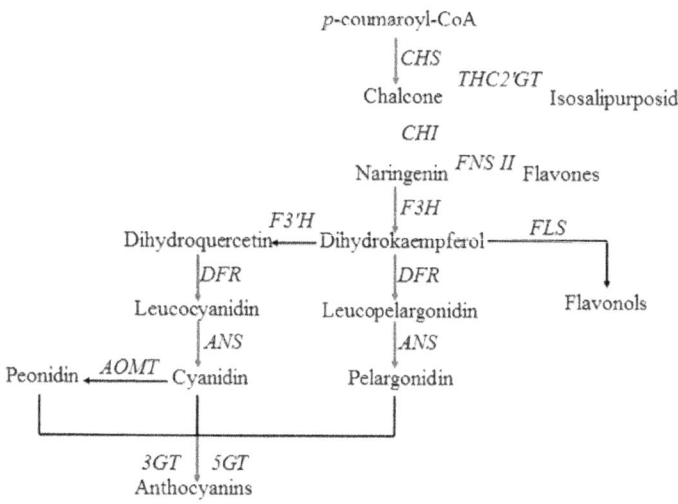

Figure 4. Simplified scheme of the flavonoid biosynthetic pathway in *P. delavayi*. Red and yellow lines indicate up-regulated genes in purple-red and yellow petals, respectively. doi:10.1371/journal.pone.0135038.g004

In the flavonoid biosynthetic pathway, the regulation of structural gene expression appears tightly organized in a spatial and temporal way during plant development, and is orchestrated by a ternary complex involving transcription factors from the R2R3-MYB, basic helix–loop–helix (bHLH), and WD40 classes [19]. Thus, besides structural genes, we also investigated the expression levels of these three transcription factor families. As shown in S4 Table, 50 differentially expressed All-unigenes were annotated as transcription factors between yellow and purple-red petals, including 21 All-unigenes encoding R2R3-MYBs, 12 encoding bHLHs and 17 encoding WD40s. Among these transcription factors, 21 were highly expressed in purple-red petals and 29 were highly expressed in yellow petals.

Table 3. Identification of 27 differentially expressed genes involved in *P. delavayi* flower pigmentation.

Unigene ID	PI_FPKM	Pd_FPKM	log2(Pd/PI)	p-value	FDR	Predicted function
Up-regulated genes (15)						
Unigene7271_All	1.0926	12.414	3.5061	1.41E-60	8.63E-59	CHS
Unigene45464_All	0	3.8864	11.9242	5.11E-07	3.18E-06	CHS
CL7622.Contig1_All	3.2381	10.0341	1.6317	3.16E-15	4.39E-14	CHI
CL2583.Contig1_All	201.7592	820.2684	2.0235	0	0	F3H
CL7561.Contig2_All	5.2308	11.3517	1.1178	4.92E-07	3.09E-06	FLS
CL1376.Contig3_All	0.2903	1.8999	2.7103	1.76E-08	1.34E-07	DFR
CL1376.Contig5_All	0.159	1.189	2.9027	5.71E-05	2.51E-04	DFR
CL1376.Contig6_All	3.0482	226.7622	6.2171	0	0	DFR
Unigene21659_All	1.8912	5.3039	1.4878	1.68E-09	1.43E-08	DFR
Unigene13390_All	62.6394	525.0521	3.0673	0	0	ANS
CL3395.Contig1_All	3.5064	8.4245	1.2646	6.32E-16	9.31E-15	3GT
CL3906.Contig2_All	0.8079	9.6622	3.5801	4.36E-72	3.15E-70	3GT
Unigene11227_All	17.9454	74.0118	2.0441	2.16E-242	5.32E-240	3GT
CL4890.Contig1_All	99.39	242.6426	1.2877	0	0	3GT
CL4890.Contig2_All	110.5854	253.0147	1.1941	0	0	3GT
Down-regulated genes (12)						
CL1498.Contig1_All	9.522	0.696	-3.7741	2.57E-54	1.41E-52	CHS
Unigene18441_All	37.3852	9.5524	-1.9683	5.21E-107	5.57E-105	THC2'GT
CL7622.Contig3_All	39.2223	19.029	-1.0435	5.65E-42	2.33E-40	CHI
Unigene30832_All	11.5703	2.6567	-2.1227	3.52E-08	2.57E-07	CHI
Unigene30833_All	9.5425	1.9625	-2.2817	6.85E-08	4.82E-07	CHI
CL8830.Contig1_All	110.819	0.5231	-7.7269	0	0	FNS II
CL8830.Contig2_All	25.5164	5.8693	-2.1202	2.31E-91	2.13E-89	FNS II
Unigene8606_All	1129.5084	318.1654	-1.8278	0	0	FLS
CL9282.Contig1_All	1.6743	0.3238	-2.3704	3.00E-05	1.39E-04	DFR
Unigene1005_All	8.1856	1.6398	-2.3196	7.78E-09	6.14E-08	3GT
Unigene33533_All	3.1914	0.5424	-2.5568	1.59E-04	6.33E-04	3GT
CL11011.Contig1_All	42.2327	7.2877	-2.5348	1.51E-83	1.26E-81	5GT

Notes: PI_FPKM, FPKM value in PI; Pd_FPKM, FPKM value in Pd.

doi:10.1371/journal.pone.0135038.t003

Expression Analysis of Candidate DEGs Probably Involved in Flavonoid Biosynthesis

To validate the unigene expression profiles from the transcriptome sequencing data, seven flavonoid biosynthesis-related All-unigenes that exhibited particularly strong expression in anthocyanin-pigmented petals compared with other flower organs, such as sepals, carpels, leaves, and stamens (data not shown), and showed the largest difference in expression level between purple-red and yellow petals, were selected for quantitative RT-PCR assays. To get more information in this analysis, we examined the expression patterns of these genes at five different developmental stages during flower opening. As shown in Fig 5, purple-red petals showed much higher gene expression levels than yellow petals for most of the selected genes, including *CHS*, *F3H*, *DFR*, *ANS* and *3GT*. Among them, *DFR* in particular was expressed at a statistically significantly higher level in purple-red petals than in yellow ones all through the flower pigmentation period, with $p < 0.01$ (Fig 5), suggesting that it might play an extremely important role in purple-red petal coloration.

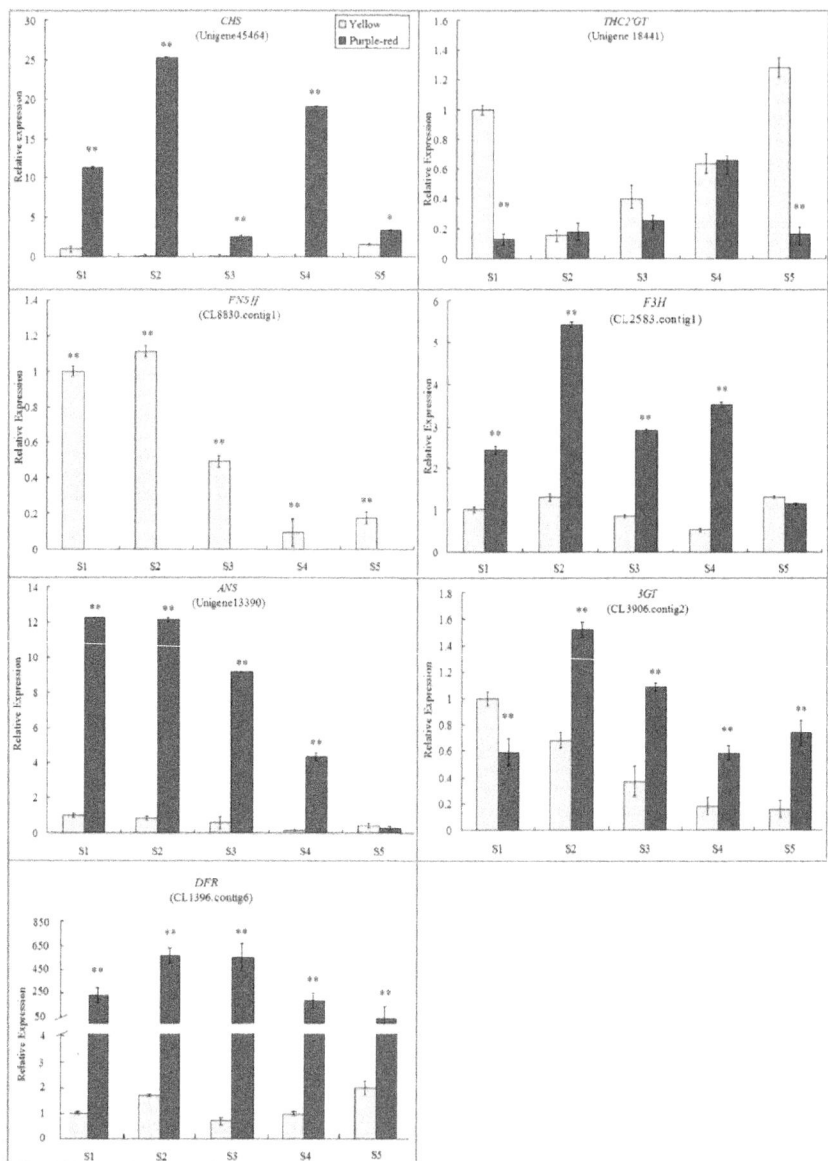

Figure 5. qRT-PCR analysis of seven flavonoid biosynthetic pathway-related candidate unigenes in Pl and Pd. qRT-PCR analysis was performed using total RNA from petals at each floral developmental stage (S1–S5). The expression in yellow petals at stage 1 was used as a calibration standard. Data from real-time PCR were normalized to the *helicase* gene expression value, and relative transcript levels are presented as means with standard errors (S.E.) of five replications. doi:10.1371/journal.pone.0135038. g005

Conversely, the expression levels of the other two selected genes were much higher in yellow petals (Fig 5). *THC2 GT* showed prominent expression at stage 1 prior to petal pigmentation and at stage 5 when the petals were fully pigmented, both with almost 8-fold greater expression than in purple-red petals at the corresponding stages, implying that a high expression level of this gene prior to petal pigmentation is essential for yellow pigment accumulation. For *FNS II*, the transcript levels were extremely high in yellow petals but hardly detected in purple-red ones throughout the entire process of flower pigmentation. These results demonstrated that the expression patterns of the seven selected DEGs analyzed by real-time RT-PCR were consistent with their respective digital expression data.

DISCUSSION

In recent years, as a cost-effective and efficient tool to discover novel genes and analyze molecular mechanisms, high-throughput mRNA sequencing based on the Illumina/Solexa platform has been widely used for transcriptome analysis of many ornamental plants without genomic sequence data, such as *P. suffruticosa* [33,34], *Gardenia jasminoides* [40],*Chrysanthemum lavandulifolium* [41] and *Zantedeschia aethiopica* [42].

Prior to this study, most sequencing efforts in wild tree peony species were based on EST sequencing; very few tags had been reported in public databases and little genetic or genomic information was available. Here, we report the results of a large-scale transcriptome analysis of petal pigmentation in an endemic species of China, *P. delavayi*, using RNA-Seq technology. Two cDNA libraries constructed from the petals of yellow (Pl) and purple-red (Pd) flowers at different opening stages were analyzed, leading to the identification of 44,811 unigenes that were annotated using known sequences in various databases and DEGs involved in flavonoid biosynthesis. The relative transcript levels of seven candidate DEGs were further validated using real-time PCR. To the best of our knowledge, this is the first attempt to use Illumina paired-end sequencing technology for the *de novo* sequencing and assembly of the *P. delavayi*flower transcriptome. We believe our data will provide important new insights and facilitate further studies of *P. delavayi* genes and their functions.

Compared with previous floral transcriptome studies in the tree peony *P. suffruticosa* [33, 34] and other woody ornamental plants, such as *Rosa* [43], *Camellia chekiangoleosa* [41] and *G.jasminoides* [40], we herein report more contigs and unigenes, suggesting that the wild tree peony species *P. delavayi* contains abundant gene resources. In previous studies of peach [44] and safflower [27] flowers, transcriptome dynamics associated with petal pigmentation were investigated. Millions of transcripts were generated from petals of different colors, and enriched functional category analysis based on GO annotations showed that genes associated with 'cell', 'cell part', 'binding', 'cellular processes', 'metabolic processes' and 'catalytic activity' were the highest-represented groups. In the present study, similar results were obtained (Fig 1), indicating that the *P. delavayi* flowers were undergoing rapid development and carrying out intensive metabolic activities with complicated regulation by many transcription factors during petal pigmentation.

The accumulation of secondary metabolites plays an important role in the formation and development of flower color, especially flavonoids [45,46]. Flavonoid biosynthesis is regulated by a series of structural and regulatory genes, and the structural genes directly control flavonoid biosynthesis and accumulation [47]. A number of researchers have demonstrated that the primary pigments related to *P. delavayi* flower color are flavonoid compounds, including chalcones, flavones and flavonols, and anthocyanins, mainly Pn3G5G, which contribute to a variety of colors, such as yellow, orange, and purple-red [8,12,13,16]. The flower coloration of yellow *P. delavayi* individuals is mainly due to the biosynthesis of the 2'-glucoside of 2',4',6',4-tetrapydroxychalcone (THC)-isosalipurposide (ISP), chrysoeriol and apigenin [8,16], whereas purple-red coloration is dependent on anthocyanin contents, particularly peonidin-3,5-glucosides (Pn3G5G) [12]. Therefore, in this work, unigenes participating in the flavonoid biosynthetic pathway were specifically selected and studied in detail, providing new insight into the regulation of flavonoid and anthocyanin biosynthesis in *P. delavayi* that should accelerate engineering of this pathway in the future.

As shown in Table 2, almost all of the structural genes in the main flavonoid biosynthesis pathway were identified, indicating that this pathway is well conserved in *P. delavayi*. We also observed that many of these genes appeared to form multi-gene families, implying that the genome of *P. delavayi*, like those of many other higher plants, has undergone one or more rounds of genome duplication during its evolution. Based on the results of unigene differential expression analysis, we predicted 27 DEGs as key structural genes involved in coloration from our sequencing data, including 15 up-regulated and 12 down-regulated genes (Table 3). In this pathway (Fig 4), CHS is the first committed enzyme, catalyzing the condensation of one molecule of 4-coumaroyl-CoA and three molecules of malonyl-CoA to generate 4,2',4',6'-tetrahydroxychalcone. We analyzed the expression patterns of 12 novel transcripts related to *CHS* and found that most of them, especially Unigene45464_All, were significantly up-regulated in purple-red petals (Fig 5). We propose that a sharp increase of transcripts encoding CHS may provide more substrate for anthocyanin biosynthesis, which is mainly responsible for petal coloration in purple-red flowers. 4,2′,4′,6′-Tetrahydroxychalcone is a key intermediate in *P.delavayi* flower coloration; it can be glycosylated by *THC2 GT* or isomerized by *CHI*, leading to the production of chalcone isosalipurposide and other flavonoids, such as flavones and flavonols, and anthocyanins, which contribute to the yellow and purple-red colors seen in *P.delavayi*, respectively. Mutations of the *CHI* gene have been shown to be necessary for isosalipurposide accumulation in carnation [48] and barley [49]. Additional mutation of the *DFR* gene confers better yellow coloration in carnation [50]. The existence of several genes encoding THC2′GT in carnation [51, 52] indicates that several non-specific enzymes may catalyze the reaction *in vivo*. In this work, only Unigene18441_ All showed sequence homology to *THC2′GT* from *Catharanthus roseus* (GenBank: BAF75901) [53], and its expression in yellow petals was almost 2-fold higher than in purple-red petals, suggesting an important role in yellow flower pigmentation. Interestingly, among the eight novel transcripts related to *CHI*, only one unigene (CL7622.Contig1_All) was down-regulated in yellow petals. Additionally, both of the transcripts (CL8830.Contig1_All and CL8830. Contig2_All) annotated as *FNS II*, which is responsible for flavone formation [54, 55], were obviously up-regulated in yellow petals. Park et al. [56] reported that in *Scutellaria baicalensis*, *SbCHI*-overexpressing hairy root lines not only showed enhanced *SbCHI* gene expression, but also produced more flavones than the control. Thus, considering that *P. delavayi* yellow flowers are pigmented by the yellow chalcone isosalipurposide and by colorless or yellow flavonol and flavone glycosides, we speculate that the high co-expression of *THC2′GT*, *CHI* and *FNS II* ensures the accumulation of pigments contributing to the yellow color.

Regarding the downstream enzymatic genes involved in the *P. delavayi* anthocyanin biosynthetic pathway, *F3H*, *DFR*, *ANS* and *3GT* were expressed at a significantly higher level in purple-red petals than in yellow petals according to both transcriptome profiling and qRT-PCR analysis (Table 3, Fig 5). We can see from Fig 4 that F3H catalyzes the formation of DHK, which can be further hydroxylated by F3›H to form DHQ. Then, DHK and DHQ are deoxidized to leucoanthocyanidins by DFR, leading to synthetic branches catalyzed by ANS and GT that produce corresponding pelargonidin-based (orange to red) and cyanidin-based (red to magenta) pigments, respectively. Cyanidin is the precursor pigment of other anthocyanidins, and can be transformed into peonidin derivatives by the action of a 3′-O-methyltransferase [57–59]. Because the purple-red color in *P. delavayi* flowers is mainly related to the synthesis of cyanidin-derived anthocyanins (cyanidin and peonidin based), our results suggest that high expression levels of *F3H*, *DFR*, *ANS* and *3GT* genes induce cyanidin and peonidin production to make flowers purple-red.

In contrast, the low transcription levels of these genes observed in yellow petals ensure the inhibition of anthocyanin formation. Thus, we propose that the molecular mechanism underlying the differences in *P. delavayi* flower pigmentation is related primarily to structural genes downstream of *CHI* in the flavonoid synthetic pathway. These results are consistent with previous studies on flower color in many other plants [60–62] and the petal pigments of yellow and purple-red *P. delavayi* individuals [8,12,16].

Transcriptional regulation of structural genes appears to be a major mechanism by which anthocyanin biosynthesis is regulated in plants. R2R3-MYB, bHLH and WD40 proteins represent the three classes of transcription factors influencing anthocyanin biosynthesis intensity and patterns, and generally controlling the expression of many different structural genes [63–65]. A large number of previous reports have demonstrated that transcription factors, especially R2R3-MYBs, play an important role in color differences in many crops, including grape [66], apple [67], cauliflower [68], oriental hybrid lily [69], and flowering peach [70]. In this study, we detected 50 differentially expressed transcription factors from these three classes, and found that some were significantly up-regulated in purple-red petals and showed positive correlation with *F3H, DFR, ANS* and *3GT* genes in the anthocyanin biosynthesis pathway (S4 Table). These results provide an informative list of candidate regulatory transcription factors involved in *P. delavayi* flower pigmentation, but further research related to these candidate regulatory genes including sequence, expression and biological function analyses needs to be carried out.

CONCLUSIONS

The Illumina HiSeq sequencing technology was used for sequencing and transcriptome analysis of the non-model plant *P. delavayi*, and provided an efficient approach for identifying critical genes involved in *P. delavayi* flower pigmentation. We identified potential candidate genes encoding key enzymes and reconstructed the flavonoid biosynthetic pathway in *P.delavayi*. The up-regulated unigenes encoding *F3H, DFR, ANS* and *3GT* might play an important role in purple-red petal pigmentation, while high co-expression of *THC2›GT, CHI* and*FNS II* ensures the accumulation of pigments contributing to the yellow color. In addition, differentially expressed transcription factors related to anthocyanin biosynthesis were detected. Our results will provide valuable resources and a substantial basis for understanding the molecular mechanisms controlling *P. delavayi* flower pigmentation, and will eventually accelerate the genetic engineering of flower color in tree peony.

METHODS

Ethics Statement

All Plant protocols were reviewed and approved by the Research Institute of Forestry, Chinese Academy of Forestry, State Forestry Administration. All necessary permits were obtained for field studies from the Diqing Forestry Bureau, Shangri-La County, Yunnan Province, China. The fieldwork conducted for sampling did not affect the local ecology and did not involve endangered species.

Plant materials

Petal samples were separately detached from purple-red-flowered and yellow-flowered individuals from a wild *P. delavayi* population in Shangri-La County (27°57′N, 99°35′E), Yunnan Province, China (Fig 6) at five different opening stages in the afternoon during the end of April to early May, 2013. The flower opening stages were described by Zhou et al. [20]: stage 1, unpigmented tight bud; stage 2, slightly pigmented soft bud; stage 3, initially opened flower; stage 4, half opened flower; stage 5, fully opened and pigmented flower with exposed anthers. The petals were immediately frozen in liquid nitrogen and after transport to the laboratory stored at −80°C until RNA extraction.

Figure 6. Fully open flowers of individuals selected for sequencing. A, yellow flowered individual; B, purple-red flowered individual. doi:10.1371/journal.pone.0135038. g006

RNA Extraction, cDNA library Construction and Transcriptome Sequencing

Total RNA was extracted from petals by the cetyltrimethylammonium bromide (CTAB) method with some modification [71]. RNase-free DNase I (Tiangen;

Beijing, China) was applied to remove contaminating genomic DNA. The RNA purity was determined with a NanoDrop 2000 spectrophotometer (NanoDrop Technologies; Wilmington, DE, USA) and 2% agarose gels were run to verify RNA integrity. Two cDNA libraries were constructed by mixing equal amounts of RNA from purple-red or yellow petals at five different opening stages, and named Pd and Pl, respectively.

Enrichment of mRNA, fragment interruption, addition of adapters, size selection and PCR amplification and RNA-Seq were performed by staff at the Beijng Genome Institute (BGI; Shenzhen, China). mRNA was isolated with magnetic oligo (dT) beads. The mRNA was mixed with fragmentation buffer, fragmented into short fragments and used as a template for first-strand and second-strand cDNA synthesis. The cDNA fragments were purified using a QiaQuick PCR extraction kit (Qiagen; Valencia, CA, USA), and resolved with elution buffer for end reparation and single nucleotide A (adenine) addition. The short fragments were connected with adapters and suitable fragments were selected for PCR amplification as templates. An Agilent 2100 Bioanalyzer (Agilent Technologies; Palo Alto, CA, USA) and an ABI StepOnePlus Real-Time PCR System (Applied Biosystems; Foster City, CA, USA) were used during the quality control steps. Finally, the two libraries were sequenced at the BGI using an Illumina HiSeq 2000 sequencing platform (Illumina Inc.; San Diego, CA, USA).

De novo Assembly and Sequence Analysis

The raw reads were first filtered by removing adaptor sequences, empty reads, repeated reads and low-quality reads to get clean reads for subsequent analysis. The sequencing data for the clean reads were deposited in the National Center for Biotechnology Information (NCBI) Sequence Read Archive (http://www. ncbi.nlm.nih.gov/Traces/sra) with accession number SRP052291. Then, *de novo* assembly of the clean reads was performed to generate non-redundant unigenes using the Trinity software with an optimized k-mer length of 25 [35].

For sequence analysis, the resulting unigenes were aligned by Blastx to several protein databases (e-value < 0.00001) such as NCBI NR, Swiss-Prot, KEGG and COG, and the best-aligned results were used to decide the sequence direction of the unigenes. If the results from different databases conflicted, a priority order of NR > Swiss-Prot > KEGG > COG was followed when deciding the sequence direction of the unigenes. If a unigene was unaligned to any of the above databases, the ESTScan software was used to decide its sequence direction [37]. For unigenes with sequence directions, we recorded their sequences from the 5′ end to the 3′ end; for those without any direction, the sequences were provided by the assembly software. The length of the sequences assembled was a criterion of assembly success. The contig and unigene length distributions were calculated.

Functional Annotation of Unigenes

To obtain functional annotations, unigene sequence-based alignments were performed against three public databases (NR, Swiss-Prot and KEGG; e-value < 0.00001), and domain-based alignments were carried out against the COG database (e-value < 0.00001). Additionally, a Blastn search was performed against the NCBI NT database with an e-value < 0.00001. The proteins with the highest sequence similarity to the unigenes were retrieved for analysis. KEGG mapping was used to study the complex biological behavior and obtain pathway annotations for the unigenes, while COG matched each annotated sequence to an ancient conserved domain, and was used to predict and classify the possible functions of the unigenes. Based on NR annotations, the Blast2GO program (version 2.3.5, http://www.blast2go.de/) was used to retrieve associated GO terms describing biological processes, molecular functions and cellular components [72]. After getting GO annotations, the WEGO (http://wego.genomics.org.cn/cgi-bin/wego/index.pl) software was used to perform GO functional classification for all unigenes and to understand the distribution of gene functions of the species at the macro level [73].

Gene Expression Analysis Using RNA-Seq Data

Unigene expression was calculated using the FPKM method [39], which eliminates the influence of different gene lengths and sequencing level on the calculation of gene expression. The false discovery rate (FDR) control is a statistical method used in multiple hypothesis testing to correct for p-value. When we obtained the FDR, we used the ratio of FPKMs from the two samples at the same time.

The smaller the FDR and the larger the ratio, the larger the difference in expression level between the two samples; therefore, for this analysis, thresholds of FDR ≤ 0.001 and ratio > 2 were chosen to judge the DEGs (Pl vs. Pd) [74]. The confirmed DEGs were subjected to GO functional enrichment analysis and KEGG pathway analysis. Then, based on NR, NT and Swiss-Prot annotations, GO functional analysis, KEGG pathway analysis and flower coloration studies in tree peony [8,11–15,20–22], the DEGs involved in petal coloration were screened for up/down-regulated unigenes, among which those DEGs with the highest absolute values for the FPKM ratio (Pl vs. Pd) were predicted as key genes for petal coloration and used for further analysis.

Quantitative Real-time PCR (qRT-PCR) Analysis

qRT-PCR was performed to verify the expression of seven genes probably involved in purple-red and yellow flower pigmentation in P $delavayi$. Total RNA was extracted from petals at five different opening stages. After treatment with RNase-free DNase I (Tiangen; Beijing, China) according to the user manual, 1 μg of total RNA was reverse-transcribed to first-strand cDNA using the PrimeScript RT reagent kit (Takara; Otsu, Japan). qRT-PCR experiments were performed in a 96-well plate with an ABI Prism 7500 Sequence Detector (Applied Biosystems, USA), using a SYBR Premix Ex TaqKit (Takara; Otsu, Japan) to monitor cDNA amplification, according to the manufacturer's protocol. As a control, parallel amplification reactions for the tree peony housekeeping gene $helicase$ (GenBank: EF608942) were also performed. Each primer set was designed based on the 3'-end cDNA sequence of the corresponding gene using the Primer Premier 5.0 software; all primer sequences are listed in S5 Table. Each 20 μL PCR reaction contained 10 μL of SYBR Premix Ex Taq (2×), 2 μL of 20× diluted RT-product, 0.8 μL of each forward and reverse primers (10 μM), 0.4 μL of ROX Reference Dye II, and 6 μL of ddH$_2$O. The thermal conditions were 95°C for 30 s and 40 cycles of 95°C for 5 s and 60°C for 35 s, and then 95°C for 15 s, 60°C for 20 s and 95°C for 15 s for the dissociation stage. After the real-time PCR, the absence of unwanted by-products was confirmed by automated melting curve analysis and agarose gel electrophoresis of the PCR products. In all experiments, five replicates for each RNA sample were included; averages were calculated and differences in the threshold cycle (Ct) were evaluated by the 7500 System Sequence Detection Software v1.3.1.

For data analysis, the relative expression ratios were calculated by the comparative $\Delta\Delta Ct$ method (ABI Prism 7500 Sequence Detection System, Applied Biosystems, USA) of relative gene quantification. To monitor the expression patterns of the seven selected transcripts during flower pigmentation, the relative quantification of gene expression was achieved by calibrating the transcription level in petals at different stages to that in yellow petals at stage 1. The expression level calculated by the formula $2^{-\Delta\Delta Ct}$ represents the x-fold difference from the calibrator.

Supporting Information

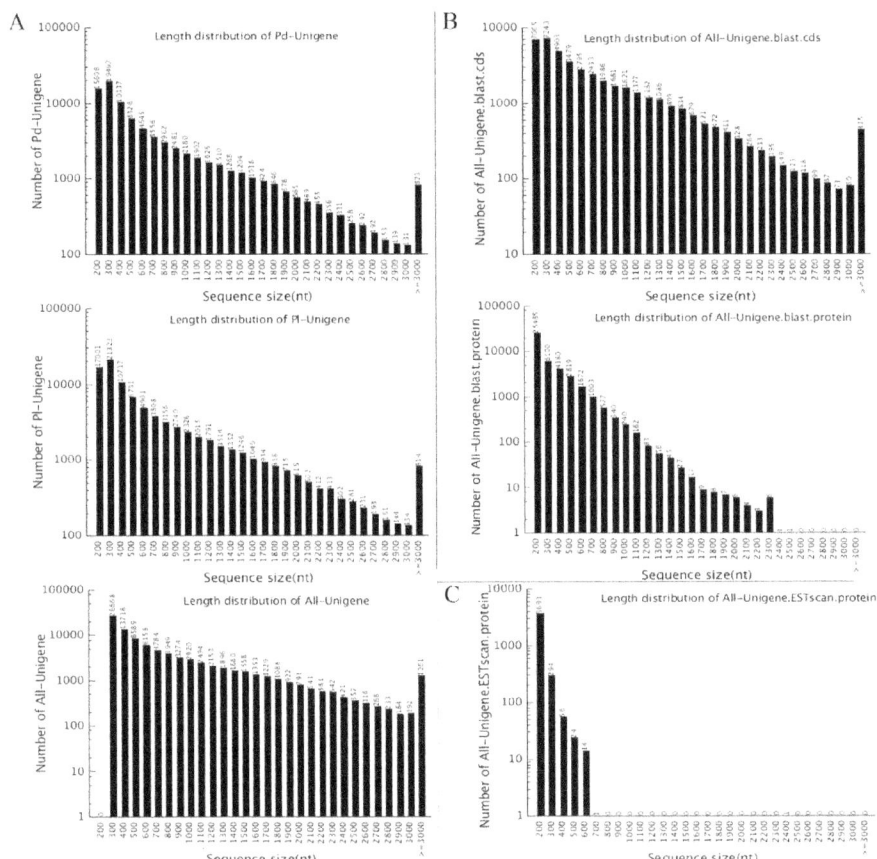

S1 Figure. Overview of Pl and Pd transcriptome assembly. A, Length distribution of the unigenes obtained from our *de novo* assembly of contigs; B, length distribution of the CDSs produced by searching unigene sequences against various protein databases and proteins predicted from the CDSs; C, length distribution of ESTs obtained from the ESTScan results.

doi:10.1371/journal.pone.0135038.s001

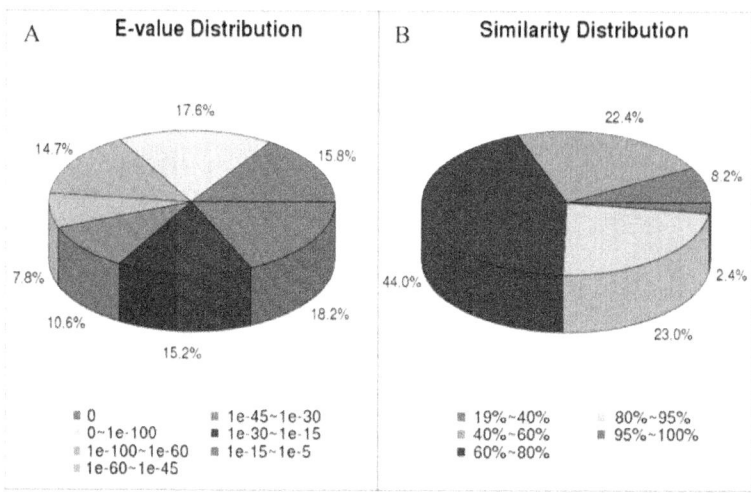

S2 Figure. Characteristics of the homology search of the unigenes. E-value (A) and similarity (B) distributions of the top Blastx hits against the NR database for each unigene.

doi:10.1371/journal.pone.0135038.s002

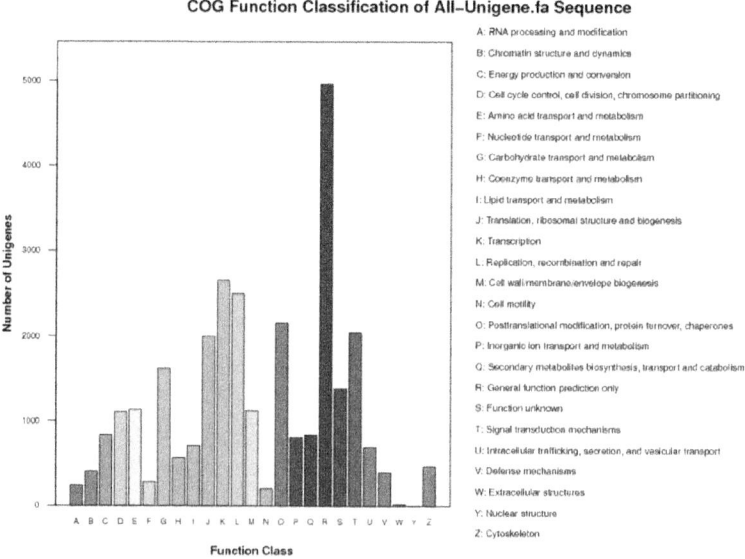

S3 Figure. Classification of the clusters of orthologous groups (COG) for the *P. delavayi* transcriptome.

The unigenes (15,073) were annotated and divided into 25 specific categories.

doi:10.1371/journal.pone.0135038.s003

S1 Table. Statistics for the annotation results.

Database	NR	NT	Swiss-Prot	KEGG	COG	GO	ALL
Annotated number	42,699	36,155	26,496	24,262	15,073	33,249	44,811
Annotated percentage	47.34%	40.08%	29.37%	26.90%	16.71%	36.86%	49.68%

S2 Table. Species distribution of the top Blastx hits in the NR database.

Species	Gene numbers	Percentage
Vitis vinifera	24,727	57.91%
Ricinus communis	4,712	11.04%
Populus trichocarpa	4,586	10.74%
Glycine max	2,127	4.98%
Medicago truncatula	1,004	2.35%
Arabidopsis thaliana	460	1.08%
Arabidopsis lyrata subsp. lyrata	335	0.78%
other	4,747	11.12%

S3 Table. Metabolic pathway analysis of the *P. delavayi* unigenes conducted using the Kyoto Encyclopedia of Genes and Genomes (KEGG) annotation system.

Code	Pathway	All genes with pathway annotation -24262	Pathway ID
1	Metabolic pathways	5288 (21.80%)	ko01100
2	Biosynthesis of secondary metabolites	2552 (10.52%)	ko01110
3	Plant-pathogen interaction	1282 (5.28%)	ko04626
4	Plant hormone signal transduction	1169 (4.82%)	ko04075
5	Spliceosome	1032 (4.25%)	ko03040
6	RNA transport	927 (3.82%)	ko03013
7	Ribosome	731 (3.01%)	ko03010
8	Protein processing in endoplasmic reticulum	669 (2.76%)	ko04141
9	Endocytosis	664 (2.74%)	ko04144

10	Glycerophospholipid metabolism	632 (2.60%)	ko00564
11	Starch and sucrose metabolism	612 (2.52%)	ko00500
12	RNA degradation	559 (2.30%)	ko03018
13	Pyrimidine metabolism	540 (2.23%)	ko00240
14	Purine metabolism	536 (2.20%)	ko00230
15	Ribosome biogenesis in eukaryotes	482 (1.99%)	ko03008
16	Ubiquitin mediated proteolysis	468 (1.92%)	ko04120
17	mRNA surveillance pathway	467 (1.93%)	ko03015
18	Ether lipid metabolism	455 (1.88%)	ko00565
19	Phenylpropanoid biosynthesis	359 (1.48%)	ko00940
20	ABC transporters	320 (1.32%)	ko02010
21	Oxidative phosphorylation	291 (1.20%)	ko00190
22	Glycolysis / Gluconeogenesis	290 (1.19%)	ko00010
23	Pentose and glucuronate interconversions	285 (1.18%)	ko00040
24	Amino sugar and nucleotide sugar metabolism	278 (1.15%)	ko00520
25	RNA polymerase	276 (1.14%)	ko03020
26	Phagosome	238 (0.98%)	ko04145
27	Pyruvate metabolism	232 (0.96%)	ko00620
28	Nucleotide excision repair	221 (0.91%)	ko03420
29	Homologous recombination	211 (0.87%)	ko03440
30	DNA replication	195 (0.80%)	ko03030
31	Stilbenoid, diarylheptanoid and gingerol biosynthesis	191 (0.79%)	ko00945
32	Zeatin biosynthesis	189 (0.78%)	ko00908

33	Circadian rhythm - plant	186 (0.77%)	ko04712
34	Terpenoid backbone biosynthesis	186 (0.77%)	ko00900
35	Flavonoid biosynthesis	179 (0.74%)	ko00941
36	Basal transcription factors	179 (0.74%)	ko03022
37	Phosphatidylinositol signaling system	178 (0.73%)	ko04070
38	Carotenoid biosynthesis	173 (0.71%)	ko00906
39	Galactose metabolism	171 (0.71%)	ko00052
40	Cyanoamino acid metabolism	170 (0.70%)	ko00460
41	Peroxisome	169 (0.70%)	ko04146
42	Inositol phosphate metabolism	169 (0.70%)	ko00562
43	Carbon fixation in photosynthetic organisms	169 (0.70%)	ko00710
44	Cysteine and methionine metabolism	155 (0.64%)	ko00270
45	Limonene and pinene degradation	149 (0.61%)	ko00903
46	Phenylalanine metabolism	146 (0.60%)	ko00360
47	Ascorbate and aldarate metabolism	145 (0.60%)	ko00053
48	Glutathione metabolism	144 (0.59%)	ko00480
49	Base excision repair	144 (0.59%)	ko03410
50	Glycine, serine and threonine metabolism	144 (0.59%)	ko00260
51	Arginine and proline metabolism	144 (0.59%)	ko00330
52	Other glycan degradation	143 (0.59%)	ko00511
53	Aminoacyl-tRNA biosynthesis	143 (0.59%)	ko00970
54	Regulation of autophagy	142 (0.59%)	ko04140

55	Glycerolipid metabolism	141 (0.58%)	ko00561
56	Photosynthesis	139 (0.57%)	ko00195
57	Propanoate metabolism	134 (0.55%)	ko00640
58	alpha-Linolenic acid metabolism	134 (0.55%)	ko00592
59	Citrate cycle (TCA cycle)	133 (0.55%)	ko00020
60	Mismatch repair	132 (0.54%)	ko03430
61	Fructose and mannose metabolism	131 (0.54%)	ko00051
62	Glycosylphosphatidy-linositol(GPI)-anchor biosynthesis	130 (0.54%)	ko00563
63	Glyoxylate and dicarboxylate metabolism	125 (0.52%)	ko00630
64	Alanine, aspartate and glutamate metabolism	122 (0.50%)	ko00250
65	Tyrosine metabolism	117 (0.48%)	ko00350
66	Pentose phosphate pathway	116 (0.48%)	ko00030
67	Proteasome	116 (0.48%)	ko03050
68	Valine, leucine and isoleucine degradation	115 (0.47%)	ko00280
69	Cutin, suberine and wax biosynthesis	111 (0.46%)	ko00073
70	Fatty acid metabolism	111 (0.46%)	ko00071
71	Nitrogen metabolism	109 (0.45%)	ko00910
72	Protein export	105 (0.43%)	ko03060
73	Porphyrin and chlorophyll metabolism	104 (0.43%)	ko00860
74	Phenylalanine, tyrosine and tryptophan biosynthesis	104 (0.43%)	ko00400
75	Flavone and flavonol biosynthesis	100 (0.41%)	ko00944
76	Steroid biosynthesis	99 (0.41%)	ko00100
77	Sphingolipid metabolism	92 (0.38%)	ko00600
78	N-Glycan biosynthesis	91 (0.38%)	ko00510

79	Pantothenate and CoA biosynthesis	91 (0.38%)	ko00770
80	SNARE interactions in vesicular transport	90 (0.37%)	ko04130
81	Natural killer cell mediated cytotoxicity	88 (0.36%)	ko04650
82	Fatty acid biosynthesis	86 (0.35%)	ko00061
83	Ubiquinone and other terpenoid-quinone biosynthesis	83 (0.34%)	ko00130
84	beta-Alanine metabolism	82 (0.34%)	ko00410
85	Tryptophan metabolism	81 (0.33%)	ko00380
86	Diterpenoid biosynthesis	79 (0.33%)	ko00904
87	Biosynthesis of unsaturated fatty acids	75 (0.31%)	ko01040
88	Sulfur metabolism	68 (0.28%)	ko00920
89	Lysine degradation	66 (0.27%)	ko00310
90	Isoflavonoid biosynthesis	65 (0.27%)	ko00943
91	Tropane, piperidine and pyridine alkaloid biosynthesis	65 (0.27%)	ko00960
92	Valine, leucine and isoleucine biosynthesis	65 (0.27%)	ko00290
93	Sesquiterpenoid and triterpenoid biosynthesis	64 (0.26%)	ko00909
94	Glycosaminoglycan degradation	56 (0.23%)	ko00531
95	Riboflavin metabolism	55 (0.23%)	ko00740
96	Butanoate metabolism	55 (0.23%)	ko00650
97	Brassinosteroid biosynthesis	53 (0.22%)	ko00905
98	Selenocompound metabolism	50 (0.21%)	ko00450
99	Circadian rhythm - mammal	50 (0.21%)	ko04710

100	One carbon pool by folate	50 (0.21%)	ko00670
101	Benzoxazinoid biosynthesis	48 (0.20%)	ko00402
102	Folate biosynthesis	47 (0.19%)	ko00790
103	Fatty acid elongation	45 (0.18%)	ko00062
104	Histidine metabolism	43 (0.18%)	ko00340
105	Glycosphingolipid biosynthesis - ganglio series	41 (0.17%)	ko00604
106	Photosynthesis - antenna proteins	40 (0.17%)	ko00196
107	Isoquinoline alkaloid biosynthesis	40 (0.17%)	ko00950
108	Linoleic acid metabolism	37 (0.15%)	ko00591
109	Non-homologous end-joining	35 (0.14%)	ko03450
110	Nicotinate and nicotinamide metabolism	35 (0.14%)	ko00760
111	Lysine biosynthesis	34 (0.14%)	ko00300
112	Vitamin B6 metabolism	34 (0.14%)	ko00750
113	Sulfur relay system	27 (0.11%)	ko04122
114	Other types of O-glycan biosynthesis	27 (0.11%)	ko00514
115	Arachidonic acid metabolism	23 (0.10%)	ko00590
116	Thiamine metabolism	23 (0.10%)	ko00730
117	Indole alkaloid biosynthesis	23 (0.10%)	ko00901
118	Taurine and hypotaurine metabolism	22 (0.09%)	ko00430
119	Monoterpenoid biosynthesis	21 (0.09%)	ko00902
120	Glucosinolate biosynthesis	21 (0.09%)	ko00966
121	Glycosphingolipid biosynthesis - globo series	19 (0.08%)	ko00603

122	Synthesis and degradation of ketone bodies	15 (0.06%)	ko00072
123	C5-Branched dibasic acid metabolism	13 (0.05%)	ko00660
124	Biotin metabolism	13 (0.05%)	ko00780
125	Lipoic acid metabolism	10 (0.04%)	ko00785
126	Anthocyanin biosynthesis	10 (0.04%)	ko00942
127	Caffeine metabolism	4 (0.02%)	ko00232
128	Betalain biosynthesis	4 (0.02%)	ko00965

S4 Table. Three transcription factors families differentially expressed between yellow and purple-red petals.

Unigene ID	Pl_FPKM	Pd_FPKM	log2(Pd/Pl)	FDR	Preicted function
Up-regulated genes(21)					
Unigene1549_All	3.4755	20.0513	2.5284	5.95E-58	R2R3-MYB transcription factor
CL3425.Contig1_All	1.6467	10.4238	2.6622	7.82E-34	R2R3-MYB transcription factor
Unigene45385_All	0.222	6.2782	4.8217	3.43E-28	R2R3-MYB transcription factor
Unigene4871_All	1.5187	6.3579	2.0657	2.75E-11	R2R3-MYB transcription factor
CL7842.Contig3_All	1.6929	5.539	1.7101	4.25E-09	R2R3-MYB transcription factor
Unigene46605_All	0.7075	3.902	2.4634	3.43E-08	R2R3-MYB transcription factor
Unigene12746_All	1.1644	5.4428	2.2248	1.81E-07	R2R3-MYB transcription factor
Unigene43869_All	0.5595	6.2769	3.4878	4.25E-07	R2R3-MYB transcription factor
Unigene46917_All	0	3.4588	11.7561	2.02E-05	R2R3-MYB transcription factor
Unigene42027_All	0	2.4331	11.2486	4.17E-04	R2R3-MYB transcription factor
CL4938.Contig2_All	5.7626	21.9523	1.9296	1.50E-62	bHLH transcription factor
Unigene9596_All	3.6358	13.5955	1.9028	2.27E-30	bHLH transcription factor

Unigene9293_All	7.6843	16.3897	1.0928	2.95E-10	bHLH transcrip-tion factor
CL9010.Con-tig1_All	1.5917	4.1257	1.3741	2.45E-04	bHLH transcrip-tion factor
Unigene20108_All	0.9557	3.5735	1.9027	4.27E-04	bHLH transcrip-tion factor
CL10265.Con-tig2_All	0.3908	4.3839	3.4877	1.28E-24	WD-repeat protein
Unigene44731_All	0	5.5337	12.434	3.07E-12	WD-repeat protein
CL2787.Con-tig1_All	2.9512	9.0466	1.6161	4.20E-11	WD-repeat protein
CL11280.Con-tig2_All	2.55	5.8947	1.2089	3.13E-07	WD-repeat protein
Unigene40185_All	0	0.7572	9.5645	4.94E-07	WD-repeat protein
CL4613.Con-tig2_All	0.1751	2.6196	3.9031	9.40E-04	WD-repeat protein
Down-regulated genes(29)					
CL1812.Con-tig1_All	556.0525	247.1279	-1.17	0	R2R3-MYB tran-scription factor
CL7988.Con-tig2_All	114.3173	54.4534	-1.0699	1.77E-170	R2R3-MYB tran-scription factor
Unigene21937_All	33.9382	4.8756	-2.7993	2.33E-73	R2R3-MYB tran-scription factor
Unigene22826_All	56.8975	25.058	-1.1831	1.05E-57	R2R3-MYB tran-scription factor
Unigene24722_All	7.0775	0	-12.789	1.99E-18	R2R3-MYB tran-scription factor
CL893.Con-tig4_All	3.3265	0.1794	-4.2128	1.80E-12	R2R3-MYB tran-scription factor
CL893.Con-tig5_All	3.5366	0.1787	-4.3068	3.93E-09	R2R3-MYB tran-scription factor
Unigene27706_All	4.2215	0.6964	-2.5998	3.31E-08	R2R3-MYB tran-scription factor
Unigene25840_All	4.4618	0	-12.1234	4.38E-05	R2R3-MYB tran-scription factor
CL964.Con-tig2_All	1.6658	0.3775	-2.1417	1.29E-04	R2R3-MYB tran-scription factor
Unigene2042_All	3.5308	1.5194	-1.2165	4.21E-04	R2R3-MYB tran-scription factor
Unigene18417_All	400.2916	171.8771	-1.2197	4.90E-266	bHLH transcrip-tion factor

CL4938.Con-tig1_All	18.6584	3.6652	-2.3479	2.94E-71	bHLH transcription factor
CL6977.Con-tig1_All	8.043	3.858	-1.0599	7.69E-12	bHLH transcription factor
Unigene18995_All	4.7929	0.974	-2.2989	1.63E-09	bHLH transcription factor
CL7910.Con-tig1_All	2.9525	0.72	-2.0359	8.70E-06	bHLH transcription factor
Unigene2947_All	4.4353	2.0981	-1.0799	2.20E-04	bHLH transcription factor
CL2997.Con-tig7_All	2.1435	0.5344	-2.004	5.30E-04	bHLH transcription factor
CL7021.Con-tig3_All	8.4055	0.6842	-3.6188	2.30E-74	WD-repeat protein
Unigene9399_All	18.7951	6.037	-1.6385	1.35E-16	WD-repeat protein
CL7021.Con-tig1_All	1.3432	0.2511	-2.4193	7.66E-09	WD-repeat protein
CL11280.Con-tig1_All	2.7106	0.489	-2.4707	1.26E-08	WD-repeat protein
CL1246.Con-tig6_All	1.1468	0.0975	-3.5561	4.79E-05	WD-repeat protein
CL10553.Con-tig2_All	3.8032	1.4858	-1.356	6.44E-05	WD-repeat protein
Unigene21299_All	8.453	4.1123	-1.0395	9.09E-05	WD-repeat protein
CL1246.Con-tig4_All	0.4181	0	-8.7077	3.25E-04	WD-repeat protein
Unigene33028_All	4.4189	1.0327	-2.0973	5.79E-04	WD-repeat protein
CL1246.Con-tig5_All	0.3895	0	-8.6055	6.28E-04	WD-repeat protein
CL1246.Con-tig3_All	0.8904	0.1998	-2.1559	9.98E-04	WD-repeat protein

S5 Table. Gene names, sequences and all primers used for qRT-PCR analysis.

Gene	Unigene ID	Forward primer(5'-3')	Reverse primer (5'-3')	Size of product (bp)
CHS	Unigene45464	CAATCATGGCAATTG-GAACA	GCTCGACCTTGTCCT-CACTG	102
THC2'GT	Unigene18441	TGTATCGGCAGTG-GAGTTCG	TATCGGAGGCAGC-GAGTGAT	131

FNS II	CL8830. contig1	ACTTCATCCCAGCACA-CACAC	TCACCCTCACAC-GACTCCAA	121
F3H	CL2583. contig1	CAGCAAC-GAAATCCCAATCA	GCGGTAGAGC-GAAGAACTCG	185
DFR	CL1376. contig6	ATGCCGAAACCGTGT-GTGT	CAAAT-GCTTCACCTTCCTCGT	134
ANS	Unigene13390	GCAAACCAGCATCAC-CAACA	GGCC-GCTTTCTTCAACTCCT	139
3GT	CL3906. contig1	TTCCAGCGAAATCC-GTCATT	ACCTCCGCAA-CAGAATCCAA	154
helicase	CL6029. Contig2	GAGTGCGGGTT-GAATCGTTG	AAGATTTCTGGATAG-GTCTGTGGC	160

Acknowledgments

This work was supported by the National High Technology Research and Development Program of China (863 Program) (2011AA10020701) and the National Natural Science Foundation of China (31201654).

Author Contributions

Conceived and designed the experiments: LZ YW QS. Performed the experiments: QS LZ. Analyzed the data: QS LZ. Contributed reagents/materials/analysis tools: KL BZ KM. Wrote the paper: QS LZ.

REFERENCES

1. Li JJ. Chinese Tree Peony and Herbaceous Peony. Beijing: China Forestry Publishing House; 1999

2. Hong DY, Pan KY. Notes on taxonomy of Paeonia sect. Moutan DC. (Paeoniaceae). Acta Phytotax Sin. 2005a; 43: 169–177.

3. Hong DY, Pan KY. Additional taxonomy motes on Paeonia sect. Moutan (Paeoniaceae). Acta Phytotax Sin. 2005b; 43: 284–287.

4. Hong DY, Pan KY. Paeonia cathayana D.Y. Hong & K. Y. Pan, a new tree peony, with revision of P.suffruticosa ssp.yinpingmudan. Acta Phytotax Sin. 2007; 43: 284–287. 5. Hong DY, Pan KY, Yu H. Taxonomy of the Paeonia delavayi complex (Paeoniaceae). Annals of the Missouri Botanical Garden. 1998; 85: 554–564.

5. Hong DY, Pan KY. Taxonomical history and revision of Paeonia sect. Moutan (Paeoniaceae). Acta Phytotax Sin. 1999, 37(4): 351–368.

6. Li K, Wang Y, Zheng BQ, Zhu XT, Wu HZ,Shi QQ. Pollen morphology of 40 Paeonia delavayi (Paeoniaceae) populations. Journal of Beijing Forestry University. 2011; 33 (1): 94–103.

7. Li CH, Du H, Wang LS, Shu QY, Zheng YR, Xu YJ. Flavonoid composition and antioxidant activity of tree peony (Paeonia section Moutan) yellow flowers. J Agric Food Chem. 2009; 57: 8496–8503. doi: 10.1021/jf902103b PMID: 19711909

8. Harborne JB, Williams CA. Advances in flavonoid research since 1992. Phytochemistry. 2000; 55 (6): 481–504. PMID: 11130659

9. Grotewold JME, Koes R. How genes paint flowers and seeds. Trends Plant Sci. 1998; 3: 212–217.

10. Hosoki T, Hamada M, Kando T, Moriwaki R, Inaba K. Comparative study of anthocyanin in tree peony flowers. J Jap Soc Hol Sci. 1991; 60: 395–403.

11. Wang LS, Hashimoto F, Shiraishi A, Aoki N, Li JJ, Shimizu K. Phenetics in tree peony species from China by flower pigment cluster analysis. J Plant Res. 2001a; 114: 213–221.

12. Wang LS, Shiraishi A, Hashimoto F, Aoki N, Shimizu K, Sakata Y. Analysis of petal anthocyanins to investigate flower coloration of Zhongyuan (Chinese) and Daikon Island (Japanese) tree peony cultivars. J Plant Res. 2001b; 114: 33–43.

13. Wang X, Cheng CG, Sun QL, Li FW, Liu JH, Zheng CC. Isolation and purification of four flavonoid constituents from the flower of Paeonia suffruticosa by high-speed counter-current chromatography. J Chromatogr A. 2005; 1075: 127–131. PMID: 15974126

14. Zhang JJ, Wang LS, Shu QY, Liu Z, Li CH, Zhang J. Comparison of anthocyanins in non-blotches and blotches of the petals of Xibei tree peony. Sci Hortic. 2007; 114: 104–111.

15. Zhou L, Wang Y, Lü CY, Peng ZH. Identification of components of flower pigments in petals of Paeonia lutea wild population in Yunnan. Journal of Northeast Forestry University. 2011; 39(8): 52–54.

16. Koes R, Verweij W, Quanttrocchio F. Flavonoids. a colorful model for the regulation and evolution of biochemical pathways. Trends Plant Sci. 2005; 10: 236–242. PMID: 15882656

17. Tanaka Y, Ohmiya A. Seeing is believing: Engineering anthocynin and carotenoid biosynthetic pathways. Curr Opin Biotechnol. 2008; 19:190–197. doi: 10.1016/j.copbio.2008.02.015 PMID: 18406131

18. Hichri I, Barrieu F, Bogs J, Kappel C, Delrot S,Lauvergeat V. Recent advances in the transcriptional regulation of the flavonoid biosynthetic pathway. J Exp Bot. 2011; 62(8): 2465–2483. doi: 10.1093/jxb/ erq442 PMID: 21278228

19. Zhou L, Wang Y, Peng Z. Molecular characterization and expression analysis of chalcone synthase gene during flower development in tree peony (Paeonia suffruticosa). Afr J Biotechnol. 2011; 10(8): 1275–1284.

20. Zhou L, Wang Y, Ren L, Shi QQ, Zheng BQ, Miao K, et al. Overexpression of Ps-CHI1, a homologue of the chalcone isomerase gene from tree peony (Paeonia suffruticosa), reduces the intensity of flower pigmentation in transgenic tobacco. Plant Cell Tiss Organ Cult. 2014; 16 (3): 285–295.

21. Zhou L, Wang Y, Peng Z. Cloning and expression analysis of dihydroflavonol 4-reductase gene PsDFR1 from tree peony (Paeonia suffruticosa). Plant Physiology Journal. 2011; 47(9): 885–892.

22. Mardis ER. Next-generation DNA sequencing methods. Annu Rev Genomics Hum Genet. 2008; 9:387–402. doi: 10.1146/annurev. genom.9.081307.164359 PMID: 18576944

23. Metzker Michael L. Sequencing technologies- the next generation. Nat Rev Genet. 2010; 11: 31–46. doi: 10.1038/nrg2626 PMID: 19997069

24. Wang Z, Gerstein M, Snyder M. RNA-Seq: a revolutionary tool for transcriptomics. Nat Rev Genet. 2009; 10: 57–63. doi: 10.1038/nrg2484 PMID: 19015660

25. Fatih Ozsolak, Milos Patrice M. RNA sequencing: advances, challenges and opportunities. Nat Rev Genet. 2011; 12: 87–98. doi: 10.1038/nrg2934 PMID: 21191423

26. Huang LL, Yang X, Sun P, Tong W, Hu SQ. The First Illumina-based de novo transcriptome sequencing and analysis of safflower flowers. PLoS ONE. 2012; 7(6): e38653. doi: 10.1371/journal.pone.0038653 PMID: 22723874

27. Han XJ, Wang YD, Chen YC, Lin LY, Wu QK. Transcriptome Sequencing and Expression Analysis of Terpenoid Biosynthesis Genes in Litsea cubeba. PLoS ONE. 2013; 8(10): e76890. doi: 10.1371/ journal. pone.0076890 PMID: 24130803

28. Wang ZY, Fang BP, Chen JY, Zhang XJ, Luo ZX, Huang LF, et al. De novo assembly and characterization of root transcriptome using Illumina paired-end sequencing and development of cSSR markers in sweet potato (Ipomoea batatas). BMC Genomics. 2010; 11: 726. doi: 10.1186/1471-2164-11-726 PMID: 21182800

29. Gai SP, Zhang YX, Mu P, Liu CY, Liu S, Dong L. Transcriptome analysis of tree peony during chilling requirement fulfillment: assembling, annotation and markers discovering. Gene. 2012; 497: 256–262. doi: 10.1016/j.gene.2011.12.013 PMID: 22197659

30. Trick M, Adamski NM, Mugford SG, Jiang CC, Febrer M, Uauy C.Combining SNP discovery from nextgeneration sequencing data with bulked segregant analysis (BSA) to finemap genes in polyploid wheat. BMC Plant Biol. 2012; 12: 14. doi: 10.1186/1471-2229-12-14 PMID: 22280551

31. Xu H, Gao Y, Wang J. Transcriptomic analysis of rice (Oryza sativa) developing embryos using the RNA-Seq technique. PLoS ONE. 2012; 7: e30646. doi: 10.1371/journal.pone.0030646 PMID: 22347394

32. Zhang C, Wang YJ, Fu JX, Dong L, Gao SL, Du DN. Transcriptomic analysis of cut tree peony with glucose supply using the RNA-Seq technique. Plant Cell Rep. 2014; 33: 111–129. doi: 10.1007/s00299- 013-1516-0 PMID: 24132406

33. Zhou H, Cheng FY, Wang R, Zhong Y, He CY. Transcriptome comparison reveals key candidate genes responsible for the unusual reblooming trait in tree peonies. PLoS ONE. 2013; 8(11): e79996. doi: 10. 1371/journal.pone.0079996 PMID: 24244590

34. Grabherr MG, Haas BJ, Yassour M, Levin JZ, Thompson DA, Amit I, et al. Full length transcriptome assembly from RNA-Seq data without a reference genome. Nat Biotechnol. 2011; 29: 644–652. doi: 10.1038/nbt.1883 PMID: 21572440

35. Natale DA, Shankavaram UT, Galperin MY, Wolf YI, Aravind L, Koonin EV. Towards understanding the first genome sequence of a crenarchaeon by genome annotation using clusters of orthologous groups of proteins (COGs). Genome Biol. 2000; 1: RESEARCH0009.

36. Iseli C, Jongeneel CV, Bucher P. ESTScan: a program for detecting, evaluating, and reconstructing potential coding regions in EST sequences. Proc Int Conf Intell Syst Mol Biol 1999; 138–148. PMID: 10786296

37. Kanehisa M, Araki M, Goto S, Hattori M, Hirakawa M, Itoh M, et al. KEGG for linking genomes to life and the environment. Nucleic Acids Res. 2008; 36: D480–484. PMID: 18077471

38. Ali M, Brian AW, Kenneth M, Lorian S, Barbara W. Mapping and quantifying mammalian transcriptomes by RNA-Seq. Nature Methods. 2008; 5: 621–628. doi: 10.1038/nmeth.1226 PMID: 18516045

39. Tsanakas GF, Manioudaki ME, Economou AS, Kalaitzis P. De novo transcriptome analysis of petal senescence in Gardenia jasminoides Ellis. BMC Genomics. 2014; 15: 554. doi: 10.1186/1471-2164- 15-554 PMID: 24993183

40. Wang Y, Huang H, Ma YP, Fu JX, Wang LL, Dai SL. Construction and de novo characterization of a transcriptome of Chrysanthemum

lavandulifolium: analysis of gene expression patterns in floral bud emergence. Plant Cell Tiss Organ Cult. 2014; 116: 297–309.

41. Cândido E de S, Fernandes Gda R, de Alencar SA, Cardoso MH, Lima SM, Miranda V de J, et al. Shedding some light over the floral metabolism by arum lily (Zantedeschia aethiopica) spathe de novo transcriptome assembly. PLoS ONE. 2014; 9(3): e90487. doi: 10.1371/journal.pone.0090487 PMID: 24614014

42. Kim J, Park JH, Lim CJ, Lin JY, Ryu J, Lee B, et al. Small RNA and transcriptome deep sequencing profiles insight into floral gene regulation in Rosa cultivars. BMC Genomics. 2012; 13: 657. doi: 10.1186/ 1471-2164-13-657 PMID: 23171001

43. Chen YN, Mao Y, Liu HL, Yu FX, Li SX, Yin TM.Transcriptome analysis of differentially expressed genes relevant to variegation in peach flowers. PLoS ONE. 2014; 9(3): e90842.

44. Nakatsuka T, Mishiba K, Kubota A, Abe Y, Yamamura S, Nakamura N, et al. Genetic engineering of novel flower color by suppression of anthocyanin modification genes in gentian. J Plant Physiol. 2010; 167:231–237. doi: 10.1016/j.jplph.2009.08.007 PMID: 19758726

45. Nishihara M, Nakatsuka T. Genetic engineering of flavonoid pigments to modify flower color in floricultural plants. Biotechnol Lett. 2011; 33: 433–441. doi: 10.1007/s10529-010-0461-z PMID: 21053046

46. Nakatsuka T, nishihara M, Mishiba K, Yamamura S. Temporal expression of flavonoid biosynthesisrelated genes regulates flower pigmentation in gentian plants. Plant Sci. 2005; 168: 1309–1318.

47. Forkmann G, Dangelmayr B. Genetic control of chalcone isomerase activity in flowers of Dianthus caryophyllus. Biochem Genet. 1980; 18 (5): 519–527.

48. Marinova K, Kleinschmidt K, Weissenbock G, Klein M. Flavonoid biosynthesis in barley primary leaves requires the presence of the vacuole and controls the activity of vacuolar flavonoid transport. Plant Physiol. 2007; 144: 432–444. PMID: 17369433

49. Itoh Y, Higeta D, Suzuki A, Yoshida H, Ozeki Y. Excision of transposable elements from the chalcone isomerase and dihydroflavonol 4-reductase genes may contribute to the varigation of the yellow-flowered carnation (Dianthus caryophyllus). Plant Cell Physiol. 2002; 43 (5): 578–585. PMID: 12040106

50. Ogata J, Itoh Y, Ishida M, Yoshida H, Ozeki Y. Cloning and heterologous expression of cDNAs encoding flavonoid glusyltransferases from Dianthus caryphyllus. Plant Biotechnol. 2004; 21: 367–375.

51. Okuhara H, Ishiguro K, Hirose C, Gao M, Togami J, Nakamura N, et al. Molecular cloning and functional expression of tetrahydroxychacone 2'-glucosyltransferase genes. Supplement to Plant and Cell Physiol. 2004; 45:420.

52. Togami J, Okuhara H, Nakamura N, Ishiguro K, Hirose C, Ochiai M, et al. Isolation of cDNAs encoding tetrahydroxychacone 2'-glucosyltransferase activity from carnation, cyclamen, and catharanthus. Plant Biotech. 2011; 28: 231–238.

53. Akashi T, Aoki T, Ayabe S. Cloning and functional expression of a cytochrome P450 cDNA encoding 2- hydroxyisoflavanone synthase involved in biosynthesis of the isoflavonoid skeleton in licorice. Plant Physiol. 1999; 121: 821–828. PMID: 10557230

54. Martens S, Mithöfer A. Flavones and flavone synthases. Phytochemistry. 2005; 66: 2399–2407. PMID: 16137727

55. Park NI, Xu H, Li X, Kim SJ, Park SU. Enhancement of flavone levels through overexpression of chalcone isomerase in hairy root cultures of Scutellaria baicalensis. Funct Integr Genomic. 2011; 11(3): 491–496.

56. Pomar F, Novo M, Masa A. Varietal differences among the anthocyanin profiles of 50 red table grape cultivars studied by high performance liquid chromatography. J Chromatogr A. 2005; 1094: 34–41. PMID: 16257286

57. Fournier-Level A, Hugueney P, Verries C, This P, Ageorges A. Genetic mechanisms underlying the methylation level of anthocyanins in grape (Vitis vinifera L.). BMC Plant Biology. 2011; 11: 179. doi: 10. 1186/1471-2229-11-179 PMID: 22171701

58. Jaakola L. New insights into the regulation of anthocyanin biosynthesis in fruits. Trends Plant Sci. 2013; 18 (9): 477–483. doi: 10.1016/j. tplants.2013.06.003 PMID: 23870661

59. Zhao DQ, Tao J, Han CX, Ge JT. Flower color diversity revealed by differential expression of flavonoid biosynthetic genes and flavonoid accumulation in herbaceous peony (Paeonia lactiflora Pall.). Mol Biol Rep. 2012; 39: 11263–11275. doi: 10.1007/s11033-012-2036-7 PMID: 23054003

60. Tan JF, Wang MJ, Tu LL, Nie YC, Lin YJ, Zhang XL. The flavonoid pathway regulates the petal colors of cotton flower. PLoS ONE. 2013; 8(8): e72364. doi: 10.1371/journal.pone.0072364 PMID: 23951318

61. Noda N, Kanno Y, Kato N, Kazuma K, Suzuki M. Regulation of gene expression involved in flavonol and anthocyanin biosynthesis during petal development in lisianthus (Eustoma grandiflorum). Physio Plantarum. 2004; 122(2): 305–313.

62. Mol J, Grotewold E, Koes R. How genes paint flowers and seeds. Trends Plant Sci. 1998; 3: 212–217.

63. Broun P. Transcriptional control of flavonoid biosynthesis: a complex network of conserved regulators involved in multiple aspects of differentiation in Arabidopsis. Curr Opin Plant Biol. 2005; 8(3): 272–279. PMID: 15860424

64. Morita Y, Saitoh M, Hoshino A, Nitasaka E, Iida S. Isolation of cDNAs for R2R3-MYB, bHLH and WDR transcriptional regulators and identification of c and ca mutations conferring white flowers in the Japanese morning glory. Plant Cell Physiol. 2006; 47: 457–470. PMID: 16446312

65. Czemmel S, Stracke R, Weisshaar B, Cordon N, Harris NN, Walker AR. The grapevine R2R3-MYB transcription factor VvMYBF1 regulates flavonol synthesis in developing grape berries. Plant Physiology. 2009; 151(3): 1513–1530. doi: 10.1104/pp.109.142059 PMID: 19741049

66. Ban Y, Honda C, Hatsuyama Y, Igarashi M, Bessho H, Moriguchi T. Isolation and functional analysis of a MYB transcription factor gene that is a key regulator for the development of red coloration in apple skin. Plant J. 2007; 48(7): 958–970.

67. Chiu LW, Zhou XJ, Burke S, Wu XL, Prior Ronald L, Li L. The purple cauliflower arises from activation of a MYB transcription factor. Plant Physiol. 2010; 154 (3): 1470–1480. doi: 10.1104/pp.110.164160 PMID: 20855520

68. Yamagishi M. Oriental hybrid lily Sorbonne homologue of LhMYB12 reguletes anthocyanin biosyntheses in flower tepals and tepal spots. Mol Breeding. 2011; 28: 381–389.

69. Uematsu C, Katayama H, Makino I, Inagaki A, Arakawa O, Martin C. Peace, a MYB-like transcription factor, regulates petal pigmentation in flowering peach 'Genpei' bearing variegated and fully pigmented flowers. J Exp Bot. 2014; doi: 10.1093/jxb/ert456

70. Chang S, Puryear J, Cairney J. A simple and efficient method for isolating RNA from pine trees. Plant Mol Biol Rep. 1993; 11: 113–116.

71. Conesa A, Götz S, García-Gómez JM, Terol J, Talón M, Robles M. Blast2GO: a universal tool for annotation, visualization and analysis in functional genomics research. Bioinformatics. 2005; 21: 3674– 3676. PMID: 16081474

72. Ye J, Fang L, Zheng H, Zhang Y, Chen J, Zhang Z, et al. WEGO: a web tool for plotting GO annotations. Nucleic Acids Res. 2006; 34: W293–7. PMID: 16845012

73. Benjamini Y, Yekutieli D. The control of the false discovery rate in multiple testing under dependency. Ann Stat. 2001; 29: 1165–1188.

Chapter 3

TRANSCRIPTOME ANALYSIS OF SYRINGA OBLATA LINDL. INFLORESCENCE IDENTIFIES GENES ASSOCIATED WITH PIGMENT BIOSYNTHESIS AND SCENT METABOLISM

Jian Zheng[1,2], Zenghui Hu[1], Xuelian Guan[1], Dequan Dou[1], Guo Bai[1], Yu Wang[1], Yingtian Guo[1], Wei Li[3], Pingsheng Leng[1]

[1] College of Landscape Architecture, Beijing University of Agriculture, Beijing, 102206, China

[2] Beijing Engineering Research Center of rural landscape planning and design, Beijing, 102206, China

[3] College of Landscape Architecture and Forestry, Qingdao Agricultural University, Qingdao, 266100, China

ABSTRACT

Syringa oblata Lindl. is a woody ornamental plant with high economic value and characteristics that include early flowering, multiple flower colors, and strong fragrance. Despite a long history of cultivation, the genetics and molecular biology of *S. oblata* are poorly understood. Transcriptome and expression profiling data are needed to identify genes and to better understand the biological mechanisms of floral pigments and scents in this species. Nine cDNA libraries were obtained from three replicates of three developmental stages: inflorescence with enlarged flower buds not protruded, inflorescence with corolla lobes not displayed, and inflorescence with flowers fully opened and emitting strong fragrance. Using the Illumina RNA-Seq technique, 319,425,972 clean reads were obtained and were assembled into 104,691 final unigenes (average length of 853 bp), 41.75% of which were annotated in the NCBI non-redundant protein database. Among the annotated unigenes, 36,967 were assigned to gene ontology categories and 19,956 were assigned to eukaryoticorthologous groups. Using the Kyoto Encyclopedia of Genes and Genomes pathway database, 12,388 unigenes were sorted into 286 pathways. Based on these transcriptomic data, we obtained a large number of candidate genes that were differentially expressed at different flower stages and that were related to floral pigment biosynthesis and fragrance metabolism. This comprehensive transcriptomic analysis provides fundamental information on the genes and pathways involved in flower secondary metabolism and development in *S. oblata*, providing a useful database for further research on *S. oblata* and other plants of genus *Syringa*.

INTRODUCTION

Color and scent are important properties of flowers and play an important role in the ecophysiology, aesthetic properties, and economic value of flowering plants. Each plant possesses a unique and distinct floral color and scent. Exploring the generation mechanism for floral color and scent is necessary to reveal their roles in plants, and to breed new varieties through regulation of color and scent.

The diversity of floral color results from the difference in flower pigments. Flower pigments vary among plant species according to the characteristics of their low-molecular-mass secondary metabolites, such as flavonoids, carotenoids, and alkaloids, of which flavonoids are the dominant compounds[1]. The flavonoid biosynthetic pathway has been thoroughly studied in model plants such as *Arabidopsis thaliana*, *petunia spp.*, *Antirrhinum majus*, *Vitis vinifera*, and *Zea mays*[2–4]. Anthocyanidins (including pelargonidin, cyanidin, delphinidin, peonidin, petunidin, and malvidin) in floral organs are the main flavonoids that determine flower color. The regulatory mechanisms of anthocyanin metabolism have been resolved in some ornamental plants, such as *Gerbera hybrida*, *Gentian scabra*, *Lilium* spp., and *P. hybrida*[5–9]. Anthocyanidins and many of the related biosynthetic enzymes [e.g., Chalcone synthase (CHS), chalcone-flavanone isomerase (CHI), flavanone 3-hydroxylase (F3H), flavonoid-3'-hydroxylase (F3'H), flavonoid 3'5'-hydroxylase (F3'5'H), dihydroflavonol 4-reductase (DFR), anthocyanidin synthase (ANS), glycosyltransferase (GT), acyltransferase (AT), and methyltransferase (MT)] have been identified and functionally characterized[3]. The different biosynthesis and regulatory mechanisms of anthocyanidins found in different plants should be explored.

Floral scents are composed of volatile compounds, of which terpenoids, phenylpropanoids/benzenoids, and fatty-acid derivatives are the main components. The floral scents of ornamental plants, such as *S. oblata*[10], *Rosa hybrida*[11], tree peony[12], *Lilium*spp.[13], and *Prunus mume*[14], have been investigated. Terpenoids have been found to be the most important components in a wide range of species[15]. Terpenoids are derived from the mevalonate (MVA) pathway in the cytosol, mainly mediating the biosynthesis of sesquiterpene, or from the plastid 2-C-methyl-d-erythritol-4-phosphate (MEP) pathway. Many of the related terpenoid biosynthetic enzymes [e.g., acetoacetyl-CoA transferase (AACT), 3-hydroxy-3-methylglutaryl-CoA synthase (HMGR), mevalonate kinase (MVK), phosphomevalonate kinase (PMK), MEP cytidylyltransferase (MCT), 4-(cytidine 5'-diphospho)-2-C-methyl-d-erythritol kinase (CMK), 2-C-methyl-d-erythritol-2,4-cyclodiphosphate synthase (MDS), 1-hydroxy-2-methyl-2-(E)-butenyl-4-diphosphate synthase (HDS), and 1-hydroxy-2-methyl-2-(E)-butenyl-4-diphosphate reductase (HDR)] have been also identified and functionally characterized[16]. However, the biosynthesis and regulatory mechanisms of floral scent, especially in woody ornamental plants, are largely unknown.

Genus *Syringa* (family Oleaceae) includes 27 wild species, most of which are distributed in China. *S. oblata*, a deciduous, hardy, and fast-growing perennial shrub, is native to China, and is distributed in northern regions such as Inner Mongolia, Qinghai, Hebei, Beijing, and Liaoning provinces[17]. *S. oblata* is widely cultivated because of its high ornamental and economic value with elegant color and unique fragrance[18]. *S. oblata* blooms in late April through May (depending on weather), one to two weeks earlier than its well-known lilac cousin *S. vulgaris*[19]. As a native and early flowering tree species in the Beijing area, *S. oblata* has been recommended as the foundation tree species for botanical gardens and afforestation. However the biosynthesis mechanisms of floral pigments and scent are largely unknown, especially on the molecular level, which limits the cultivation of new varieties of *S. oblata*.

Because of the lack of genomic evidence, the regulatory mechanisms of floral pigments and scents biosynthesis in *S. oblata* are difficult to investigate on the molecular level. RNA sequencing (RNA-Seq), one of the most recent Illumina sequencing techniques, has dramatically improved the efficiency of genomic exploration and gene discovery in non-model plant species for which reference genome sequence data are not available[20]. RNA-Seq generates absolute gene expression measurements and thus provides greater insight and accuracy than microarray data[20, 21]. RNA-Seq has been used to examine flower development in several garden plants, such as *Chimonanthus praecox*[22], *Cymbidium sinense*[23], *Cymbidium ensifolium*[24], and *Salvia splendens*[25], and genes related to flowering, signal transduction, and flower development have been identified. RNA-Seq provides a feasible method to investigate the metabiosynthesis and regulatory mechanisms of floral pigments and scents through transcriptomic analysis.

In this study, we used RNA-seq technology to analyze the transcriptome of *S. oblata* flowers in different developmental stages. Using DEGseq software [26, 27], the differential gene expression in the flower buds and in open flowers was examined. We identified genes associated with pathways of important secondary metabolic processes that were differentially expressed during flower development, which provides comprehensive information about gene expression at the transcriptional level and increase the understanding of the molecular mechanisms of flower pigment biosynthesis and floral scent metabolism in *S. oblata*. The data also provide an important bioinformatic resource for investigating the flowering pathway and other biological mechanisms in this and other lilac species.

MATERIALS AND METHODS

Plant Materials

Three *S. oblata* plants were randomly selected from the lilac resource nursery at Beijing Agricultural University, a resource conservation unit of National Forest Genetic Resources Platform (NFGRP), Beijing, China. Three distinct stages of *S. oblata* flower development were defined: i) flower bud stage (SOFB), with flower buds enlarged and bud scales not protruded; ii) bud stage (SOB), with inflorescence emerged and corolla lobes not displayed; and iii) flowering stage (SOF), with flowers fully opened and emitting fragrance (Fig 1). Materials at each stage were dissected from the plants and immediately frozen and stored in liquid nitrogen prior to further analysis, with three replicates per stage.

Figure 1. Flower development stages in *Syringa oblata*. Images of *Syringa oblata* flowers: a, at flower bud stage (SOFB); b, bud stage (SOB): c, flowering stage (SOF). doi:10.1371/journal.pone.0142542.g001

RNA Extraction

Total RNA was extracted from the plant materials using an RNAsimple plant RNA purification kit (Tiangen Biotech, Beijing, China). The quality and quantity of purified RNA was examined using a NanoDrop ND-1000 UV/ Visible spectrophotometer (Wilmington, DE) and Agilent Bioanalyzer 2100 (Agilent Technologies, Santa Clara, CA) for gel electrophoresis. For each developmental stage, equal amounts of high-quality RNA from the three plant samples were used in cDNA library construction and Illumina deep sequencing.

Construction of cDNA library for Illumina Sequencing

Nine cDNA libraries were prepared using an Illumina RNA-Seq sample preparation kit (RNeasy Micro kit, Cat.#74004, Qiagen, China) using 10 μg of total RNA. The mRNA was isolated by polyA selection with oligo (dT) beads and fragmented using fragmentation buffer. Synthesis of cDNA, end repair, A-base addition, and ligation of the Illumina-indexed adaptors were then performed according to the Illumina protocol. Libraries were size-selected on 2% Low-Range Ultra Agarose (Bio-Rad Laboratories (Shanghai) Co., Ltd) for cDNA target fragments of 300–500 bp; this was followed by PCR amplification using Phusion DNA polymerase (NEB) for 15 PCR cycles. Following quantification by TBS380, paired-end libraries were sequenced using the Illumina HiSeq 2500 system (2×100 nt multiplex) at Shanghai Biotechnology Co., Ltd. (Shanghai, China). Data analysis and base calling were performed with the Illumina instrument software.

Sequence Data Analysis and *de novo* Assembly

The raw reads were first filtered by removing the adapter sequences, fragments <20 bp in length, and low-quality sequences, which included reads with N percentage (the percentage of nucleotides in a read that could not be sequenced) >5% and reads containing more than 20% nucleotides with Q-value≤10. The Q-value represents the sequencing quality of related nucleotides. Using the CLC Genomics Workbench software (V6.0.4,http://www.clcbio.com/products/clc-genomics-workbench/) [28–30], clean reads from the nine libraries obtained from the three flower stages were used in *de novo* splicing and read-mapping to the transcriptome in the absence of a reference genome. The spliced sequence was called the primary unigene. The primary unigene was further assembled to the final unigene with the CAP3 EST software[31]. The final unigene was used for further analysis.

Sequence Annotation and Classification

For annotation, the unigene sequences were searched against the NCBI non-redundant (NR) protein database[32] using BlastX (http://www.ncbi.nlm.nih.gov/BLAST), with a cut-off E-value of 10–5. Gene ontology (GO) terms were extracted from the annotation of high-score BLAST matches in the NCBI NR proteins database (E-value ≤ 1.0×10–5) using Blast2GO (http://www.blast2go.com), and then were sorted into categories using in-house perl scripts[33]. Functional annotation of the proteome was carried out by a BlastP homology search against the NCBI eukaryotic Orthologous Groups (KOG) database (http://www.ncbi.nlm.nih.gov/COG/). KEGG pathway annotations were performed using the online KEGG Automatic Server (KAAS,http://www.genome.jp/kegg/kaas/) [34].

Expression Analysis

Final unigenes with differential expression among the three stages of *S. oblata* flower development were detected with DEGseq software [26, 27] using three replicates per stage. A general Chi-squared test of statistical significance was used, and the false discovery rate (FDR) of results was controlled. If FDR was lower than 0.05 and the highest RPKM (reads per kilobase per million reads) of the unigene was twice that of the lowest one, the unigene was considered as differentially expressed.

GO enrichment analysis of differentially expressed genes (DEGs) was performed using the GOseq R package[35], which corrected for gene length bias. GO terms with corrected P-value less than 0.05 were considered significantly enriched by DEGs. We used the KOBAS software to test the statistical enrichment of DEGs in KEGG pathways[36].

Significantly altered genes among SOFB, SOB and SOF were described using heatmap analysis with unsupervised hierarchical clustering. The raw intensity (RPKM) was log2 transformed and then used for the calculation of Z scores[37].

Quantitative real-time PCR (qRT-PCR) Validation

Real-time PCR was performed on an ABI Power SYBR Green PCR Master Mix and 7900 HT Sequence Detection System (Applied Biosystems). Sequences of the specific primer sets are listed in the S1 File. The ubiquitin domain-containing protein gene (*DSK2B*, Acc. No.Q9SII8) and actin gene (*Actin-7*, Acc. No. P53492) of *S. oblata* were used as internal markers. qRT-PCR was performed using the SYBR Premix Ex Taq Kit (TaKaRa) according to the manufacturer's protocol. The results were normalized to the expression level of the constitutive ubiquitin and actin genes. A comparative Ct method ($2^{-\Delta\Delta Ct}$) of relative quantification was used to evaluate the quantitative variation. All quantitative PCR for each gene used three biological replicates, with three technical replicates per experiment. The RNA samples for qRT-PCR were the same as those for Illumina sequencing.

RESULTS

Illumina Sequencing and Assembly of Sequence Reads

In this study, nine cDNA libraries were constructed and subjected to Illumina deep sequencing. We obtained 337,669,638 raw reads and 319,425,972 clean reads from the three developmental-stage libraries after eliminating primer and adapter sequences and filtering out low-quality reads. We obtained 130,981 contigs from these reads. Based on the paired-end reads, these contigs were further assembled into primary unigenes using CLC Genomics Workbench (version: 6.0.4; Table 1). Finally, using CAP3 EST, we spliced the primary unigenes a second time and obtained a transcriptome database for *S. oblata*. The *de novo* assembly generated 104,691 final unigenes (Table 1). The distribution of unigene sequence length is shown in Fig 2. Approximately 30% of the unigenes were 401–600 bp in length, 23% were 201–400 bp, and 47% were >600 bp (Fig 2).

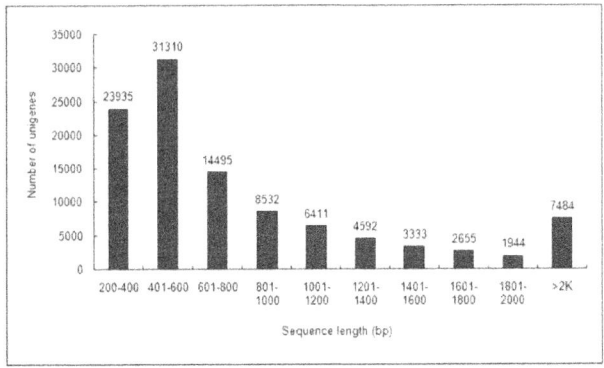

Figure 2. Length distribution of assembled *Syringa oblata* **unigenes.** All clean reads for each flower development stage (see Fig 1) were combined and resulted in 104,691 unigenes. Horizontal and vertical axes show the size and the number of unigenes, respectively. doi:10.1371/journal.pone.0142542.g002

Table 1. Summary of assembly statistics for *Syringa oblata* **transcriptome.**

Statistics	Counts	Total length (bp)	N50 (bp)	Average length (bp)	Longest (bp)
Contigs	130,981	89,118,182	858	680	15,223
Primary unigenes	106,708	89,735,247	1,077	841	15,756
Final unigenes	104,691	89,306,170	1,099	853	15,756

doi:10.1371/journal.pone.0142542.t001

Unigene Annotation

All unigenes for *S. oblata* were searched against the NCBI non-redundant (NR) database using BlastX with a cut-off E-value of 10–5. Among these unigenes, approximately 42% showed significant similarity to known proteins in the NR database, and 29% were matched to the SwissProt database (Table 2). In further analysis of the matching unigenes, we found that *S.oblata* unigenes were most closely matched with gene sequences from *Vitis vinifera*, followed by *Solanum tuberosum*, *S. lycopersicum*, *Theobroma cacao*, and *Populus trichocarpa* (Fig 3).

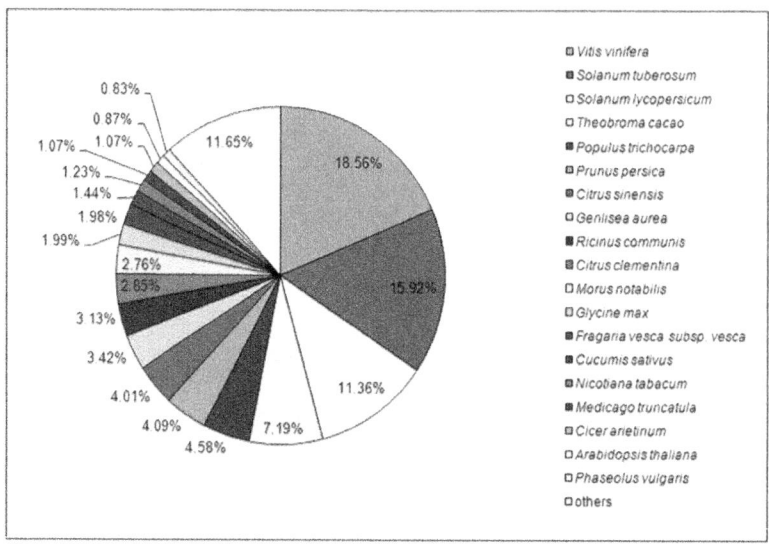

Figure 3. Species-based distribution of BLASTX matches for unigenes against NCBI NR database.

We used all plant proteins in the NCBI NR database in performing the homology search; for each sequence, we selected the closest match for analysis. doi:10.1371/journal.pone.0142542.g003

Table 2. Statistics of annotation results for *Syringa oblata* **unigenes.**

Database	NR	SwissProt	GO	KOG	KEGG	All
Number annotated	43,712	30,262	36,967	19,956	12,388	44,164

doi:10.1371/journal.pone.0142542.t002

In addition to the NCBI NR and SwissProt databases, *S. oblata* unigenes were aligned with Gene Ontology (GO), eukaryoticorthologous groups of proteins (KOG), and the Kyoto Encyclopedia of Genes and Genomes (KEGG). In total, 42% of the unigenes provided significant BLAST results; of these, approximately 84% had matches in the GO database, 45% had matches in the KOG database, and 25% had matches in KEGG (Table 2). The significance of a BLAST search depends on the length of the query sequence; thus, short reads obtained from sequencing are rarely matched to known genes [38]. Approximately 23% of the unigenes we identified were short (200–400 bp), which might help to explain the proportion that did not match. In other words, these short reads generated by sequencing or assembly might have resulted in the low significance[25].

GO Classification of Unigenes

We used Blast2GO to classify the functions of the predicted *S. oblata* unigenes. Of the 104,691 final unigenes, 36,967 were successfully annotated by GO assignments. The annotated unigenes were assigned into three main GO categories and 62 subcategories, and some belonged to more than one of the three categories (Fig 4). Based on annotation against the NR database, 33,208 GO terms were assigned. The greatest proportion (~62%) was assigned to biological processes (GO: 0008150); the others were assigned to molecular functions (GO: 0005575; ~24%) or cellular components (GO: 0003674; ~14%) (Fig 4).

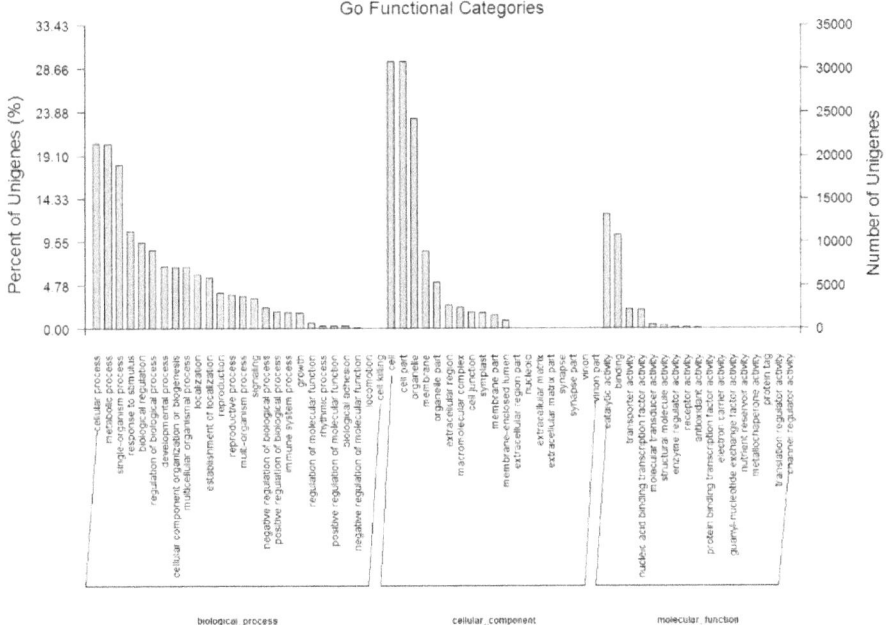

Figure 4. Gene ontology (GO) classification of unigenes of *Syringa oblata* inflorescence.

Left y-axis indicates the percentage of unigenes in subcategories of each main category. Right y-axis indicates the number of unigenes in each subcategory. doi:10.1371/journal.pone.0142542.g004

Among the biological processes, dominant subcategories included cellular, metabolic, and single-organism processes and response to stimulus and biological regulation were also important (Fig 4). The large proportion of annotated unigenes involved in metabolic processes suggested that novel genes involved in pathways of secondary metabolite synthesis could be identified. Dominant subcategories of cellular components included cells, cell parts, organelles, and membranes. Molecular function subcategories with the largest numbers of annotated unigenes were catalytic activity, binding, transporter activity, and nucleic acid binding transcription factor activity (Fig 4, S2 File).

KOG Classification of Unigenes

The KOG database is a useful platform for functional annotation of newly sequenced genomes. Based on sequence homology, 19,956 unigenes were assigned to the KOG functional classification into 25 categories denoting their involvement in cellular processes, biochemistry and metabolism, signal transduction, and other functions (Fig 5). Signal transduction mechanisms were dominant, and general functional prediction and posttranslational modification–protein turnover–chaperones shared a large proportion of genes (Fig 5, S3 File). However, 924 unigenes (4.6%) were assigned to the category of secondary metabolite biosynthesis, transport, and catabolism, which suggests that identification of novel genes involved in secondary metabolism pathways is a promising area of study (S3 File).

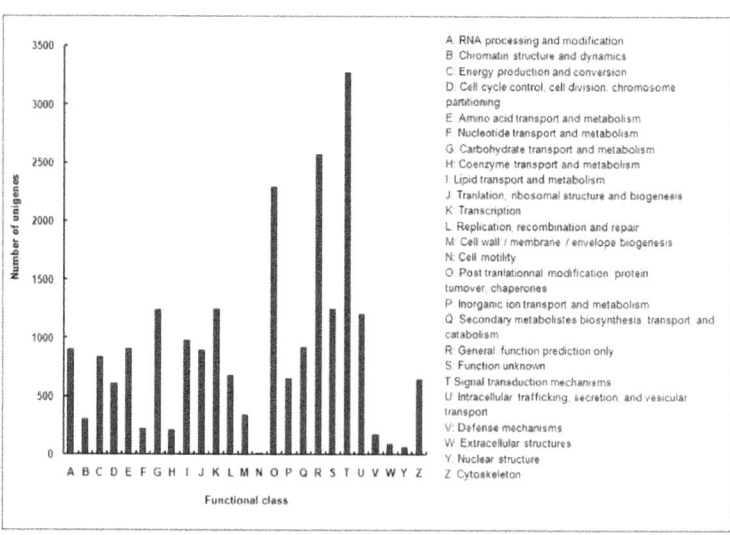

Figure 5. KOG functional classification of unigenes of *Syringa oblata* **inflorescence.** From a total of 104,691 final unigenes, 19,956 annotated unigenes with significant homology in the KOG database (E-value ≤1.0 E⁻5) were classified into 25 KOG categories. doi:10.1371/journal.pone.0142542.g005

Metabolic Pathway Assignment by KEGG

The KEGG database places emphasis on biochemical pathways and provides an alternative approach to the categorization of gene function. Here, 12,388 annotated unigenes were assigned to 286 KEGG pathways (S4 File). Metabolic pathways had the most representation (~23%), followed by biosynthesis of secondary metabolites (~13%). Several signaling pathways (AMPK, MAPK, p53) and biosynthesis pathways of important secondary metabolites (e.g. phenylpropanoid, carotenoid, flavonoid, flavone and flavonol, anthocyanin, volatile terpenoids) were represented (S4 File). These data provide a valuable resource for further mining pathways of interest in *S. oblata*.

We focused on the metabolic pathways involved in flower pigments and scents biosynthesis. Flavonoids are primary compounds affecting the formation of flower or fruit color, and three pathways are involved in flavonoid metabolism: the anthocyanin, flavonoid, and flavone/flavonol biosynthetic pathways [39]. Among these pathways, 55 unigenes (average length,1130 bp) were associated with flower pigment synthesis, including 16 unigenes annotated to anthocyanin biosynthesis, 31 unigenes annotated to flavonoid biosynthesis, and 14 unigenes annotated to the flavone/flavonol biosynthetic pathway (S5 File). Further analysis revealed that nine unigenes were between 300 and 500 bp, 20 unigenes were between 500 and 1000 bp, and 26 unigenes were >1000 bp; moreover, all unigenes had very high homology with the NR database (E <10^{-12}) (S5 File). In addition, six unigenes were annotated to both the flavonoid and the flavone/flavonol biosynthetic pathway, which suggested that these genes could be key nodes in the three flavonoid metabolic pathways (S5 File).

Volatile terpenoids are important components of the floral scent in *S. oblata*[40]. Three metabolic pathways: terpenoid backbone biosynthesis, monoterpenoid biosynthesis, and limonene and pinene degradation contributed to the biosynthesis of terpenoid volatiles, and 213 unigenes (average length,1069 bp) were annotated (S5 File). Among these, 45% of the unigenes were involved in limonene and pinene degradation, 43% were involved in terpenoid backbone biosynthesis, and only 25 unigenes were clustered in monoterpenoid biosynthesis (S5 File).

This analysis of the metabolic pathways of flower pigment and floral scent helps to provide a basis for cloning and functional analysis of the key genes involved these important characteristics of *S. oblata*.

Variation in Gene Expression among Flower Development Stages

Among the annotated unigenes, we found that 83,343 genes were expressed in samples of all three developmental stages, and 2,203, 1,510, and 1,682 genes were found to be expressed specifically in the SOFB, SOB, and SOF stages, respectively (Fig 6A). To identify differentially expressed genes (DEGs) during flower development, we performed genome-wide expression analysis for the SOFB, SOB, and SOF developmental stages (Fig 1). Following Anders and Huber [27], we identified genes that were differentially expressed between two samples by comparing SOB and SOFB, SOF and SOB, and SOF and SOFB. In total, 1,068 differentially regulated genes, including 645 upregulated and 423 downregulated genes, were identified in SOB compared to SOFB (Fig 6B). The functional annotations and statistics for all DEGs between SOB and SOFB are in the . We also classified the functions of the 1,068 DEGs by Blast2GO. Among biological processes, dominant subcategories included metabolic processes, cellular processes, and single-organism processes. In cellular components, the two largest subcategories were cell and cell parts. Catalytic activity was dominant in the molecular functions section (S6 File). GO enrichment analysis showed that most of the upregulated DEG sets were related to biological processes, cellular components, molecules, and DNA binding. The KEGG enrichment classification showed that 23 groups were significantly enriched, and most of these were upregulated primarily in the SOB library, which was correlated with metabolic and biosynthesis pathways of secondary metabolites. Based on sequence homology, the 1,068 DEGs were assigned to the KOG functional classification into 23 categories (S1 Fig).

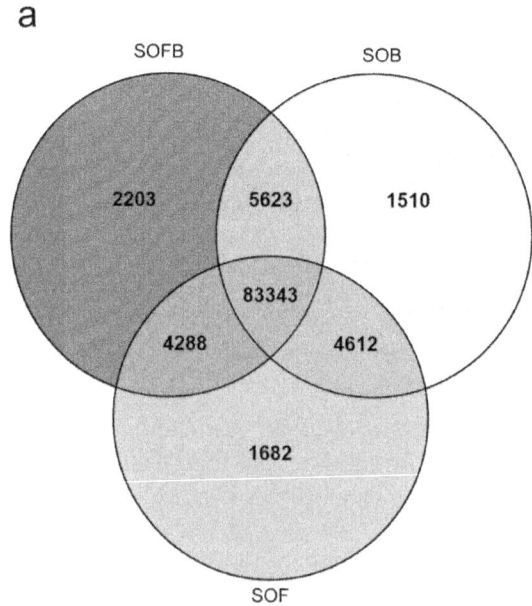

a

b

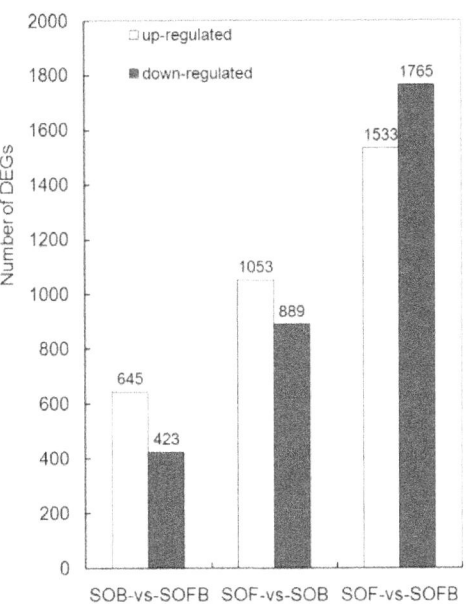

Figure 6. Distribution of genes among each flower developmental stage. a: Venn diagram illustrating the expression patterns of genes among flower developmental stages. b: Number of DEGs in each flower developmental stage. DEGs are compared between developmental stages. SOFB, flower bud stage; SOB, bud stage; SOF, flowering stage. doi:10.1371/journal.pone.0142542.g006

Compared with SOB, more DEGs were identified in SOF (Fig 6B), and the statistical information for these DEGs is summarized in the . The results of the GO functional classification between the SOF and SOB stages were similar to those for the comparison of SOB to SOFB (S7 File). GO enrichment analysis indicated that approximately half of the gene sets were downregulated and half were upregulated in the SOF library. The downregulated genes were mainly related to responses, chloroplasts, membranes, and cytosol, and the upregulated genes were related to molecular functions, kinase activity, nuclei, and biological processes. In the KEGG classification, 32 groups were significantly enriched, and about half of these genes were upregulated and half were downregulated in the SOF library. These enriched genes were mainly related to metabolic pathways and biosynthesis of secondary metabolites. The KOG functional classification showed that the DEGs between the two stages were also assigned to 23 categories (S2 Fig).

We also compared gene expression between SOF and SOFB, and found 3298 genes with differential expression (Fig 6B). The functional annotations and statistical information of all 3298 DEGs are summarized in the . GO functional classification indicated that the most of the 3298 DEGs were involved in metabolic processes, cellular processes, and single-organism processes in the biological process section; the dominant subcategories were cell and cell parts in the cellular components section; and catalytic activity was the dominant subcategory in the molecular functions section (S8 File). GO enrichment analysis showed that most of the gene sets were downregulated in the SOF library, and these genes were mainly related to epigenetics, flower development, cytosols, and binding (S9 File). KEGG classification showed that 14 groups were significantly enriched, and most of these genes were downregulated in the SOF library and were related to ribosomes and DNA replication (S9 File). In KOG functional classification, the DEGs between the two stages were also assigned to 24 categories (S3 Fig).

Genes related to Flower Pigment Biosynthesis and Floral Scent Metabolism

During development, the flowers of *S. oblata* underwent rapid changes in color and scent, from green and unscented (SOFB) to amaranth and slightly scented (SOB), and then to lilac with very strong scent (SOF). We focused on genes related to flower pigment biosynthesis and scent metabolism, and chose some genes that showed significant differential expression among the three developmental stages (Fig 7). The functional annotation for these unigenes was provided in S13 File. For example, contig_2561 showed lower expression levels in SOFB and SOF than in SOB and was homologous with *Catharathus roseus* cinnamic acid 4-hydrolase (C4H), related to flavonoid biosynthesis. Other genes associated with anthocyanins biosynthesis showed higher expression levels in SOB and SOF than in SOFB. These genes included contigs_52106 (homologous with *Perilla frutescens* chalcone synthase, CHS), 14647 (homologous with *Camellia chekiangoleosa* flavanone 3-hydroxylase, F3H), 9238 (homologous with *Petunia × hybrida* flavonol synthase, FLS), 22308 (homologous with *Eustoma grandiflorum*anthocyanidin synthase, ANS), and 32424 (homologous with the *Olea europaea* flavonoid 3-O-glucosyltransferase, 3GT). In addition, some genes of anthocyanin biosynthesis pathways, such as contig_38186 (homologous with COMT1), contig_27568 (homologous with DOMT-like) and contig_33752 (homologous with GT) displayed the highest expression in the SOF stage, which indicated that these genes located downstream of anthocyanin biosynthesis. In particular, the mean FPKMs of contig_27568 were 1.08, 1.43, and 4295 in each SOFB, SOB, and SOF stage, respectively (P = 0), which means that the *DOMT-like* gene might be specifically and strongly expressed at the SOF stage.

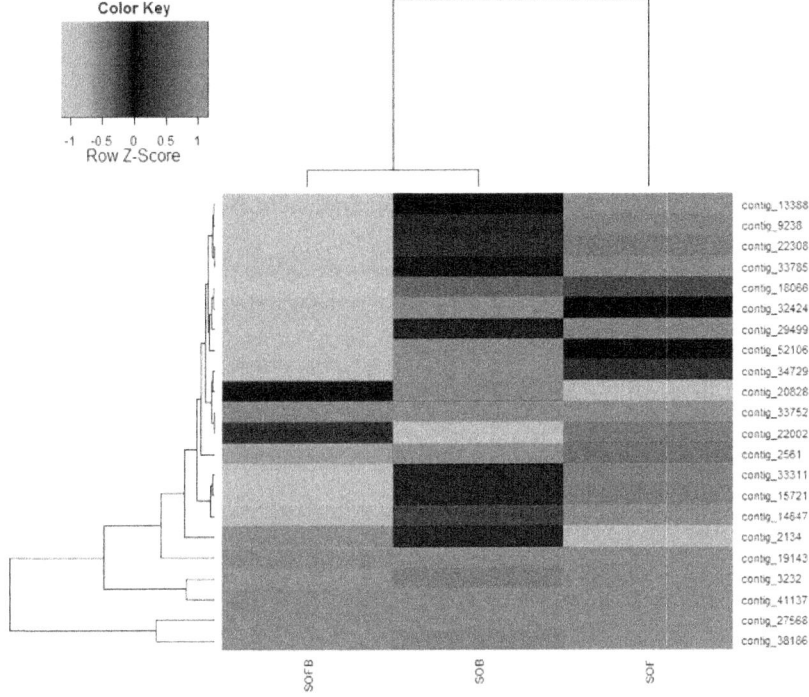

Figure 7. Differential expression of genes related to flower pigmentation and floral scent biosynthesis in *Syringa oblata*. Each column represents an experimental sample (SOFB, SOB, and SOF), and each row presents a gene. Expression differences are shown by different colors. Red indicates high expression and green indicates low expression. doi:10.1371/journal.pone.0142542.g007

Some key genes related to fragrance biosynthesis showed a significant difference in expression among the three flower developmental stages. The functional annotation for these unigenes is shown in the S13 File. The contigs_3232 (homologous with *Salvia miltiorrhiza*1-deoxy-d-xylulose 5-phosphate synthase, DXS), 33311 (homologous with *Nicotiana tabacum* hydroxy-methylglutaryl-coenzyme A reductase, HMGR), 13388 (homologous with *S. miltiorrhiza*geranyldiphosphatesynthase, GPS), and 33785 (homologous with *Oryza sativa* subsp.*Japonica*1-deoxy-d-xylulose 5-phosphate reductoisomerase, DXR) had the highest levels of expression in SOF and encoded key enzymes involved in biosynthesis of terpenoid volatiles; however, the three terpenoid synthase genes with contigs_22002 (homologous with camphene/tricyclene synthase [TPS3] of *S. lycopersicum*),

19143 (terpene synthase[TPS] of*O. europaea*), and 41137 ([3S]-linalool/[E]-nerolidol synthase [LIS] of *V. vinifera*) showed the highest expression levels in SOF. In contrast, contigs_2134 and 18066 from terpene degradation genes showed the highest expression in SOFB and SOB, respectively. In addition, we found that some cytochrome P450 genes involved in metabolism of terpenoid volatiles, such as contigs_2134 (homologous with *V. vinifera* cytochrome P450 76A2, CYP76A2), 819 (homologous with *A. thaliana* cytochrome P450 78A7, CYP78A7), and 17171 (homologous with *V. vinifera* cytochrome P450 77A3, CYP77A3), showed the highest expression in SOFB stage. However, contig_18066 (homologous with *V. vinifera* cytochrome P450 77A2, CYP77A2) displayed higher expression in SOB and SOF than in the SOFB stage. Finally, we found that three genes, *DXS* (contig_3232), *TPS* (contig_19143), and *LIS* (contig_41137), were induced specifically and strongly (>100-fold change in expression) at the SOF stage.

Apart from the structural genes, nine transcription factors regulating anthocyanin and fragrance biosynthesis were found in this study. We found that the six genes encoding the transcription factors [e.g., contigs 28146 (homologous with *Prunus avium* transcription factor R2R3-MYB), 28147 (homologous with *Theobroma cacao* transcription factor MYB), 38962 (homologous with *Camellia sinensis* transcription factor R2R3-MYB1), 25616 (homologous with *Citrus sinensis* transcription factor bHLH79), 9153 (homologous with *S. tuberosum* transcription factor bHLH147), and 64997 (homologous with *Fragaria vesca* bHLH79-like transcription factor)] showed higher expression in SOB and SOF than in the SOFB stage. However, the other three genes [e.g., contigs_11688 (homologous with *A. thaliana* transcription factors MYB44), 21735 (homologous with *S. tuberosum* transcription factor bHLH3), and 20570 (homologous with *A. thaliana* transcription factor bHLH48)] displayed higher expression in SOFB than in the other two stages.

Real-time Quantitative PCR Validation of RNA-seq Results

To validate the sequencing data, 22 genes involved in flower pigment biosynthesis and fragrance metabolism were selected for real-time qPCR with gene-specific primers designed using Primer version 5.0 (S1 File). The expression patterns of these genes in each flower developmental stage are shown in Fig 8.

Using *S. oblata DSK2B* as a reference gene, expression patterns determined by real-time qPCR were consistent with those obtained by RNA-Seq (Fig 7, Fig 8), confirming the accuracy of the RNA-Seq results. We also obtained identical real-time qPCR results using *S. oblata ACT-7* as a reference gene (S4 Fig).Thus, the data generated here can be used to investigate specific flowering genes that show comparative expression levels among different developmental phases.

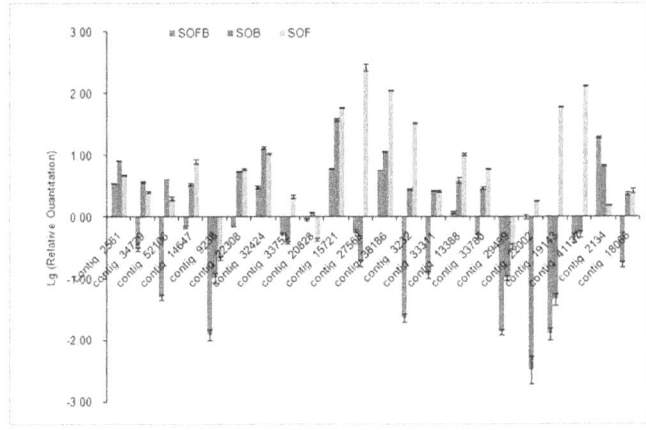

Figure 8. Real-time qPCR validation of genes involved in flower pigmentation and scent biosynthesis in *Syringa oblata*. The y-axis indicates fold changes in expression among flowering stages (SOFB, SOB, and SOF) using results from RT-qPCR. Data were normalized against *Syringa oblata* reference gene *DSK2B*. Quantitative PCR for each gene used three biological replicates, with three technical replicates per experiment. Bars indicate SD. doi:10.1371/journal.pone.0142542.g008

DISCUSSION

Illumina Sequencing and Sequence Annotation

S. oblata is a popular and traditional garden plant in many countries, including China. However, little is known about the mechanisms responsible for floral color and fragrance, and genomic information for this species is currently unavailable. The aims of this project were to generate a large amount of cDNA sequence data that would facilitate more detailed studies of *S. oblata* and to identify genes that control flower pigments and floral scents compounds. The availability of transcriptomic data for *S. oblata* will meet the initial informational needs for functional studies (e.g., molecular genetics, variety breeding, biochemical characterization) of this species and its relatives. Here, RNA-seq was performed using Illumina sequencing and generated 319,425,972 high-quality reads that were assembled into 104,691 final unigenes with an average sequence length of 853 base pairs. The average length of the assembled unigenes was greater than that obtained for *C. sinense* (612 bp)[23], *Myrica rubra* (437 bp)[41], bamboo (736 bp)[42], and *Hevea brasiliensis* (485 bp) [43] using similar sequencing technologies. For gene annotation, the 104,691 final unigenes were used for BLASTX and annotation against NCBI protein databases including NR, SwissProt, KOG, KEGG, and GO. Because of the limited availability of genetic information, 44,164 unigenes were identified through BLAST searches, and 57.81% of unigenes had no homologues in the NCBI databases. Interestingly, similar to the results for *C. praecox*[22], the annotated unigenes of *S. oblata* showed higher homology to those of *V. vinifera* (Fig 3), which may reflect a closer evolutionary relationship between the two species.

Anthocyanin Biosynthesis Genes in *S. oblata*

A set of genes that includes *CHS, CHI, F3H, F3'H, F3',5'H, DFR, ANS, GT, AT*, and *MT* affect anthocyanin biosynthesis[44–50], and these genes are divided into upstream and downstream structural genes. Most of the upstream genes belong to the family of early biosynthesis genes (*EBGs*), and the downstream genes belong to the family of late biosynthesis genes (*LBGs*)[51,52]. All genes in the anthocyanin biosynthesis pathway showed differential expression patterns during plant growth and development, consistent with findings in other species. In *Lilium* 'Asiatic lightning', the expression of *CHS* and *DFR* increases with flower development and reaches a maximum in the flowering period, which suggests that the gene expression is coordinated with the production of anthocyanins[53]. Three patterns of gene expression are observed in *Gentiana triflora*: *CHS* and *CHI* are expressed throughout flower development, *F3'H* is highly expressed in the early stage of flower development, and *F3H, F3 5 H*, and *DFR* are expressed only in the late stage of flower development[54]. In *Pericallis hybrida, CHS, CHI, F3H, F3'H, DFR, F3'5'H*, and *3MaT* are expressed at high levels in early stages, then the expression decreases moderately until there is low expression of these genes in petals in the late flowering stage[55]. In this study, we found that the upstream genes, *C4H* and *CHS*, and the downstream genes, *3GT/UDGT, F3H, FLS, ANS, coAOMT, DOMT-like*, and *COMT-1*, were upregulated in SOF relative to SOFB. *C4H, CHS*, and *3GT/UDGT* were expressed at a high level in the SOB stage (Fig 7, Fig 8). Interestingly, the *MTs* (including *coAOMTs, DOMT-like, COMT-1*) were expressed at a low level in the SOFB stage and at the highest level in the SOF stage, which indicated that the production of pigments in *S. oblata* was coordinated with the methylation of anthocyanidins derived from leucoanthocyanidins catalyzed by ANS. These results showed that the expression patterns of genes related to anthocyanin biosynthesis in *S. oblata* were similar to those of *Lilium* 'Asiatic lightning,' *G. triflora*, and *P. hybrida*, but were yet unique.

Volatile Terpenoid Metabolism Genes in *S. oblata*

The biosynthesis and emission of terpenes have been investigated in many plants, including snapdragon[56, 57], *Clarkia breweri*[58], *A. thaliana*[59], and *Lavandula angustifolia*[60]. In *S.oblata*, terpenoid biosynthesis genes involved in the MVA and MEP pathways were identified as*DXS, DXR, HMGR, GPS, TPS3, TPS,* and *LIS*. These genes showed similar expression patterns across the development stages and had the strongest expression during the SOF stage, which may result in the emission of volatile terpenoids in a large quantity, which we found in the full flowering stage (in press). The *DXS, DXR,* and *LIS* genes isolated from *R. rugosa*flowers show consistent expression during development, from budding to the withering stage[61]. In *L. angustifolia*, the expression of two *TPS* genes is positively correlated with the emission of volatile terpenoids during development of the inflorescence [60]. In snapdragon flowers, transcript levels of genes including *DXS, DXR, MCT, CMK,* and *TPS* in the MEP pathway (leading to formation of volatile terpenoids) are upregulated during petal development, implying that transcriptional induction of the MEP pathway precedes scent formation[57]. However, the expression levels of genes in the MVA pathway are not significantly upregulated with petal development, with the exception of *TPS* downstream[57]. The MEP and MVA pathways were both activated with flower development in *S. oblata*, in contrast to the findings in snapdragon. Thus, the distinct expression patterns of MEP and MVA pathway genes of different plants might result in differences in biosynthesis and emission of floral terpenoids. Interestingly, when the MEP and MVA pathways were stimulated, genes for terpene-degrading enzymes (contigs 2134 and 18066) showed lower expression levels, thus maintaining the release of strong floral scent during blooming (Fig 7, Fig 8).

Cytochrome P450s are important in the oxidative, peroxidative, and reductive metabolism of numerous and diverse endogenous compounds including terpenoids. In this study, we found one cytochrome P450 gene (*CYP77A2*) that was up-regulated significantly at the SOB stage and down-regulated at the SOF stage, which was similar to a previous report of expression of a cytochrome P450 gene in winter sweet[22]. However, the other cytochrome P450 genes (*CYP76A2, CYP78A7,* and *CYP77A3*) were down-regulated at the SOB and SOF stage in *S. oblata*. Cytochrome P450 genes are found in all kingdoms and show extraordinary diversity in their chemical reactions. Although cytochrome P450 genes are one of the largest gene families in plants, their functions in flowers are as yet largely unknown.

Transcriptional Regulation of Anthocyanin Biosynthesis in *S. oblata*

Transcription factors (e.g., R2R3-MYB, bHLH, and WD40 proteins) could activate or inhibit expression of structural genes to regulate the biosynthesis of anthocyanins. Previous studies have shown that anthocyanin biosynthesis depends on regulation of MYB-bHLH compounds in some plants, e.g., in *G. hybrida*, interaction of GhMYB10 (R2R3-MYB) and GhbHLH encoded by *GhMYC1* activates the expression of *GhDFR* [7]. However, GtbHLH1 interacting with GtMYB3 jointly promote the expression of *GtF3'5'H* and *Gt5AT*, but were found not to influence the activity of *GtCHS* in *G. triflora*[8]. LhbHLH2 was shown to interact with LhMYB6 and LhMYB12, which activate the expression of *LhDFR*, *LhCHSa* and *LhCHSb* in *Lilium* spp. [9]. In addition, anthocyanin biosynthesis is also regulated by MYB-bHLH-WD40 compounds (MBW compounds) in some plants, e.g., in *A. thaliana*, anthocyanin biosynthesis in seeds and vegetative organs is regulated by MBW compounds such as TT2 (R2R3-MYB protein)-TT8 (bHLH042)-TTG1 (WD40 protein) compounds that promote the biosynthesis of procyanidins in seeds [6]. In petunias, the structural genes, e.g., LBGs (*DFR, ANS13, GT, AOMT*) and *CHSJ*, for anthocyanin biosynthesis are also regulated by MBW compounds; however, the EBGs (*CHSA, CHI, F3H*) cannot be regulated by MBW compounds[5]. In this study, we found that three MYB (R2R3-MYB, MYB, and R2R3-MYB1) and three bHLH (bHLH79, bHLH147, and bHLH79-like) transcription factors were up-regulated at the SOB and SOF stages, and another three transcription factors (MYB44, bHLH3, and bHLH48) were down-regulated at the SOB and SOF stages in *S. oblata* . Yet, the molecular regulation mechanism for flower pigment biosynthesis depending on MYB-bHLH or MBW compounds still needs further research. MYB and bHLH transcription factors involved in anthocyanin biosynthesis and their roles in the regulation of structural gene expression have been studied in model and some ornamental plants, which could provide a good reference for relative studies on anthocyanin synthesis in *S.oblata*.

Transcriptional Regulation of Floral Scent Biosynthesis in *S. oblata*

In contrast to the rapid progress in recent years in characterizing and mapping the reactions leading to the formation of floral scent, little is known about regulation of floral scent production at the molecular level. The genetic basis and functional significance of scent production have also been investigated in petunia. To date, ODORANT1 (ODO1) is the first transcription factor that has been characterized as a regulator of scent production in flowers[62]. ODO1 belongs to the R2R3-MYB transcription factor family (subgroup with AtMYB42 and AtMYB85), and its suppression in *P. hybrida* leads to decreased levels of emitted volatile phenylpropanoids[62,63]. The importance of regulating flux toward phenylpropanoids is also demonstrated in *P. hybrida* flowers over-expressing the *A. thaliana* MYB factor PRODUCTION OF ANTHOCYANIN PIGMENT1 (PAP1)[64]. Recently, another two transcription factors, EMISSION OF BENZENOIDS I and II (EOBI and EOBII), belonging to subgroup of R2R3-MYB family (subgroup 19), are also demonstrated to regulate benzenoid biosynthesis in petunias[65, 66]. Both *EOBI* and *EOBII* positively regulate *ODO1*, which determines the floral scent production in*P. hybrida*. However, the regulating mechanism of transcription factors the terpene has not yet been reported. In this study, we found that two R2R3-MYB (contigs_28146 and 38962) transcription factors were up-regulated at the SOB and SOF stages in *S. oblata* . However, the molecular regulation mechanism of floral scent biosynthesis depending on MYBs still needs further research.

CONCLUSION

The combination of RNA-seq and DEGs analysis based on Illumina sequencing technology provided comprehensive information on gene expression in *S. oblata*. In this study, we revealed numerous differentially expressed genes between different flowering stages. Candidate genes for biosynthesis of anthocyanin, flavonoids, flavones and flavonols, terpenoids, and monoterpenoids and for degradation of limonene and pinene were rapidly identified by this approach. In summary, this comprehensive database provides essential information for investigating flowering and other biological pathways in *S. oblata* and will be useful for improving the horticultural and ornamental quality of this species.

Supporting Information

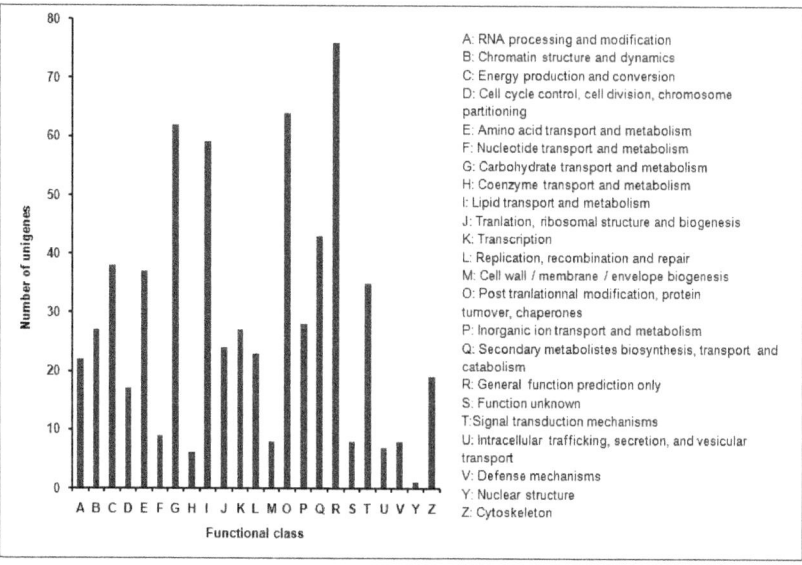

S1 Figure. KOG classification of DEGs between SOB and SOFB.

doi:10.1371/journal.pone.0142542.s001

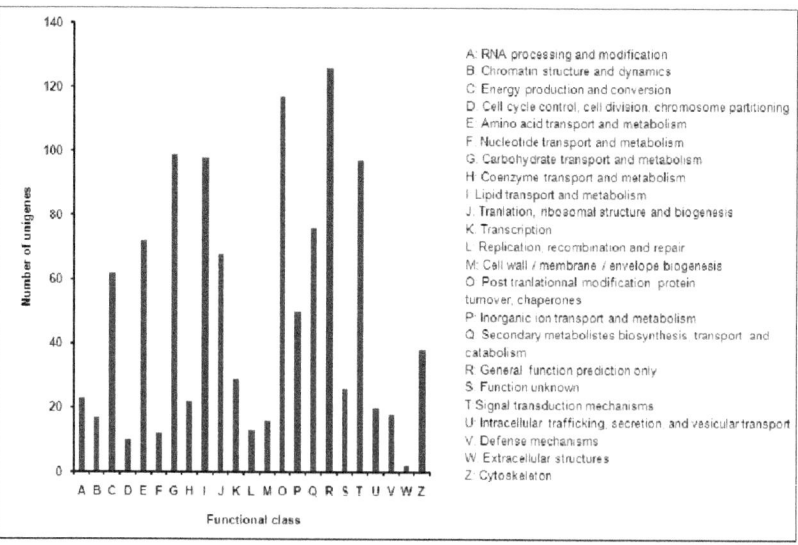

S2 Figure. KOG classification of DEGs between SOF and SOB. doi:10.1371/journal.pone.0142542.s002

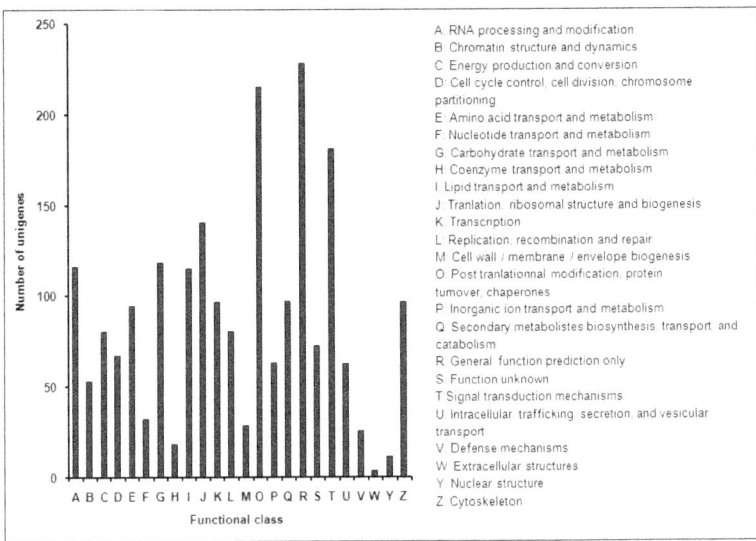

S3 Figure. KOG classification of DEGs between SOF and SOFB. doi:10.1371/journal.pone.0142542.s003

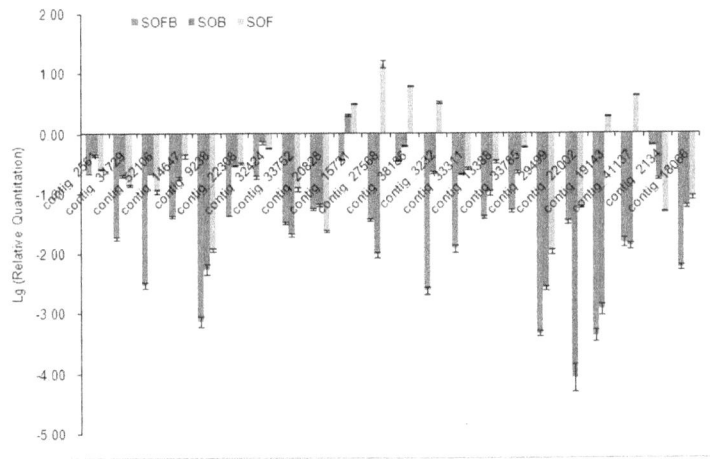

S4 Figure. Real-time qPCR validation of genes involved in flower pigmentation and scent biosynthesis in *S. oblata*. The y-axis indicates fold change in expression among SOFB (blue bars), SOB (red bars) and SOF (green bars) samples using the results of RT-qPCR. Data were normalized against the reference S. oblata ACT-7 gene. Quantitative PCR for each gene was performed with three biological replicates, with three technical replicates per experiment. Bars indicate SD.

doi:10.1371/journal.pone.0142542.s004

S1 File. Primers used for analysis of gene expression by qRT-PCR.

Primer Name	Sequence (5'to3')
contig_2561Forward Primer	CCTCCACGACGCCAAACTT
contig_2561Reverse Primer	GCCACCAAGCATTGACCAA
contig_9238Forward Primer	TGCGAATTTAGGGACACCAT
contig_9238Reverse Primer	CCAACTGGCTCAAACTCGAT
contig_14647Forward Primer	TTCCACCGGACATATCAAACC
contig_14647Reverse Primer	TGGCTCGTGATTTCTTTGCTT
contig_15721Forward Primer	GGCTACGACAACACCCTATGG
contig_15721Reverse Primer	TTCCTCATCGGCGCATCT
contig_20828Forward Primer	CAGTTCAGCCGTGCTCTCAA
contig_20828Reverse Primer	ACCCAAATAATCCCTCTCACTATCC
contig_22308Forward Primer	GGCTAACTTGCTACCATAACCTTGA
contig_22308Reverse Primer	CCGGTCGAAGAGAAGGAGAA
contig_27568Forward Primer	CATCTGCAGAAGGAATTGACTCA
contig_27568Reverse Primer	TGCTGGCTTGGAAGGGACTA
contig_32424Forward Primer	GCACGCATCAGTTGTATACATCAG
contig_32424Reverse Primer	CTTCTGCTAATGCAGCCAGTTCT
contig_33752Forward Primer	AAGCGCGAAGAGTTTGTGATC
contig_33752Reverse Primer	TTCCATTGCTAGCTAAGCGTGAT
contig_34729Forward Primer	GCTTGGCAAATGCCTCGTTA
contig_34729Reverse Primer	CCATGGCCGAGCACATC
contig_38186Forward Primer	ATGTTGGTGGTGGAACTGGT
contig_38186Reverse Primer	ATATCTCCGCCAACATGCTC
contig_52106Forward Primer	GGGCTGACCCCATTCCTTT
contig_52106Reverse Primer	GTGGTGGTTGAAGTGCCTAAATT
contig_426Forward Primer	TGCCCTGGGAGCATCATC
contig_426Reverse Primer	CAGCCCCTTGTTTGTGATAATG
contig_4430Forward Primer	GAGCAGAACCACGCACCATA
contig_4430Reverse Primer	GGACGACCAAACCCTAGTTAGCT
contig_3232Forward Primer	CGTCGGGAATATGTTATGGG
contig_3232Reverse Primer	GAAATGGCATTGGAGCAGTT
contig_33311Forward Primer	GAGAATCCGGCGAATTATGA

contig_33311Reverse Primer	AACATTTTGCACTCCCTTGG
contig_33785Forward Primer	GATGTCATCATATTCCGCCC
contig_33785Reverse Primer	TTTTAGGGATTGGCCTGTTG
contig_13388Forward Primer	ATTGCCATATCAACCCCAAA
contig_13388Reverse Primer	TTTAGCCAAAGGAGCCAAGA
contig_29499Forward Primer	GGCAGCTTCAATCCGTAGAG
contig_29499Reverse Primer	GAATCTGCACGTACCCCACT
contig_22002Forward Primer	TGGAAATTGCACGGATAACA
contig_22002Reverse Primer	TTGCCAACCAACGAATTACA
contig_19143Forward Primer	GGTCAGGTCATCCTTCCTCA
contig_19143Reverse Primer	ACAAGAAGAAGCGGTCCAGA
contig_41137Forward Primer	GCCCTGAAGCAAACCATTTA
contig_41137Reverse Primer	CAAGGCACTTCTTGACACCA
contig_2134Forward Primer	AGAGGCTCGACAACAGGAAA
contig_2134Reverse Primer	GTAGCAGCTCTGCAAAACCC
contig_18066Forward Primer	CGGCAAGAAAATCTCCACAT
contig_18066Reverse Primer	TACTGCGTAAACACCCTCCC

S2 File. Summary of GO term assignment for the *S. oblata* inflorescence transcriptome (Level2).

doi:10.1371/journal.pone.0142542.s006

Go type	Term	Unigene counts	Percent (%)	GO_id
biological_ process	cellular process	21398	65.87%	GO:0009987
	metabolic process	21249	65.41%	GO:0008152
	single-organism process	18869	58.08%	GO:0044699
	response to stimulus	11276	34.71%	GO:0050896
	biological regulation	9888	30.44%	GO:0065007
	regulation of biological process	9018	27.76%	GO:0050789
	developmental process	7158	22.03%	GO:0032502
	cellular component organization or biogenesis	7133	21.96%	GO:0071840
	multicellular organismal process	7051	21.70%	GO:0032501
	localization	6244	19.22%	GO:0051179
	establishment of localization	5889	18.13%	GO:0051234
	reproduction	4130	12.71%	GO:0000003
	reproductive process	3880	11.94%	GO:0022414
	multi-organism process	3737	11.50%	GO:0051704
	signaling	3503	10.78%	GO:0023052
	negative regulation of biological process	2413	7.43%	GO:0048519
	positive regulation of biological process	1970	6.06%	GO:0048518
	immune system process	1924	5.92%	GO:0002376
	growth	1798	5.53%	GO:0040007
	regulation of molecular function	661	2.03%	GO:0065009
	rhythmic process	337	1.04%	GO:0048511
	positive regulation of molecular function	280	0.86%	GO:0044093
	biological adhesion	278	0.86%	GO:0022610
	negative regulation of molecular function	154	0.47%	GO:0044092
	locomotion	56	0.17%	GO:0040011
	cell killing	2	0.01%	GO:0001906

cellular_ component	cell	30837	90.98%	GO:0005623
	cell part	30834	90.97%	GO:0044464
	organelle	24228	71.48%	GO:0043226
	membrane	8928	26.34%	GO:0016020
	organelle part	5342	15.76%	GO:0044422
	extracellular region	2716	8.01%	GO:0005576
	macromolecular complex	2414	7.12%	GO:0032991
	cell junction	1854	5.47%	GO:0030054
	symplast	1836	5.42%	GO:0055044
	membrane part	1540	4.54%	GO:0044425
	membrane-enclosed lumen	936	2.76%	GO:0031974
	extracellular region part	69	0.20%	GO:0044421
	nucleoid	55	0.16%	GO:0009295
	extracellular matrix	25	0.07%	GO:0031012
	extracellular matrix part	13	0.04%	GO:0044420
	synapse	10	0.03%	GO:0045202
	synapse part	6	0.02%	GO:0044456
	virion	1	0.00%	GO:0019012
	virion part	1	0.00%	GO:0044423

molecular_ function	catalytic activity	13162	45.56%	GO:0003824
	binding	10793	37.36%	GO:0005488
	transporter activity	2204	7.63%	GO:0005215
	nucleic acid binding transcription factor activity	2188	7.57%	GO:0001071
	molecular transducer activity	502	1.74%	GO:0060089
	structural molecule activity	392	1.36%	GO:0005198
	enzyme regulator activity	261	0.90%	GO:0030234
	receptor activity	216	0.75%	GO:0004872
	antioxidant activity	141	0.49%	GO:0016209
	protein binding transcription factor activity	74	0.26%	GO:0000988
	electron carrier activity	63	0.22%	GO:0009055
	guanyl-nucleotide exchange factor activity	52	0.18%	GO:0005085
	nutrient reservoir activity	31	0.11%	GO:0045735
	metallochaperone activity	9	0.03%	GO:0016530
	protein tag	8	0.03%	GO:0031386
	translation regulator activity	6	0.02%	GO:0045182
	channel regulator activity	4	0.01%	GO:0016247
	Total number of unigenes by GO assignments	**36967**		

S3 File. KOG annotation of *S. oblata* **unigenes.** doi:10.1371/journal.pone.0142542. s007

KOG catego- ries	KOG description	Number	Percent (%)
A	RNA processing and modification	902	4.52%
B	Chromatin structure and dynamics	303	1.52%
C	Energy production and conversion	835	4.18%
D	Cell cycle control, cell division, chromosome partitioning	610	3.06%
E	Amino acid transport and metabolism	906	4.54%

F	Nucleotide transport and metabolism	225	1.13%
G	Carbohydrate transport and metabolism	1240	6.21%
H	Coenzyme transport and metabolism	208	1.04%
I	Lipid transport and metabolism	983	4.93%
J	Translation, ribosomal structure and biogenesis	892	4.47%
K	Transcription	1249	6.26%
L	Replication, recombination and repair	676	3.39%
M	Cell wall/membrane/envelope biogenesis	336	1.68%
N	Cell motility	15	0.08%
O	Posttranslational modification, protein turnover, chaperones	2294	11.50%
P	Inorganic ion transport and metabolism	658	3.30%
Q	Secondary metabolites biosynthesis, transport and catabolism	924	4.63%
R	General function prediction only	2566	12.86%
S	Function unknown	1252	6.27%
T	Signal transduction mechanisms	3276	16.42%
U	Intracellular trafficking, secretion, and vesicular transport	1203	6.03%
V	Defense mechanisms	174	0.87%
W	Extracellular structures	94	0.47%
Y	Nuclear structure	65	0.33%
Z	Cytoskeleton	648	3.25%
Total		19,956	100.00%

S4 File. Summary of KEGG pathways involved in the *S. oblata* **inflorescence transcriptome.** doi:10.1371/journal.pone.0142542.s008

Number	Pathway_ID	Pathway_description	Unigene_number	Percent (%)
1	ko01100	Metabolic pathways	2906	23.46%
2	ko01110	Biosynthesis of secondary metabolites	1574	12.71%
3	ko01120	Microbial metabolism in diverse environments	619	5.00%
4	ko04141	Protein processing in endo-plasmic reticulum	521	4.21%
5	ko05169	Epstein-Barr virus infection	400	3.23%
6	ko01230	Biosynthesis of amino acids	396	3.20%
7	ko01200	Carbon metabolism	395	3.19%
8	ko04075	Plant hormone signal trans-duction	378	3.05%
9	ko00240	Pyrimidine metabolism	352	2.84%
10	ko04626	Plant-pathogen interaction	328	2.65%
11	ko00500	Starch and sucrose metabolism	318	2.57%
12	ko03040	Spliceosome	275	2.22%
13	ko03013	RNA transport	251	2.03%
14	ko00230	Purine metabolism	245	1.98%
15	ko00020	Citrate cycle (TCA cycle)	245	1.98%
16	ko04721	Synaptic vesicle cycle	240	1.94%
17	ko03010	Ribosome	230	1.86%
18	ko03015	mRNA surveillance pathway	224	1.81%
19	ko00190	Oxidative phosphorylation	220	1.78%
20	ko05203	Viral carcinogenesis	217	1.75%
21	ko00520	Amino sugar and nucleotide sugar metabolism	208	1.68%
22	ko04144	Endocytosis	207	1.67%
23	ko03018	RNA degradation	206	1.66%
24	ko00010	Glycolysis / Gluconeogenesis	206	1.66%
25	ko04110	Cell cycle	205	1.65%

26 ko05164		Influenza A	205	1.65%
27	ko00940	Phenylpropanoid biosynthesis	201	1.62%
28	ko04120	Ubiquitin mediated proteolysis	195	1.57%
29	ko04151	PI3K-Akt signaling pathway	190	1.53%
30	ko04114	Oocyte meiosis	189	1.53%
31	ko04910	Insulin signaling pathway	180	1.45%
32	ko00564	Glycerophospholipid metabolism	168	1.36%
33	ko00270	Cysteine and methionine metabolism	167	1.35%
34	ko05205	Proteoglycans in cancer	164	1.32%
35	ko04145	Phagosome	158	1.28%
36	ko00620	Pyruvate metabolism	157	1.27%
37	ko05168	Herpes simplex infection	157	1.27%
38	ko05166	HTLV-I infection	155	1.25%
39	ko04152	AMPK signaling pathway	154	1.24%
40	ko04722	Neurotrophin signaling pathway	147	1.19%
41	ko00590	Arachidonic acid metabolism	143	1.15%
42	ko04111	Cell cycle - yeast	139	1.12%
43	ko00970	Aminoacyl-tRNA biosynthesis	133	1.07%
44	ko03008	Ribosome biogenesis in eukaryotes	132	1.07%
45	ko04810	Regulation of actin cytoskeleton	132	1.07%
46	ko04010	MAPK signaling pathway	132	1.07%
47	ko04068	FoxO signaling pathway	130	1.05%
48	ko00360	Phenylalanine metabolism	129	1.04%
49	ko01212	Fatty acid metabolism	128	1.03%
50	ko05206	MicroRNAs in cancer	127	1.03%
51	ko00983	Drug metabolism - other enzymes	127	1.03%
52	ko04150	mTOR signaling pathway	124	1.00%
53	ko04146	Peroxisome	123	0.99%

54	ko04113	Meiosis - yeast	122	0.98%
55	ko04666	Fc gamma R-mediated phagocytosis	121	0.98%
56	ko04070	Phosphatidylinositol signaling system	120	0.97%
57	ko03420	Nucleotide excision repair	119	0.96%
58	ko04261	Adrenergic signaling in cardiomyocytes	118	0.95%
59	ko00260	Glycine, serine and threonine metabolism	117	0.94%
60	ko00040	Pentose and glucuronate interconversions	117	0.94%
61	ko04720	Long-term potentiation	117	0.94%
62	ko00052	Galactose metabolism	116	0.94%
63	ko04728	Dopaminergic synapse	116	0.94%
64	ko00710	Carbon fixation in photosynthetic organisms	115	0.93%
65	ko00480	Glutathione metabolism	115	0.93%
66	ko05034	Alcoholism	114	0.92%
67	ko00561	Glycerolipid metabolism	113	0.91%
68	ko04932	Non-alcoholic fatty liver disease (NAFLD)	112	0.90%
69	ko00330	Arginine and proline metabolism	112	0.90%
70	ko04919	Thyroid hormone signaling pathway	112	0.90%
71	ko03030	DNA replication	111	0.90%
72	ko00680	Methane metabolism	111	0.90%
73	ko04921	Oxytocin signaling pathway	110	0.89%
74	ko00562	Inositol phosphate metabolism	108	0.87%
75	ko05132	Salmonella infection	108	0.87%
76	ko04310	Wnt signaling pathway	107	0.86%
77	ko04014	Ras signaling pathway	106	0.86%
78	ko04020	Calcium signaling pathway	102	0.82%
79	ko04390	Hippo signaling pathway	101	0.82%
80	ko04142	Lysosome	100	0.81%
81	ko00630	Glyoxylate and dicarboxylate metabolism	100	0.81%

82	ko04022	cGMP-PKG signaling pathway	98	0.79%
83	ko04066	HIF-1 signaling pathway	96	0.77%
84	ko04724	Glutamatergic synapse	96	0.77%
85	ko00909	Sesquiterpenoid and triterpenoid biosynthesis	96	0.77%
86	ko00945	Stilbenoid, diarylheptanoid and gingerol biosynthesis	96	0.77%
87	ko00903	Limonene and pinene degradation	96	0.77%
88	ko03440	Homologous recombination	95	0.77%
89	ko00624	Polycyclic aromatic hydrocarbon degradation	95	0.77%
90	ko01210	2-Oxocarboxylic acid metabolism	94	0.76%
91	ko00960	Tropane, piperidine and pyridine alkaloid biosynthesis	93	0.75%
92	ko00900	Terpenoid backbone biosynthesis	92	0.74%
93	ko03460	Fanconi anemia pathway	91	0.73%
94	ko03050	Proteasome	90	0.73%
95	ko04510	Focal adhesion	90	0.73%
96	ko00051	Fructose and mannose metabolism	87	0.70%
97	ko00627	Aminobenzoate degradation	87	0.70%
98	ko05134	Legionellosis	86	0.69%
99	ko04915	Estrogen signaling pathway	86	0.69%
100	ko04914	Progesterone-mediated oocyte maturation	85	0.69%
101	ko04623	Cytosolic DNA-sensing pathway	85	0.69%
102	ko00363	Bisphenol degradation	85	0.69%
103	ko04650	Natural killer cell mediated cytotoxicity	84	0.68%
104	ko04912	GnRH signaling pathway	83	0.67%
105	ko00860	Porphyrin and chlorophyll metabolism	82	0.66%
106	ko00030	Pentose phosphate pathway	82	0.66%

107	ko00250	Alanine, aspartate and gluta-mate metabolism	81	0.65%
108	ko04370	VEGF signaling pathway	81	0.65%
109	ko00300	Lysine biosynthesis	79	0.64%
110	ko03430	Mismatch repair	78	0.63%
111	ko00460	Cyanoamino acid metabo-lism	78	0.63%
112	ko00195	Photosynthesis	77	0.62%
113	ko04612	Antigen processing and presentation	77	0.62%
114	ko04360	Axon guidance	77	0.62%
115	ko05161	Hepatitis B	76	0.61%
116	ko00400	Phenylalanine, tyrosine and tryptophan biosynthesis	75	0.61%
117	ko04350	TGF-beta signaling pathway	75	0.61%
118	ko00280	Valine, leucine and isoleu-cine degradation	74	0.60%
119	ko04530	Tight junction	71	0.57%
120	ko04712	Circadian rhythm - plant	70	0.57%
121	ko04115	p53 signaling pathway	70	0.57%
122	ko00071	Fatty acid degradation	70	0.57%
123	ko00510	N-Glycan biosynthesis	69	0.56%
124	ko00053	Ascorbate and aldarate me-tabolism	68	0.55%
125	ko04976	Bile secretion	68	0.55%
126	ko03022	Basal transcription factors	67	0.54%
127	ko00720	Carbon fixation pathways in prokaryotes	67	0.54%
128	ko04391	Hippo signaling pathway - fly	67	0.54%
129	ko04015	Rap1 signaling pathway	67	0.54%
130	ko00410	beta-Alanine metabolism	66	0.53%
131	ko00565	Ether lipid metabolism	66	0.53%
132	ko04620	Toll-like receptor signaling pathway	66	0.53%
133	ko04270	Vascular smooth muscle contraction	65	0.52%
134	ko03060	Protein export	63	0.51%

135	ko04916	Melanogenesis	63	0.51%
136	ko04520	Adherens junction	63	0.51%
137	ko00830	Retinol metabolism	63	0.51%
138	ko00350	Tyrosine metabolism	61	0.49%
139	ko00592	alpha-Linolenic acid metabolism	61	0.49%
140	ko00640	Propanoate metabolism	60	0.48%
141	ko03020	RNA polymerase	59	0.48%
142	ko02010	ABC transporters	59	0.48%
143	ko00906	Carotenoid biosynthesis	58	0.47%
144	ko00061	Fatty acid biosynthesis	58	0.47%
145	ko05202	Transcriptional misregulation in cancer	58	0.47%
146	ko05133	Pertussis	58	0.47%
147	ko00100	Steroid biosynthesis	57	0.46%
148	ko01040	Biosynthesis of unsaturated fatty acids	57	0.46%
149	ko03320	PPAR signaling pathway	57	0.46%
150	ko04662	B cell receptor signaling pathway	57	0.46%
151	ko04540	Gap junction	57	0.46%
152	ko05031	Amphetamine addiction	57	0.46%
153	ko04611	Platelet activation	54	0.44%
154	ko03410	Base excision repair	53	0.43%
155	ko04130	SNARE interactions in vesicular transport	53	0.43%
156	ko00910	Nitrogen metabolism	53	0.43%
157	ko00310	Lysine degradation	53	0.43%
158	ko04730	Long-term depression	53	0.43%
159	ko00513	Various types of N-glycan biosynthesis	51	0.41%
160	ko00600	Sphingolipid metabolism	51	0.41%
161	ko00380	Tryptophan metabolism	51	0.41%
162	ko04660	T cell receptor signaling pathway	51	0.41%
163	ko00130	Ubiquinone and other terpenoid-quinone biosynthesis	50	0.40%

164	ko04664	Fc epsilon RI signaling pathway	49	0.40%
165	ko00920	Sulfur metabolism	48	0.39%
166	ko04727	GABAergic synapse	48	0.39%
167	ko04917	Prolactin signaling pathway	48	0.39%
168	ko04380	Osteoclast differentiation	47	0.38%
169	ko04062	Chemokine signaling pathway	46	0.37%
170	ko00780	Biotin metabolism	45	0.36%
171	ko00670	One carbon pool by folate	44	0.36%
172	ko01220	Degradation of aromatic compounds	44	0.36%
173	ko04742	Taste transduction	44	0.36%
174	ko04210	Apoptosis	43	0.35%
175	ko04140	Regulation of autophagy	42	0.34%
176	ko00450	Selenocompound metabolism	42	0.34%
177	ko00062	Fatty acid elongation	42	0.34%
178	ko00253	Tetracycline biosynthesis	42	0.34%
179	ko00511	Other glycan degradation	41	0.33%
180	ko00073	Cutin, suberine and wax biosynthesis	41	0.33%
181	ko04920	Adipocytokine signaling pathway	41	0.33%
182	ko04012	ErbB signaling pathway	41	0.33%
183	ko04750	Inflammatory mediator regulation of TRP channels	41	0.33%
184	ko00770	Pantothenate and CoA biosynthesis	40	0.32%
185	ko04961	Endocrine and other factor-regulated calcium reabsorption	39	0.31%
186	ko04713	Circadian entrainment	39	0.31%
187	ko00430	Taurine and hypotaurine metabolism	39	0.31%
188	ko04966	Collecting duct acid secretion	37	0.30%
189	ko04621	NOD-like receptor signaling pathway	37	0.30%
190	ko00982	Drug metabolism - cytochrome P450	37	0.30%

191	ko04670	Leukocyte transendothelial migration	37	0.30%
192	ko00980	Metabolism of xenobiotics by cytochrome P450	37	0.30%
193	ko00950	Isoquinoline alkaloid biosynthesis	36	0.29%
194	ko04710	Circadian rhythm	36	0.29%
195	ko00290	Valine, leucine and isoleucine biosynthesis	35	0.28%
196	ko04972	Pancreatic secretion	35	0.28%
197	ko04064	NF-kappa B signaling pathway	35	0.28%
198	ko04320	Dorso-ventral axis formation	35	0.28%
199	ko04723	Retrograde endocannabinoid signaling	34	0.27%
200	ko00650	Butanoate metabolism	33	0.27%
201	ko00904	Diterpenoid biosynthesis	33	0.27%
202	ko04918	Thyroid hormone synthesis	33	0.27%
203	ko00340	Histidine metabolism	32	0.26%
204	ko04330	Notch signaling pathway	32	0.26%
205	ko00941	Flavonoid biosynthesis	31	0.25%
206	ko00908	Zeatin biosynthesis	31	0.25%
207	ko04668	TNF signaling pathway	30	0.24%
208	ko00440	Phosphonate and phosphinate metabolism	30	0.24%
209	ko04745	Phototransduction - fly	30	0.24%
210	ko05204	Chemical carcinogenesis	30	0.24%
211	ko04630	Jak-STAT signaling pathway	29	0.23%
212	ko00563	Glycosylphosphatidylinositol(GPI)-anchor biosynthesis	27	0.22%
213	ko02020	Two-component system	27	0.22%
214	ko00790	Folate biosynthesis	26	0.21%
215	ko04726	Serotonergic synapse	26	0.21%
216	ko00625	Chloroalkane and chloroalkene degradation	26	0.21%
217	ko04260	Cardiac muscle contraction	25	0.20%
218	ko00760	Nicotinate and nicotinamide metabolism	25	0.20%

219	ko04962	Vasopressin-regulated water reabsorption	25	0.20%
220	ko00591	Linoleic acid metabolism	25	0.20%
221	ko00902	Monoterpenoid biosynthesis	25	0.20%
222	ko04970	Salivary secretion	24	0.19%
223	ko04960	Aldosterone-regulated sodium reabsorption	24	0.19%
224	ko04725	Cholinergic synapse	24	0.19%
225	ko04011	MAPK signaling pathway - yeast	24	0.19%
226	ko04978	Mineral absorption	23	0.19%
227	ko00750	Vitamin B6 metabolism	23	0.19%
228	ko04112	Cell cycle - Caulobacter	23	0.19%
229	ko04744	Phototransduction	23	0.19%
230	ko00730	Thiamine metabolism	22	0.18%
231	ko04340	Hedgehog signaling pathway	22	0.18%
232	ko04971	Gastric acid secretion	21	0.17%
233	ko00905	Brassinosteroid biosynthesis	20	0.16%
234	ko04974	Protein digestion and absorption	20	0.16%
235	ko04973	Carbohydrate digestion and absorption	20	0.16%
236	ko00521	Streptomycin biosynthesis	20	0.16%
237	ko05032	Morphine addiction	20	0.16%
238	ko00603	Glycosphingolipid biosynthesis - globo series	20	0.16%
239	ko04013	MAPK signaling pathway - fly	20	0.16%
240	ko04740	Olfactory transduction	18	0.15%
241	ko03070	Bacterial secretion system	17	0.14%
242	ko00626	Naphthalene degradation	17	0.14%
243	ko04122	Sulfur relay system	16	0.13%
244	ko00362	Benzoate degradation	16	0.13%
245	ko00942	Anthocyanin biosynthesis	16	0.13%
246	ko00660	C5-Branched dibasic acid metabolism	15	0.12%
247	ko04975	Fat digestion and absorption	15	0.12%
248	ko00740	Riboflavin metabolism	14	0.11%

249	ko04622	RIG-I-like receptor signaling pathway	14	0.11%
250	ko00531	Glycosaminoglycan degradation	14	0.11%
251	ko00944	Flavone and flavonol biosynthesis	14	0.11%
252	ko05033	Nicotine addiction	14	0.11%
253	ko03450	Non-homologous end-joining	13	0.10%
254	ko00524	Butirosin and neomycin biosynthesis	13	0.10%
255	ko00072	Synthesis and degradation of ketone bodies	12	0.10%
256	ko04711	Circadian rhythm - fly	12	0.10%
257	ko00196	Photosynthesis - antenna proteins	11	0.09%
258	ko00540	Lipopolysaccharide biosynthesis	11	0.09%
259	ko04911	Insulin secretion	10	0.08%
260	ko00401	Novobiocin biosynthesis	9	0.07%
261	ko00514	Other types of O-glycan biosynthesis	9	0.07%
262	ko00361	Chlorocyclohexane and chlorobenzene degradation	9	0.07%
263	ko00604	Glycosphingolipid biosynthesis - ganglio series	9	0.07%
264	ko04964	Proximal tubule bicarbonate reclamation	8	0.06%
265	ko00785	Lipoic acid metabolism	7	0.06%
266	ko00351	DDT degradation	7	0.06%
267	ko00966	Glucosinolate biosynthesis	6	0.05%
268	ko00901	Indole alkaloid biosynthesis	5	0.04%
269	ko01051	Biosynthesis of ansamycins	5	0.04%
270	ko00643	Styrene degradation	4	0.03%
271	ko05030	Cocaine addiction	4	0.03%
272	ko00254	Aflatoxin biosynthesis	4	0.03%
273	ko00232	Caffeine metabolism	3	0.02%
274	ko00140	Steroid hormone biosynthesis	3	0.02%
275	ko01053	Biosynthesis of siderophore group nonribosomal peptides	3	0.02%

276	ko04977	Vitamin digestion and absorption	3	0.02%
277	ko00550	Peptidoglycan biosynthesis	2	0.02%
278	ko00281	Geraniol degradation	2	0.02%
279	ko00364	Fluorobenzoate degradation	2	0.02%
280	ko00965	Betalain biosynthesis	2	0.02%
281	ko00623	Toluene degradation	2	0.02%
282	ko04913	Ovarian steroidogenesis	2	0.02%
283	ko00471	D-Glutamine and D-glutamate metabolism	2	0.02%
284	ko04514	Cell adhesion molecules (CAMs)	1	0.01%
285	ko00523	Polyketide sugar unit biosynthesis	1	0.01%
286	ko04512	ECM-receptor interaction	1	0.01%

S5 File. Summary of flower pigment biosynthesis and floral scent metabolism in *S. oblata*. doi:10.1371/journal.pone.0142542.s009

S6 File. GO classification of DEGs between SOB and SOFB.

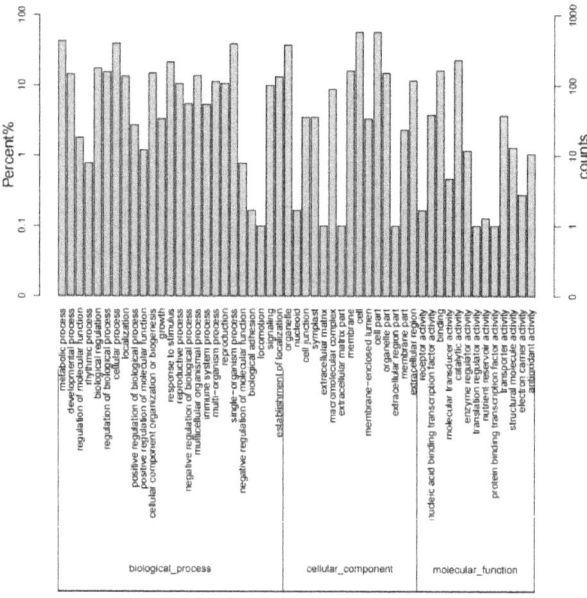

S7 File. GO classification of DEGs between SOF and SOB.

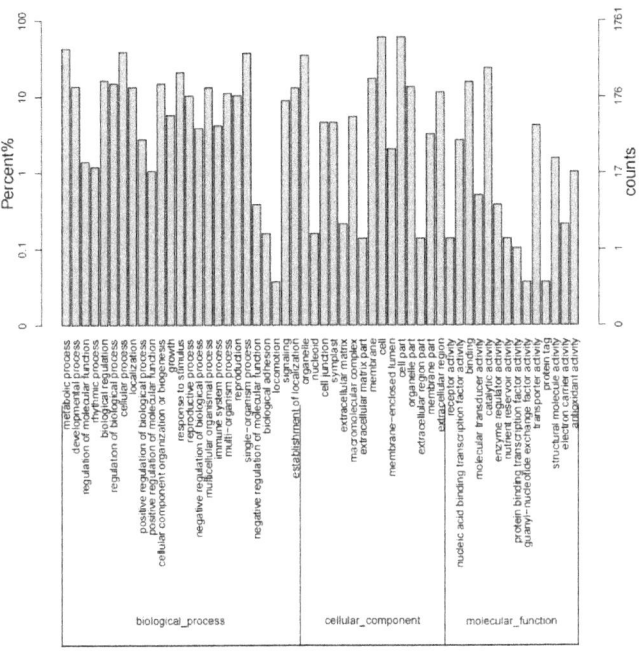

S8 File. GO classification of DEGs between SOF and SOFB.

doi:10.1371/journal.pone.0142542.s015

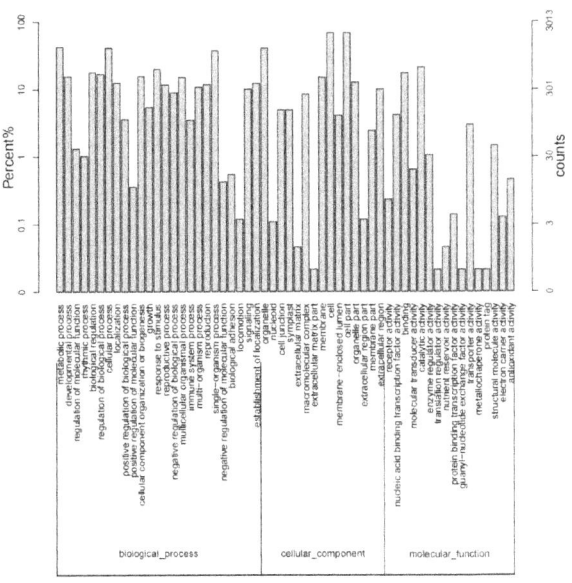

S9 File. Comparison of gene set enrichment between SOF and SOFB by GO and KEGG.

doi:10.1371/journal.pone.0142542.s016

Pathway ID	KEGG pathway category	DEGs with pathway annotation (3298)	Up-reg-ulation	Down-regula-tion	Qvalue
ko03030	**DNA replication**	35(1.06%)	0	35	2.19E-06
ko00040	Pentose and glucuro-nate interconversions	36(1.09%)	32	4	2.79E-06
ko00362	Benzoate degradation	9(0.27%)	9	0	1.15E-04
ko04020	Calcium signaling pathway	28(0.85%)	2	26	2.11E-04
ko00900	Terpenoid backbone biosynthesis	26(0.79%)	22	4	2.16E-04
ko04918	Thyroid hormone synthesis	13(0.39%)	3	10	2.74E-04
ko00071	Fatty acid degradation	21(0.64%)	15	6	3.49E-04
ko03008	Ribosome biogenesis in eukaryotes	33(1.00%)	0	33	3.66E-04
ko03010	**Ribosome**	50(1.52%)	0	50	4.71E-04
ko04914	Progesterone-mediated oocyte maturation	23(0.70%)	1	22	8.47E-04
ko00592	alpha-Linolenic acid metabolism	18(0.55%)	17	1	1.02E-03
ko04910	Insulin signaling pathway	9(0.27%)	8	1	1.78E-03
ko04728	Dopaminergic synapse	4(0.12%)	3	1	2.07E-03
ko05134	Legionellosis	22(0.67%)	4	18	2.21E-03

Acknowledgments

We thank the National Forest Genetic Resources Platform (NFGRP) for providing the *Syringa oblata.* resources. We also thank Professor Yongxiu Liu (Institute of Botany, Chinese Academy of Sciences) and Shengqing Shi (Research Institute of Forestry, Chinese Academy of Forestry) for their excellent advice on this paper.

Author Contributions

Conceived and designed the experiments: PL. Performed the experiments: JZ

XG DD GB YW YG. Analyzed the data: JZ ZH WL. Contributed reagents/ materials/analysis tools: JZ ZH. Wrote the paper: JZ.

REFERENCES

1. Huang JX, Wang LS, Li XM, Lu YQ. Advances in molecular basis and evolution of floral color variation. Chin Bull Bot. 2006, 23(4): 321–333.

2. Shirley BW, Kubasek WL, Storz G, Bruggemann E, Koornneef M, Ausubel FM, et al. Analysis of Arabidopsis mutants deficient in flavonoid biosynthesis. Plant J. 2002, 8(5): 659–671.

3. Grotewold E. The genetics and biochemistry of floral pigments. Annu Rev Plant Biol. 2006, 57: 761– 780. PMID: 16669781

4. Tanaka Y, Sasaki N, Ohmiya A. Biosynthesis of plant pigments: anthocyanins, betalains and carotenoids. Plant J. 2008, 54(4): 733–749. doi: 10.1111/j.1365-313X.2008.03447.x PMID: 18476875

5. Albert NW, Lewis DH, Zhang H, Schwinn KE, Jameson PE, Davies KM. Members of an R2R3-MYB transcription factor family in Petunia are developmentally and environmentally regulated to control complex floral and vegetative pigmentation patterning. Plant J. 2011, 65(5): 771–784. doi: 10.1111/j.1365- 313X.2010.04465.x PMID: 21235651

6. Baudry A, Heim MA, Dubreucq B, Caboche M, Weisshaar B, Lepiniec L. TT2, TT8, and TTG1 synergistically specify the expression of BANYULS and proanthocyanidin biosynthesis in Arabidopsis thaliana. Plant J. 2004, 39(3): 366–380. PMID: 15255866

7. Elomaa P, Uimari A, Mehto M, Albert VA, Laitinen RAE, Teeri TH. Activation of anthocyanin biosynthesis in Gerbera hybrida (Asteraceae) suggests conserved protein-protein and protein-promoter interactions between the anciently diverged monocots and eudicots. Plant Physiol. 2003, 133(4): 1831–1842. PMID: 14605235

8. Nakatsuka T, SanaeHaruta K, Pitaksutheepong C, Abe Y, Kakizaki Y, Yamamoto K, et al. Identification and characterization of R2R3-MYB and bHLH transcription factors regulating anthocyanin biosynthesis in gentian flowers. Plant Cell Physiol. 2008, 49(12): 1818–1829. doi: 10.1093/pcp/pcn163 PMID: 18974195

9. Yamagishi M, Shimoyamada Y, Nakatsuka T, Masuda K. Two R2R3-MYB genes, homologs of petunia AN2, regulate anthocyanin biosyntheses in flower tepals, tepal spots and leaves of Asiatic hybrid lily. Plant Cell Physiol. 2010, 51(3): 463–474. doi:

10. 10.1093/pcp/pcq011 PMID: 20118109 10. LI Z, CAO H, LIU L, LI

B. Chemical constituents of aroma in fresh Syringa oblata flowers. J Zhejiang For Coll. 2006, 23(2): 159–162.

11. Hendel-Rahmanim K, Masci T, Vainstein A, Weiss D. Diurnal regulation of scent emission in rose flowers. Planta. 2007, 226(6): 1491–1499. PMID: 17636322

12. Zhao J, Hu ZH, Leng PS, Zhang HX, Cheng FY. Fragrance composition in six tree peony cultivars. Kor J Hortic Sci Technol. 2012, 30(6): 617–625.

13. Zhang HX, Hu ZH, Leng PS, Wang WH, Xu F, Zhao J. Qualitative and quantitative analysis of floral volatile components from different varieties of Lilium spp. Scientia Agri Sin. 2013, 46(4): 790–799.

14. Hao RJ, Zhang Q, Yang WR, Wang J, Cheng TR, Pan HT, et al. Emitted and endogenous floral scent compounds of Prunus mume and hybrids. Biochem System Ecol. 2014, 54: 23–30.

15. Knudsen JT, Roger E, Jonathan G, Stahl B. Diversity and distribution of floral scent. Bot Rev. 2006, 72 (1): 1–120.

16. Nagegowda DA. Plant volatile terpenoid metabolism: Biosynthetic genes, transcriptional regulation and subcellular compartmentation. FEBS Lett.

17. 2010, 584: 2965–2973. doi: 10.1016/j.febslet.2010.05.045 PMID: 20553718

18. Cui HX, Jiang GM, Zang SY. The distribution, origin and evolution of Syringa. Bull Bot Res. 2004, 24 (2): 141–145.

19. Zang SY, Cui HX. lilac. Shanghai: Shanghai Science and Technology Press. 2000.

20. Li YN, Svenson SE. Syringa oblata. Am Nurserym. 2014, 214(2): 38.

21. Wang Z, Gerstein M, Snyder M. RNA-Seq: a revolutionary tool for transcriptomics. Nature Rev Genet. 2009, 10(1): 57–63. doi: 10.1038/nrg2484 PMID: 19015660

22. 'tHoen PAC, Ariyurek Y, Thygesen HH, Vreugdenhil E, Vossen RHAM, Menezes ReXd, et al. Deep sequencing-based expression analysis shows major advances in robustness, resolution and inter-lab portability over five microarray platforms. Nucleic Acids Res. 2008, 36(21): e141. doi: 10.1093/nar/ gkn705 PMID: 18927111

23. Liu DF, Sui SZ, Ma J, Li ZN, Guo YL, Luo DP, et al. Transcriptomic analysis of flower development in wintersweet (Chimonanthus praecox). PLoS ONE. 2013, 9(1): e86976.

24. Zhang JX, Wu KL, Zeng SJ, Silva JATd, Zhao XL, Tian CE, et al. Transcriptome analysis of Cymbidium sinense and its application to

the identification of genes associated with floral development. BMC Genomics. 2013, 14: 279. doi: 10.1186/1471-2164-14-279 PMID: 23617896

25. Li XB, Luo J, Yan TL, Xiang L, Jin F, Qin DH, et al. Deep sequencing-based analysis of the Cymbidium ensifolium floral transcriptome. PLoS ONE. 2013, 8(12): e85480. doi: 10.1371/journal.pone.0085480 PMID: 24392013

26. Ge XX, Chen HW, Wang HL, Shi AP, Liu KF. De Novo assembly and annotation of Salvia splendens transcriptome using the Illumina platform. PLoS ONE. 2014, 9(3): e87693. doi: 10.1371/journal.pone. 0087693 PMID: 24622329

27. Robinson MD, McCarthy DJ, Smyth GK. edgeR: a Bioconductor package for differential expression analysis of digital gene expression data. Bioinformatics. 2010, 26(1): 139–140. doi: 10.1093/ bioinformatics/ btp616 PMID: 19910308

28. Anders S, Huber W. Differential expression analysis for sequence count data. Genome Biol. 2010, 11: R106. doi: 10.1186/gb-2010-11-10-r106 PMID: 20979621

29. Su CL, Chao YT, Alex-Chang YC, Chen WC, Chen CY, Lee AY, et al. De novo assembly of expressed transcripts and global analysis of the phalaenopsis aphrodite transcriptome. Plant Cell Physiol. 2011, 52 (9): 1501–1514. doi: 10.1093/pcp/pcr097 PMID: 21771864

30. Garg R, Patel RK, Tyagi AK, Jain M. De novo assembly of chickpea transcriptome using short reads for gene discovery and marker identification. DNA Res. 2011, 18(1): 53–63. doi: 10.1093/dnares/dsq028 PMID: 21217129

31. Bräutigam A, Mullick T, Schliesky S, Weber APM. Critical assessment of assembly strategies for nonmodel species mRNA-Seq data and application of next-generation sequencing to the comparison of C3 and C4 species. J Exp Bot. 2011, 62(9): 3093–3012. doi: 10.1093/jxb/err029 PMID: 21398430

32. Huang XQ, Madan A. CAP3: A DNA sequence assembly program. Genome Res. 1999, 9: 868–877. PMID: 10508846

33. Pruitt KD, Tatusova T, Maglott DR. NCBI reference sequences (RefSeq): a curated non-redundant sequence database of genomes, transcripts and proteins. Nucleic Acids Res. 2007, 35(suppl. 1): D61– D65.

34. Ashburner M, Ball CA, Blake JA, Botstein D, Butler H, Cherry JM, et al. Gene ontology: tool for the unification of biology. The Gene Ontology Consortium. Nat Genet. 2000, 25(1): 25–34. PMID: 10802651

35. Moriya Y, Itoh M, Okuda S, Yoshizawa AC, Kanehisa M. KAAS: an automatic genome annotation and pathway reconstruction server. Nucleic Acids Res. 2007, 35(suppl. 2): 182–185.

36. Young MD, Wakefield MJ, Smyth GK, Oshlack A. Gene ontology analysis for RNA-seq: accounting for selection bias. Genome Biol. 2010, 11: R14. doi: 10.1186/gb-2010-11-2-r14 PMID: 20132535

37. Mao XZ, Cai T, Olyarchuk JG, Wei LP. Automated genome annotation and pathway identification using the KEGG Orthology (KO) as a controlled vocabulary. Bioinformatics. 2005, 21(19): 3787–3793. PMID: 15817693

38. Cheadle C, Vawter MP, Freed WJ, Becker KG. Analysis of microarray data using Z score transformation. J Mol Diagn. 2003, 5(2): 73–81. PMID: 12707371

39. Novaes E, Drost DR, Farmerie WG, Pappas GJ, Grattapaglia D, Sederoff RR, et al. High-throughput gene and SNP discovery in Eucalyptus grandis, an uncharacterized genome. BMC Genomics. 2008, 9: 312. doi: 10.1186/1471-2164-9-312 PMID: 18590545

40. Iwashina T, Konta F, Kitajima J. Anthocyanins and flavonols of Chimonanthus praecox (Calycanthaceae) as flower pigments. J Jap Bot. 2001, 76(3): 166–172.

41. Hui RH, Li TC, Hou DY. Analysis of volatile componentsfrom flower and leaf of syringa oblata Lindl. By GC/MS. J Chin Mass Spectr Soc. 2002, 23(4): 210–213.

42. Feng C, Chen M, Xu CJ, Bai L, Yin XR, Li X, et al. Transcriptomic analysis of Chinese bayberry (Myrica rubra) fruit development and ripening using RNA-Seq. BMC Genomics. 2012, 13:19. doi: 10.1186/1471-2164-13-19 PMID: 22244270

43. Liu MY, Qiao GR, Jiang J, Yang HQ, Xie LH, Xie JZ, et al. Transcriptome sequencing and de novo analysis for Ma Bamboo (Dendrocalamus latiflorus Munro) using the Illumina platform. PLoS ONE. 2012, 7 (10): e46766. doi: 10.1371/journal.pone.0046766 PMID: 23056442

44. Li DJ, Deng Z, Qin B, Liu XH, Men ZH. De novo assembly and characterization of bark transcriptome using Illumina sequencing and development of EST-SSR markers in rubber tree (Hevea brasiliensis Muell. Arg.). BMC Genomics. 2012, 13: 192. doi: 10.1186/1471-2164-13-192 PMID: 22607098

45. Deng MH, Wen JF, Huo JL, Zhu HS, Dai XZ, Zhang ZQ, et al. Cloning and characterization of two novel purple pepper genes (CHS and F3H). Afr J Biotechnol. 2012, 11(9): 2389–2397.

46. Kunu W, Thanonkeo S, Thanonkeo P. Cloning and expression analysis of dihydroxyflavonol 4-reductase (DFR) in Ascocenda spp. Afr J Biotechnol. 2012, 11(64): 12702–12709.

47. Ogata J, Itoh Y, Ishida M, Yoshida H, Ozeki Y. Cloning and heterologous expression of cDNAs encoding flavonoid glucosyltransferases from Dianthus caryophyllus. Plant Biotechnol. 2004, 21(5): 367– 375.

48. Hatayama M, Ono E, Yonekura-Sakakibara K, Tanaka Y, Nishino T, Nakayama T. Biochemical characterization and mutational studies of a chalcone synthase from yellow snapdragon (Antirrhinum majus) flowers. Plant Biotechnol. 2006, 23: 273–378.

49. Zhou L, Wang Y, Peng Z. Molecular characterization and expression analysis of chalcone synthase gene during flower development in tree peony (Paeonia suffruticosa). Afr J Biotechnol. 2011, 10(8): 1275–1284.

50. Cardoso S, Lau W, Dias JE, Fevereiro P, Maniatis N. A Candidate-gene association study for berry colour and anthocyanin content in Vitis vinifera L. PLoS ONE. 2012, 7(9): e46021. doi: 10.1371/journal. pone.0046021 PMID: 23029369

51. Togami J, Okuhara H, Nakamura N, Ishiguro K, Hirose C, Ochiai M, et al. Isolation of cDNAs encoding tetrahydroxychalcone 2'-glucosyltransferase activity from carnation, cyclamen, and catharanthus. Plant Biotechnol. 2011, 28: 231–238.

52. Pelletier MK, Murrell JR, Shirley BW. Characterization of flavonol synthase and leucoanthocyanidin dioxygenase genes in Arabidopsis. Further evidence for differential regulation of "early" and "late" genes. Plant Physiol. 1997, 113(4): 1437–1445. PMID: 9112784

53. Pelletier MK, Shirley BW. Analysis of flavanone 3-hydroxylase in Arabidopsis seedlings-Coordinate regulation with chalcone synthase and chalcone isomerase. Plant physiol. 1996, 111(1): 339–345. PMID: 8685272

54. Nakatsuka A, Lzumi Y, Yamagishi M. Spatial and temporal expression of chalcone synthase and dihydroflavonol 4-reductase genes in the Asiatic hybrid lily. Plant Sci. 2003, 165(4): 759–767.

55. Nakatsuka T, Nishihara M, Mishiba K, Yamamura S. Temporal expression of flavonoid biosynthesisrelated genes regulates flower pigmentation in gentian plants. Plant Sci. 2005, 168(5): 1309–1318.

56. Hu K. Expression of genes on anthocyanin biosythesis pathway control flower coloration in Chrysanthemum and Cineraria (PhD thesis). Beijing: Beijing Forestry University. 2010, 63–74.

57. Dudareva N, Martin D, Kish CM, Kolosova N, Gorenstein N, Fäldt J, et al. (E)-β-ocimene and myrcene synthase genes of floral scent biosynthesis in snapdragon: function and expression of three terpene synthase genes of a new terpene synthase subfamily. Plant Cell. 2003, 15(5): 1227–1241. PMID: 12724546

58. Muhlemann JK, Maeda H, Chang C-Y, Miguel PS, Baxter I, Cooper bB, et al. Developmental changes in the metabolic network of snapdragon flowers. PloS ONE. 2012, 7(7): e40381. doi: 10.1371/journal.pone.0040381 PMID: 22808147

59. Dudareva N, Cseke L, Blanc VM, Pichersky E. Evolution of floral scent in Clarkia: novel patterns of SLinalool synthase gene expression in the C. breweri flower. Plant Cell. 1996, 8(7): 1137–1148. PMID: 8768373

60. Chen F, Tholl D, D'Auria JC, Farooq A, Pichersky E, Gershenzon J. Biosynthesis and emission of terpenoid volatiles from Arabidopsis flowers. Plant Cell. 2003, 15(2): 481–494. PMID: 12566586

61. Guitton Y, Nicolè F, Moja S, Valot N, Legrand S, Jullien F, et al. Differential accumulation of volatile terpene and terpene synthase mRNAs during lavender (Lavandula angustifolia and L. x intermedia) inflorescence development. Physiol Plantarum. 2010, 138(2): 150–163.

62. Feng LG, Chen C, Li TL, Wang M, Tao J, Zhao DQ, et al. Flowery odor formation revealed by differential expression of monoterpene biosynthetic genes and monoterpene accumulation in rose (Rosa rugosa Thunb.). Plant Physiol Biochem. 2014, 75: 80–88. doi: 10.1016/j.plaphy.2013.12.006 PMID: 24384414

63. Verdonk JC, Haring MA, Tunen AJv, Schuurink RC. ODORANT1 regulates fragrance biosynthesis in petunia flowers. Plant Cell. 2005, 17(5): 1612–1624. PMID: 15805488

64. Yuan YW, Byers KJRP, HDB Jr. The genetic control of flower-pollinator specificity. Curr Opin Plant Biol. 2013, 16(4): 422–428. doi: 10.1016/j.pbi.2013.05.004 PMID: 23763819

65. Ben ZM, Negre-Zakharov F, Masci T, Ovadis M, Shklarman E, Ben-Meir H, et al. Interlinking showy traits: co-engineering of scent and colour biosynthesis in flowers. Plant Biotechnol J. 2008, 6(4): 403– 415. doi: 10.1111/j.1467-7652.2008.00329.x PMID: 18346094

66. Spitzer-Rimon B, Marhevka E, Barkai O, Marton I, Edelbaum O, Masci T, et al. EOBII, a gene encoding a flower-specific regulator of phenylpropanoid volatiles' biosynthesis in petunia. Plant Cell. 2010, 22 (6): 1961–1976. doi: 10.1105/tpc.109.067280 PMID: 20543029

67. Spitzer-Rimon B, Farhi M, Albo B, Cna'ani A, Zvi MMB, Masci T, et

al. The R2R3-MYB-like regulatory factor EOBI, acting downstream of EOBII, regulates scent production by activating ODO1 and structural scent-related genes in petunia. Plant Cell. 2012, 24(12): 5089–5105. doi: 10.1105/tpc.112.105247 PMID: 23275577

Chapter 4

GENETIC ENGINEERING OF YELLOW BETALAIN PIGMENTS BEYOND THE SPECIES BARRIER

Takashi Nakatsuka[1], Eri Yamada[1], Hideyuki Takahashi[1], Tomohiro Imamura[1], Mariko Suzuki[2], Yoshihiro Ozeki[2], Ikuko Tsujimura[1], Misa Saito[1], Yuichi Sakamoto[1], Nobuhiro Sasaki[2] & Masahiro Nishihara[1]

[1] Iwate Biotechnology Research Center, 22-174-4 Narita, Kitakami, Iwate 024-0003, Japan

[2] Department of Biotechnology and Life Science, Faculty of Engineering, Tokyo University of Agriculture and Technology, 2-24-16 Naka-cho, Koganei, Tokyo 184-8588, Japan

ABSTRACT

Betalains are one of the major plant pigment groups found in some higher plants and higher fungi. They are not produced naturally in any plant species outside of the order Caryophyllales, nor are they produced by anthocyanin-accumulating Caryophyllales. Here, we attempted to reconstruct the betalain biosynthetic pathway as a self-contained system in an anthocyanin-producing plant species. The combined expressions of a tyrosinase gene from shiitake mushroom and a DOPA 4,5-dioxygenase gene from the four-o'clock plant resulted in successful betalain production in cultured cells of tobacco BY2 and *Arabidopsis* T87. Transgenic tobacco BY2 cells were bright yellow because of the accumulation of betaxanthins. LC-TOF-MS analyses showed that proline-betaxanthin (Pro-Bx) accumulated as the major betaxanthin in these transgenic BY2 cells. Transgenic*Arabidopsis* T87 cells also produced betaxanthins, but produced lower levels than transgenic BY2 cells. These results illustrate the success of a novel genetic engineering strategy for betalain biosynthesis.

INTRODUCTION

The coloration of flowers and fruits is due to the accumulation of pigments such as flavonoids (including anthocyanins), carotenoids, and betalains[1,2]. Anthocyanins and carotenoids are widely distributed in angiosperms, whereas betalains are water-soluble nitrogenous pigments found only in some plants in the order Caryophyllales (but not in the families Caryophyllaceae and Molluginaceae)[3] and in some higher fungi, such as fly agaric (*Amanita muscaria*)[4]. Although betalains have functions analogous to those of anthocyanins as pigments, anthocyanins and betalains are mutually exclusive pigments in plants[3]. This raises the taxonomic mystery of why betalains cannot be produced in any plants outside of the order Caryophyllales[5,6].

Betalains, which include the red-violet betacyanins and the yellow betaxanthins, are biosynthesized from the amino acid tyrosine by several enzymatic and spontaneous chemical steps (Fig. 1)[7]. The first step is the conversion of tyrosine to dihydroxyphenylalanine (L-DOPA) by uncharacterized enzyme(s) with tyrosine-hydroxylating activity. L-DOPA is converted to betalamic acid (BA), which is an important precursor of betalain biosynthesis, by DOPA 4,5-dioxygenase (DOD)[8]. In addition, L-DOPA is also converted to *cyclo*-DOPA by a cytochrome P450 enzyme[9]. BA conjugates with an amino acid or amine to form yellow betaxanthins. This condensation of BA with amino acids is considered as a spontaneous non-enzymatic reaction[10]. There are two pathways postulated for betacyanin biosynthesis (Fig. 1). Namely, glycosylation steps occur at the precursor molecule, *cyclo*-DOPA, or after the condensation of *cyclo*-DOPA with BA to form betanidin. Both betanidin 5-*O*-glucosyltransferase (betanidin 5GT) and *cyclo*-DOPA 5-*O*-glucosyltransferase (*c*DOPA 5GT) have been identified[11,12]. Both betacyanins and betaxanthins accumulate in the vacuole, like anthocyanins[2]. As described above, the first enzymatic step in betalain biosynthesis is the formation of L-DOPA from tyrosine, which involves a hydroxylation step (Fig. 1). In plants, tyrosine-hydroxylating activity has mainly been attributed to tyrosinase, but has also been attributed to tyrosine hydroxylase[13,14,15,16]. However, the involvement of either enzyme has not been shown conclusively, nor has any gene encoding a product with this activity been yet identified in betalain-producing plants. In contrast, the tyrosinase involved in the formation of melanin, the brown pigment produced during development and post-harvest storage of mushrooms, has been well studied and characterized[17].

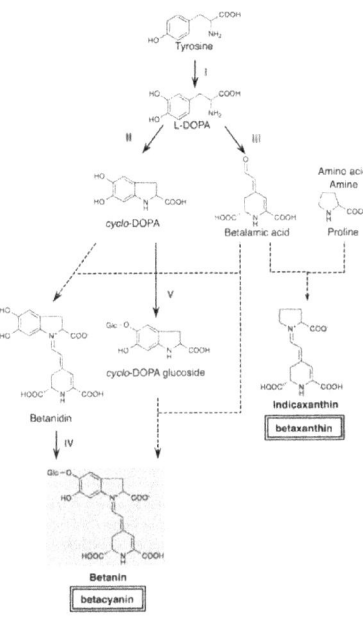

Figure 1: Betalain biosynthetic pathway. Thick arrows show enzymatic reactions, dotted arrows show spontaneous reactions. I, enzyme with tyrosine-hydroxylating activity (LeTYR in this study); II, L-DOPA oxidase; III, DOPA 4,5-dioxygenase (MjDOD in this study); IV, betanidin 5-*O*-glucosyltransferase; V,*cyclo*-DOPA 5-*O*-glucosyltransferase. Enzymes used for betaxanthin engineering in this study are shown in red font.

Schliemann et al.18 proposed that the concerted action of two enzymes, DOD from *A. muscaria* and DOPA oxidase from *Portulaca grandiflora*, synthesized betanidin, one of the betacyanins, *in vitro*. Feeding of betalamic acid to broad bean (*Vicia faba* L.) seedlings, which do not synthesize betaxanthins, resulted in the formation of dopaxanthin10. More recently, Harris et al.19 reported that transgenic *Arabidopsis* expressing the *DOD* gene from the fungus *A. muscaria* produced betaxanthin and betacyanin when supplied with the substrate L-DOPA. However, there are no reports on production of betalain pigments in non-betalain producing plants *in vivo* without feeding of the pathway intermediates.

In this study, we have shown that production of betaxanthin pigments without exogenous substrates is possible in suspension-cultured cells of both tobacco and *Arabidopsis*. We demonstrated that the combined expressions of a fungal tyrosinase gene and a plant *DOD* gene made a successful heterologous expression system for production of betaxanthins in these cultured cells. This is the first report of a self-contained betalain biosynthetic system in plant species outside of the order Caryophyllales.

RESULTS

Establishment of Transgenic BY2 Suspension cells for Production of Betaxanthins

To produce betalain pigments in anthocyanin-producing plant species outside of the order Caryophyllales, we first selected tobacco (*Nicotiana tabacum*), which belongs to the family Solanaceae, as the transformation host. The tobacco BY-2 cell line is the most popular suspension cell line in plant research and its rapid growth characteristics make it an ideal experimental material20. We produced transgenic tobacco BY2 cells expressing tyrosinase from shiitake mushroom (*Lentinula edodes*, *LeTYR*)21 and/or *MjDOD* from the four-o'clock plant (*Mirabilis jalapa*)22 under the control of the cauliflower mosaic virus 35S promoter (35Sp). Calli expressing both 35Sp-LeTYR and 35Sp-MjDOD showed bright yellow coloration (Fig. 2a). Transgenic calli expressing 35Sp-LeTYR or 35Sp-MjDOD showed no color change compared with those of the vector control and untransformed wild-type (Fig. 2a).

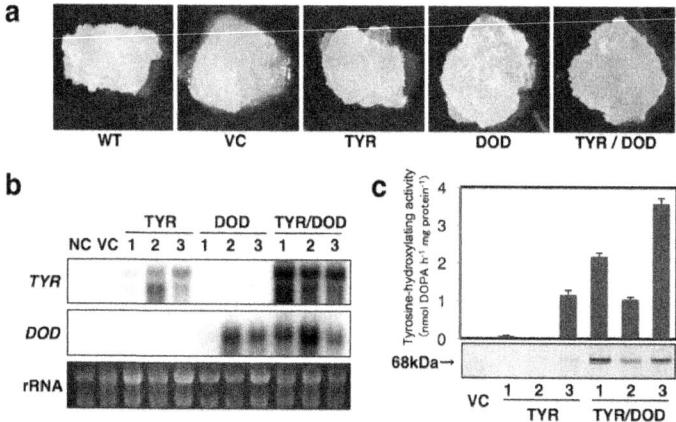

Figure 2: Analyses of transgenic tobacco BY2 calli. (a) Images of typical transgenic BY2 calli. WT, wild-type; VC, vector control (pIG121-Hm); TYR, 35Sp-LeTYR line no. 3; DOD, 35Sp-MjDOD line no. 3; TYR/DOD, 35Sp-LeTYR/35Sp-MjDOD line no. 3. (b) Northern blot analysis. Total RNAs were isolated from wild-type (WT) and transgenic BY2 including vector control (VC), and three independent transgenic lines 35Sp-LeTYR (TYR), 35Sp-MjDOD (DOD), and 35Sp-LeTYR/35Sp-MjDOD (TYR/DOD). Total RNAs (5 μg) were separated on an agarose gel and then transferred to a nylon membrane and hybridized with *LeTYR* or *MjDOD* probes. Ethidium bromide staining of rRNA bands shows quality and loading controls. (c) Tyrosine-hydroxylating activity and accumulation of LeTYR proteins in transgenic BY2 calli. The activity was determined by CE-MS analysis as described in Materials and methods. Western blot analysis was performed using anti-LeTYR antibody21.

Expression and Enzymatic Activity Analysis of Transgenic Tobacco BY2 cells

We analyzed transgene expressions in each of the three transgenic BY2 lines by northern blot analysis (Fig. 2b). The expressions of *LeTYR* were detected in both 35Sp-LeTYR and 35Sp-LeTYR/35Sp-MjDOD transgenic BY2 calli. However, the expression levels of *LeTYR* in 35Sp-LeTYR/35Sp-MjDOD lines were higher than those in 35Sp-LeTYR lines (Fig. 2b). In addition, we detected degraded and aberrant *LeTYR* transcripts in transgenic BY2 calli. These degraded and aberrant transcripts of *LeTYR* were also observed in the fruiting body of shiitake mushroom21. There were similar expression levels of *MjDOD* in 35Sp-MjDOD and 35Sp-LeTYR/35Sp-MjDOD lines, except for 35Sp-MjDOD line no. 1. Western blot analysis of the tyrosinase protein with an anti-LeTYR polyclonal antibody was performed for transgenic BY2 calli (Fig. 2c). In the crude protein extracts from shiitake mushroom, there were two bands corresponding to the premature 68-kDa and mature 45-kDa proteins, as reported by Sato et al.21. In transgenic BY2 calli with 35Sp-LeTYR or 35Sp-LeTYR/35Sp-MjDOD, premature 68-kDa signals were detected, but mature 45-kDa signals were scarcely detected. The 68-kDa signals in 35Sp-LeTYR lines were much weaker than those in 35Sp-LeTYR/35Sp-MjDOD co-expressing lines (Fig. 2c). The results of western blot analysis were largely consistent with the expression levels determined by northern blot analysis (Figs. 2b, c).

Next, we investigated tyrosine-hydroxylating activities in 35Sp-LeTYR and 35Sp-LeTYR/35Sp-MjDOD transgenic BY2 cell lines. The crude proteins extracted from transgenic BY2 lines showed no DOPA oxidase activity in the 3-methyl-2-benzothiazolinone hydrazone hydrochloride (MBTH) assay (data not shown). Thus, the tyrosine-hydroxylating activity, which synthesizes L-DOPA from tyrosine, was compared among the transgenic BY2 cell lines (Fig. 2c). No tyrosine-hydroxylating activity was detected in wild-type, the vector control, or 35Sp-MjDOD lines. Among the three 35Sp-LeTYR lines, line no. 3, which showed the strongest 68-kDa signal in the western blot analysis, showed tyrosine-hydroxylating activity of 1.2 nmol L-DOPA h^{-1} mg protein^{-1}. The 35Sp-LeTYR/35Sp-MjDOD lines showed tyrosine-hydroxylating activities comparable to, or 3-fold higher than, that of 35Sp-LeTYR line no. 3. These tyrosine-hydroxylating activities were largely consistent with the expression levels determined by northern blot and western blot analyses.

LC-TOF-MS Analyses of Betaxanthins in Transgenic BY2 cells

Pigments accumulated in transgenic BY2 calli were investigated by LC-TOF-MS. We detected BA (m/z = 212.056, λmax = 425 nm), the precursor of betaxanthin, in 35Sp-LeTYR/35Sp-MjDOD transgenic cell lines, but not in the vector control, 35Sp-LeTYR, or 35Sp-MjDOD lines. Therefore, the combined expression of *LeTYR* and *MjDOD* was essential for accumulation of betalamic acid in tobacco BY2 cells.

35Sp-LeTYR/35Sp-MjDOD transgenic BY2 lines nos. 1, 2, and 3 accumulated 1.16 ± 0.13, 1.69 ± 0.18, and 1.67 ± 0.05 μmole betaxanthins/g dry weight, respectively, as calculated from the molar extinction coefficient for betaxanthins. When 4-week-old calli were compared, transgenic line no. 3 showed the most stable proliferation and vivid yellow color among the three transgenic lines (Fig. 3a,Supplementary Fig. S1). The accumulation of betaxanthins started to increase at 4 weeks after subculture and reached the maximum at 8 weeks, thereafter decreasing rapidly (Fig. 3b). However, their calli turned brown after 5 weeks subculture, and became stunted (Fig. 3a,Supplementary Fig. S1). Therefore, 4-week-old calli of 35Sp-LeTYR/35Sp-MjDOD line no. 3 were used for further analyses.

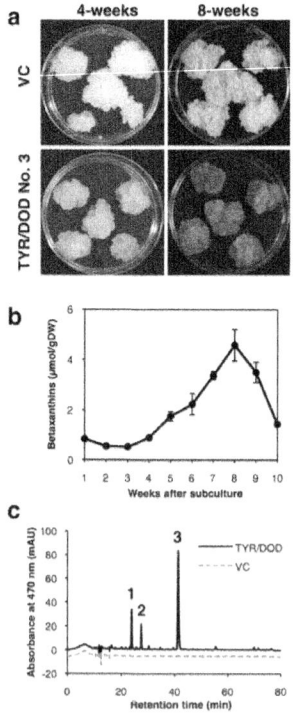

Figure 3: Betaxanthin accumulation in 35Sp-LeTYR/35Sp-MjDOD transgenic BY2 cells. (a) Typical phenotypes of vector control (VC) and 35Sp-LeTYR/35Sp-MjDOD line no. 3 (TYR/DOD no. 3) calli after 4- and 8-weeks subculture. (b) Change in betaxanthin concentration over time in transgenic calli line no. 3 after subculture. Betaxan-

thins were extracted in 80% (v/v) methanol containing 0.04% (v/v) HCl, and concentration was calculated from absorbance at 480 nm. (c) Identification of betaxanthins by LC-TOF-MS in 35Sp-LeTYR/35Sp-MjDOD transgenic BY2 cells. Chromatography profiles of extracts from vector control (dashed line) and 35Sp-LeTYR/35Sp-MjDOD line no.3 (thick line). Peak 1, retention time 23.8 min, $m/z = 340.114$, λmax $= 470$ nm, Glutamine-Bx (Vulgaxanthin I); Peak 2, 27.4 min, $m/z = 313.104$, λmax $= 470$ nm, Threonine-Bx; Peak 3, 41.3 min, $m/z = 309.1084$, λmax $= 480$ nm, Proline-Bx (Indicaxanthin).

HPLC analysis detected one major and some minor peaks at 470 nm (Fig. 3c). There was no clear peak in the vector control, 35Sp-LeTYR, or 35Sp-MjDOD transgenic BY2 calli. The major peak (retention time 41.3 min, $m/z = 309.108$, λmax $= 480$ nm) was assigned to proline-Bx (Pro-Bx, indicaxanthin; Fig. 1) by comparison with synthesized betaxanthin standards. The 35Sp-LeTYR/35Sp-MjDOD transgenic BY2 line no. 3 accumulated a maximum of 3.3 μmoles Pro-Bx/g dry weight (Fig. 3c, peak 3), based on synthetic authentic Pro-Bx as the standard. The other minor peaks 1 (23.8 min, $m/z = 340.114$, λmax $= 470$ nm) and 2 (27.4 min, $m/z = 313.104$, λmax $= 470$ nm) were assigned to glutamine-Bx (Gln-Bx, vulgaxanthin I) and threonine-Bx (Thr-Bx), respectively (Fig. 3c). Accumulation of betaxanthins was also detected in other 35Sp-LeTYR/35Sp-MjDOD transgenic BY2 lines.

Production of Betaxanthins in Arabidopsis T87 cells

In addition to tobacco BY2 cells, we produced 35Sp-LeTYR/35Sp-MjDOD transgenic calli of *Arabidopsis* T87 cells. We obtained 40 independent transgenic calli with 35SpLeTYR/35SpMjDOD, and selected two transgenic calli lines that showed high expression levels of both *LeTYR* and *MjDOD* genes in a northern blot analysis (Fig. 4a). Western blot analysis detected 68-kDa signals of the LeTYR protein in both transgenic lines, like in transgenic BY2 calli (Fig. 4a). The wild-type *Arabidopsis* T87 cells were also yellow when cultured in darkness; therefore, we did not observe a marked difference in color phenotype between the vector control and the transgenic lines (Fig. 4b). However, the aqueous extracts from transgenic lines were slightly yellow, whereas that from the vector control was colorless (Fig. 4c). HPLC analysis detected some very weak peaks in *Arabidopsis* transgenic line no. 1, but we could not identify specific betaxanthins.

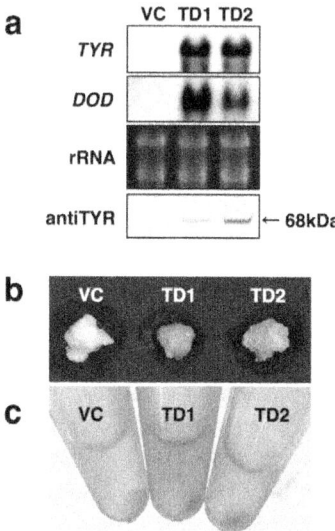

Figure 4: Analysis of 35Sp-LeTYR/35Sp-MjDOD transgenic *Arabidopsis* **T87 cells.** (a) Analyses of *LeTYR* and *MjDOD* transcripts and LeTYR protein accumulation. VC, vector control pIG121Hm; TD1 and TD2, independent 35Sp-LeTYR/35Sp-MjDOD transgenic T87 cell lines. (b) Images of typical transgenic T87 calli. (c) Aqueous extracts from transgenic T87 calli.

These results indicated that the combined expressions of *LeTYR* and *MjDOD* could induce the accumulation of betaxanthins in cultured cells of non-Caryophyllales plants, but the specific components and amounts of betalains produced differed between tobacco and *Arabidopsis*.

DISCUSSION

Betalains are Caryophyllales-specific plant pigments that are responsible for the red-violet and yellow coloration instead of anthocyanins in some plants in the order Caryophyllales. The betalain biosynthetic pathway in plants was deduced from genetic and metabolite analysis of *P. grandiflora* mutants23, and then the genes encoding DOD, betanidin 5GT, and *c*DOPA5GT were identified from plants in the Caryophyllales8·11·22·24. More recently, *CYP76AD1*encoding L-DOPA oxidase was identified from *Beta vulgaris*, and it was revealed that the oxidation of L-DOPA to *cyclo*-DOPA was catalyzed by this enzyme in betacyanin biosynthesis9. It was suggested that this cytochrome P450, CYP76AD1, may have tyrosine-hydroxylating activity to synthesize L-DOPA from tyrosine9, although further confirmation is necessary. Thus, the gene encoding an enzyme with tyrosine-hydroxylating activity, which is responsible for the crucial first step in betalain biosynthesis, was not identified, although the enzyme had been partially purified in earlier works13·14. The results of those previous studies strongly suggested that a tyrosinase belonging to the copper protein polyphenol oxidase (PPO) family is the tyrosine-hydroxylating enzyme involved in betalain biosynthesis, although it has never been proven conclusively. Most plant PPOs have a plastid transit peptide at the N-terminal17. Although PPO genes have been cloned from pokeweed (*Phytolacca americana*), it is not certain whether they actually function in betalain biosynthesis25. Müeller et al.26 reported the purification of tyrosinase from the betalain-containing pileus of the mushroom *Amanita muscaria*, but no additional experiments had been conducted until now.

In previous studies, broad bean and *DOD*-expressing *Arabidopsis* produced betalains when fed with betalamic acid and L-DOPA, respectively10·19. These results suggested that anthocyanin-producing plant species have the potential to produce betalains if they are provided with suitable precursors for betalain biosynthesis. However, a betalain-specific gene that encodes an enzyme with tyrosine-hydroxylating activity, whether from higher plants or fungi, was not available for combining with *DOD* to complete the biosynthetic pathway. We considered that mushroom tyrosinases linked to the synthesis of melanin pigments have been well characterized17 and might be used to reconstruct betalain biosynthesis. In fact, LeTYR from shiitake mushroom resulted in accumulation of L-DOPA in BY2 cells; that is, the foreign LeTYR protein showed tyrosine-hydroxylating activity in cultured plant cells (Fig. 2c). Thereafter, MjDOD protein could convert L-DOPA produced by LeTYR into betalamic acid, leading to the formation of yellow betaxanthins in transgenic tobacco BY2 cells expressing both *MjDOD* and *LeTYR* genes (Fig. 2a; Fig. 3). The betaxanthin accumulation peaked at 8 weeks after subculture, even though the transgenic calli turned brown in prolonged culture after 5 weeks (Figs. 3a,b, Supplemental Fig. S1). This suggested that there was accumulation of other metabolites as a result of the foreign tyrosinase activity. Since the browning also occurred in transgenic BY2 cells expressing *LeTYR* only, these were not betalain-derived compounds. This browning was considered to represent accumulation of L-DOPA-derived (but not betaxanthin-derived) metabolites via dopaquinone that might be toxic to plant cells. One strategy to reduce or eliminate browning is to convert the metabolites to nontoxic compounds by introducing other detoxifying enzymes. If successful, this may result in more effective accumulation of betaxanthins in transgenic BY2 cells. We also produced a transgenic *Arabidopsis* T87 cell line by introducing the same constructs (Fig. 4). However, betaxanthin accumulated to lower levels in T87 transgenic cells than in BY2 transgenic cells, and the coloration of *Arabidopsis* transgenic cells could not be observed visually (Fig. 4b). It is unknown why the amount of heterologous betaxanthin produced differed among the plant species used as transformation hosts, but there are several possible explanations: the transport efficiency of BA may differ between tobacco and *Arabidopsis*, and/or there may be differences in tyrosine availability or in competing endogenous reactions that use tyrosine and/or the L-DOPA substrate. Further research is required to elucidate the precise reason(s).

The condensation of BA with amino acids *in planta* is a spontaneous reaction10. Pro-Bx accumulated preferentially in transgenic BY2 cells, reaching a concentration of 3.3 µmoles/g dry weight (Fig. 3c). Pro-Bx (indicaxanthin) accumulated as a major pigment in *Opuntia ficus-indica*27 and *Mirabilis jalapa*28. The yellow callus from *B. vulgaris* produced 4.2 µmoles/g dry weight of total betaxanthin29. Therefore, the amounts of betaxanthins accumulated in transgenic BY2 cells were similar to those reported in calli and roots from plants in the Caryophyllales. When BA was fed to broad bean, the major betaxanthin was dopaxanthin10. In that case, amino acid analysis of the hypocotyl extract of broad bean seedlings revealed that L-DOPA was the most abundant amino acid present. However, the accumulation levels of amino acids are not the only factors influencing condensation reactions with BA30. Further studies are necessary to clarify the mechanisms of accumulation of specific betaxanthins in plant cells.

We transformed 35Sp-LeTYR/35Sp-MjDOD transgenic BY2 calli with a *cDOPA5GT* gene from *M. jalapa*24. However, none of the transgenic cell lines accumulated betacyanins. This indicated that *cyclo*-DOPA, which is the substrate of cDOPA5GT, was not abundantly accumulated, probably because of deficient or low L-DOPA oxidase activity. It is noteworthy that C-terminal processing of tyrosinase has been reported in several fungi31,32,33. Processing of tyrosinase is catalyzed by endogenous proteases such as the serine proteases, and this proteolytic cleavage plays a critical role in enzyme activity. For example, analysis of a recombinant tyrosinase from *Pholiota microspora* revealed that the latent 67-kDa premature form did not have catalytic activity but the C-terminal cleaved 44-kDa mature form was catalytically active33, although the analysis did not distinguish between monophenolase and diphenolase activities. In transgenic BY2 calli expressing *LeTYR*, the premature 68-kDa LeTYR was detected from the early growth stages but the mature 45-kDa LeTYR appeared at later growth stages (Fig. 2c, Supplementary Fig. S2). It is unknown why LeTYR was not efficiently processed, but it is likely that plant proteases could not efficiently process the fungal tyrosinase protein because of its low expression level at the early growth stages of BY2 cells. In light of these facts, it is most likely that the unprocessed LeTYR accumulated in BY2 cells had only tyrosine-hydroxylating activity, and not DOPA oxidase activity.

However, betacyanins were not accumulated even at later growth stages when the mature 45-kDa LeTYR appeared, suggesting that the enzymatic activity of the shiitake tyrosinase expressed in BY2 calli was insufficient for efficient production of *cyclo*-DOPA. Therefore, the additional expression of *CYP76AD1* might induce betacyanin accumulation in transgenic BY2 calli. Moreover, Goldfeder et al.34demonstrated that the amino acid residue at position 218 influenced the monophenolase/diphenolase activity ratio of a bacterial tyrosinase by affecting the positioning of substrates within the active site. Protein engineering of a fungal tyrosinase involved in enzymatic activity or C-terminal processing might be also useful for the effective production of betalains in plant cells.

In conclusion, our results demonstrated that it is possible to engineer betalain biosynthesis in a heterologous expression system without extraneous substrate feeding. A fungal tyrosinase was able to substitute for the tyrosine-hydroxylating enzyme of betaxanthin-accumulating plants to accumulate L-DOPA in plant cells. Betalains are naturally limited to the order Caryophyllales in higher plants, but genetic engineering can now overcome the species barrier. As betalain pigments have beneficial properties for human nutrition and plant coloration, our approach may contribute to the development of genetically modified crops in the future.

METHODS

Vector Construction and BY2 Transformation

The coding region sequences (CDS) of *L. edodes* tyrosinase (*LeTYR,*accession number AB027512)21 and *M. jalapa* DOPA 4,5-dioxygenase (*MjDOD*, AB435372)22 were used to replace the *GUS* gene in the binary vector pBI121 (kanamycin resistance) and pEBisHR (hygromycin resistance), respectively. pBI121-35Sp-LeTYR and pEBisHR-35Sp-MjDOD binary vectors were transformed into*Agrobacterium tumefaciens* EHA105 or EHA101.

The transformation of BY2 cells was performed following the procedure of An35. Transgenic calli were selected on medium containing 200 mg l^{-1} kanamycin or 20 mg l^{-1} hygromycin B, and 500 mg l^{-1} carbenicillin, and were subcultured every 2 weeks. *A. thaliana*ecotype Columbia suspension-cultured T87 cells36 were obtained from the RIKEN BioResource Center. T87 cells were transformed by an *Agrobacterium*-mediated method as described by Ogawa et al.37.

Expression Analysis

Total RNAs were isolated from 200 mg (fresh weight) transgenic calli using a Fast RNA pro GREEN Kit (MP Bio Japan, Tokyo, Japan). Total RNAs (5 µg) were separated on a 1% agarose gel, and then transferred onto a Hybond N$^+$ membrane (GE healthcare, Little Chalfont, UK). The probes for *LeTYR* and *MjDOD* genes were labeled using a DIG-PCR labeling Kit (Roche Applied Sciences, Indianapolis, IN, USA). The hybridization and detection procedures were performed using the DIG Luminescent Detection Kit for Nucleic Acids (Roche) according to manufacturer›s instructions.

Western blot Analysis of LeTYR Protein

Tissues (1 g fresh weight) of the transgenic BY2 and *Arabidopsis* T87 cell lines were lyophilized, and crude proteins were extracted using 400 µl 50 mM HEPES buffer [pH 7.0]. Protein concentration was measured as described by Bradford38. For each transgenic callus, 10 µg crude protein was separated by SDS-PAGE (10% acrylamide) and transferred onto a polyvinylidene difluoride (PVDF) membrane. The membrane was blocked by blocking buffer (phosphate buffered saline [PBS] plus 5% skimmed milk and 0.3% Tween-20) for at least 12 h and then probed with a rabbit polyclonal antibody raised against LeTYR21. After extensive washing, membranes were incubated with the goat anti-rabbit IgG secondary antibody conjugated to alkaline phosphatase (Bio-Rad Laboratories, Hercules, CA, USA) and detected using an Immun-Blot Assay Kit (Bio-Rad).

Tyrosine-Hydroxylating Activity Assay

To investigate tyrosine-hydroxylating activity, crude proteins were extracted from freeze-dried transgenic BY2 cells at 4 weeks after subculture using 400 µl 50 mM HEPES buffer [pH 7.0]. After centrifugation at 20,000 × g at 4°C for 10 min, the supernatants were collected and protein concentrations were measured as described above. Then, 25 µl crude extract containing 150 µg protein was mixed with 25 µl 2 mM tyrosine and incubated at 30°C for 30 min. The reaction was terminated by incubating the mixture at 95°C for 3 min. Boiled crude extracts were also assayed as the control. The mixture was centrifuged at 20,000 × g for 5 min, and the supernatants were subjected to further analysis. The synthesized L-DOPA was quantified by capillary electrophoresis-mass spectrometry (CE-MS; Agilent Technologies, Santa Clara, CA, USA) according to the method of Takahashi et al39. The activities were represented by the amount of newly synthesized L-DOPA (nmol

h^{-1} mg^{-1} protein). The MBTH assay was also performed to detect diphenolase activity as described by Rodriguez-Lopez et al.40.

Metabolite Analysis

Transgenic BY2 cells (100 mg fresh weight) were powdered using a Micro Smash M100 (TOMY) and homogenized with 200 μl 1% (v/v) formic acid. After centrifugation, the supernatant was filtered through a Millipore 5-kDa-cutoff filter (Millipore) and the filtrates were used for analysis. Betaxanthins were analyzed by Agilent 6520 Accurate-Mass quadrupole time-of-flight liquid chromatography/mass spectrometry (Q-TOF LC/MS; Agilent Technologies). Separation was performed on a Develosil PRAQUEOUS-AR-5 column (4.6 mm × 250 mm; Numura Chemical Co., Seto, Japan) with a gradient elution of 3–20% (v/v) acetonitrile containing 0.1% (v/v) formic acid for 60 min, followed by 20–60% (v/v) for 30 min at a flow rate of 0.25 ml min 1^{-1} at 30°C. Absorbance at 470 nm was monitored by an Agilent 1200 Diode-Array Detector SL (Agilent Technologies). The total concentration of betaxanthin was calculated using the molar extinction coefficient (480 nm, = 48,000)29. Pro-Bx was quantified based on the area on the HPLC chromatogram monitored at 470 nm using synthetic authentic Pro-Bx as the standard. The HPLC separation was performed as described previously41.

ACKNOWLEDGMENTS

We thank Dr. Toshitsugu Sato, Kitami Institute of Technology, for providing the shiitake tyrosinase gene and for helpful discussions. The authors thank Dr. Ikuo Nakamura, Chiba University, for providing the pEBis binary vector. We thank the RIKEN BioResource Center for providing Arabidopsis thaliana T87 cells.

REFERENCES

1.   Grotewold, E. The genetics and biochemistry of floral pigments. Annu. Rev. Plant Biol. 57, 761–780 (2006).

2.   Tanaka, Y., Sasaki, N. & Ohmiya, A. Biosynthesis of plant pigments: anthocyanins, betalains and carotenoids. Plant J. 54, 733–749 (2008).

3.   Stafford, H. A. Anthocyanins and betalains: evolution of the mutually exclusive pathways. Plant Sci. 101, 91–98 (1994).

4.   Hinz, U. G., Fivaz, J., Girod, P. A. & Zyrd, J. P. The gene coding for the DOPA dioxygenase involved in betalain biosynthesis in Amanita muscaria and its regulation. Mol. Gen. Genet. 256, 1–6 (1997).

5.   Clement, J. S. & Mabry, T. J. Pigment evolution in the Caryophyllales: a systematic overview. Bot. Acta 109, 360–367 (1996).

6.   Brockington, S. F., Walker, R. H., Glover, B. J., Soltis, P. S. & Soltis, D. E. Complex pigment evolution in the Caryophyllales. New Phytol. 190, 854–864 (2011).

7.   Strack, D., Vogt, T. & Schliemann, W. Recent advances in betalain research. Phytochemistry 62, 247–269 (2003).

8.   Christinet, L., Burdet, F. X., Zaiko, M., Hinz, U. & Zryd, J. P. Characterization and functional identification of a novel plant 4,5-extradiol dioxygenase involved in betalain pigment biosynthesis in Portulaca grandiflora. Plant Physiol. 134, 265–274 (2004).

9.   Hatlestad, G. J.et al. The beet R locus encodes a new cytochrome P450 required for red betalain production. Nat. Genet. 44, 816–820 (2012).

10.   Schliemann, W., Kobayashi, N. & Strack, D. The decisive step in betaxanthin biosynthesis is a spontaneous reaction. Plant Physiol. 119, 1217–1232 (1999).

11.   Vogt, T., Grimm, R. & Strack, D. Cloning and expression of a cDNA encoding betanidin 5-O-glucosyltransferase, a betanidin- and flavonoid-specific enzyme with high homology to inducible glucosyltransferases from the Solanaceae. Plant J. 19, 509–519 (1999).

12.   Sasaki, N., Adachi, T., Koda, T. & Ozeki, Y. Detection of UDP-glucose:cycloDOPA 5-O-glucosyltransferase activity in four o'clocks (Mirabilis jalapa L.). FEBS Lett. 568, 159–162 (2004).

13.   Steiner, U., Schliemann, W., Bo"hm, H. & Strack, D. Tyrosinase involved in betalain biosynthesis of higher plants Planta 208, 114–124 (1999).

14.   Steiner, U., Schliemann, W. & Strack, D. Assay for tyrosine hydroxylation activity of tyrosinase from betalain-forming plants and cell cultures. Anal. Biochem. 238, 72–75 (1996).

15.   Yamamoto, K., Kobayashi, N., Yoshitama, K., Teramoto, S. & Komamine, A. Isolation and purification of tyrosine hydroxylase from callus cultures of Portulaca grandiflora. Plant Cell Physiol. 42, 969–975 (2001).

16.   Luthra, P. M. & Singh, S. Identification and optimization of tyrosine hydroxylase activity in Mucuna pruriens DC. var. utilis. Planta 231, 1361–1369 (2010).

17.   van Gelder, C. W. G., Flurkeya, W. H. & Wichers, H. J. Sequence and structural features of plant and fungal tyrosinases. Phytochemistry 45, 1309–1323 (1997).

18.   Schliemann, W., Steiner, U. & Strack, D. Betanidin formation from

dihydroxyphenylalanine in a model assay system. Phytochemistry 49, 1593–1598 (1998).

19. Harris, N. N. et al. Betalain production is possible in anthocyanin-producing plant species given the presence of DOPA-dioxygenase and L-DOPA. BMC Plant Biol. 12, 34 (2012).

20. Nagata, T., Okada, K., Takebe, I. & Matsui, C. Delivery of tobacco mosaic virus RNA into protoplasts mediated by reverse-phase evaporation vesicles (liposomes). Mol. Gen. Genet. 184, 161–165 (1981).

21. Sato, T. et al. The tyrosinase-encoding gene of Lentinula edodes, Letyr, is abundantly expressed in the gills of the fruit-body during post-harvest preservation. Biosci. Biotechnol. Biochem. 73, 1042–1047 (2009).

22. Sasaki, N. et al. Detection of DOPA 4,5-dioxygenase (DOD) activity using recombinant protein prepared from Escherichia coli cells harboring cDNA encoding DOD from Mirabilis jalapa. Plant Cell Physiol. 50, 1012–1016 (2009).

23. Trezzini, G. F. & Zryd, J.-P. Portulaca grandiflora: a model system for the study of the biochemistry and genetics of betalain synthesis. Acta Horticul. 280, 581–585 (1990).

24. Sasaki, N. et al. Isolation and characterization of cDNAs encoding an enzyme with glucosyltransferase activity for cyclo-DOPA from four o'clocks and feather cockscombs. Plant Cell Physiol. 46, 666–670 (2005).

25. Joy, R. W., Sugiyama, M., Fukuda, H. & Komamine, A. Cloning and characterization of polyphenol oxidase cDNAs of Phytolacca americana. Plant Physiol. 107, 1083–1089 (1995).

26. Mu¨eller, L. A., Hinz, U. & Zry¨d, J.-P. Characterization of a tyrosinase from Amanitamuscaria involved in betalain biosynthesis. Phytochemistry 42, 1511– 1515 (1996).

27. Stintzing, F. C., Schieber, A. & Carle, R. Identification of betalains from yellow beet (Beta vulgaris L.) and cactus pear [Opuntia ficus-indica (L.) Mill.] by highperformance liquid chromatography-electrospray ionization mass spectrometry. J. Agric. Food Chem. 50, 2302–2307 (2002).

28. Piattelli, M., Minale, L. & Nicolaus, R. A. Pigments of centrospermae—V. : Betaxanthins from Mirabilis jalapa L. Phytochemistry 4, 817–823 (1965).

29. Girod, P. A. & Zryd, J. P. Secondary metabolism in cultured red beet (Beta vulgris L.) cells: Differential regulation of betaxanthin and betacyanin biosynthesis. Plant Cell Tiss. Org. Cult. 25, 1–12 (1991).

30. Bo¨hm, H. & Ma¨ck, G. Betaxanthin formation and free amino acids in

hairy roots of Beta vulgaris var. lutea depending on nutrient medium and glutamate or glutamine feeding. Phytochemistry 65, 1361–1368 (2004).

31. Wichers, H. J. et al. Cloning, expression and characterisation of two tyrosinase cDNAs from Agaricus bisporus. Appl. Microbiol. Biotechnol. 61, 336–341 (2003).

32. Selinheimo, E. et al. Production and characterization of a secreted, C-terminally processed tyrosinase from the filamentous fungus Trichoderma reesei. FEBS J. 273, 4322–4335 (2006).

33. Kawamura-Konishi, Y., Maekawa, S., Tsuji, M. & Goto, H. C-terminal processing of tyrosinase is responsible for activation of Pholiota microspora proenzyme. Appl. Microbiol. Biotechnol. 90, 227–234 (2011).

34. Goldfeder, M., Kanteev, M., Adir, N. & Fishman, A. Influencing the monophenolase/diphenolase activity ratio in tyrosinase. Biochim. Biophys. Acta 1834, 629–633 (2013).

35. An, G. High efficiency transformation of cultured tobacco cells. Plant Physiol. 79, 568–570 (1985).

36. Axelos, M., Curie, C., Mazzolini, L., Bardet, C. & Lescure, B. A protocol for transient gene expression in Arabidopsis thaliana protoplasts isolated from cell suspension cultures. Plant Physiol. Biochem. 30, 123–128 (1992).

37. Ogawa, Y. et al. Efficient and high-throughput vector construction and Agrobacterium-mediated transformation of Arabidopsis thaliana suspensioncultured cells for functional genomics. Plant Cell Physiol. 49, 242–250 (2008).

38. Bradford, M. M. A rapid and sensitive method for the quantitation of microgram quantities of protein utilizing the principle of protein-dye binding. Anal. Biochem. 72, 248–254 (1976).

39. Takahashi, H., Uchimiya, H. & Hihara, Y. Difference in metabolite levels between photoautotrophic and photomixotrophic cultures of Synechocystis sp. PCC 6803 examined by capillary electrophoresis electrospray ionization mass spectrometry. J. Exp. Bot. 59, 3009–3018 (2008).

40. Rodriguez-Lopez, J. N., Escribano, J. & Garcia-Canovas, F. A continuous spectrophotometric method for the determination of monophenolase activity of tyrosinase using 3-methyl-2-benzothiazolinone hydrazone. Anal. Biochem. 216, 205–212 (1994).

41. Sekiguchi, H., Ozeki, Y. & Sasaki, N. In vitro synthesis of betaxanthins

using recombinant DOPA 4,5-dioxygenase and evaluation of their radical-scavenging activities. J. Agric. Food Chem. 58, 12504–12509 (2010).

Chapter 5

PEACE, A MYB-LIKE TRANSCRIPTION FACTOR, REGULATES PETAL PIGMENTATION IN FLOWERING PEACH 'GENPEI' BEARING VARIEGATED AND FULLY PIGMENTED FLOWERS

Chiyomi Uematsu[1], Hironori Katayama[2], Izumi Makino[1], Azusa Inagaki[1], Osamu Arakawa[3] and Cathie Martin[4]

[1] Botanical Gardens, Graduate School of Science, Osaka City University, Osaka 576-0004, Japan

[2] Food Resources Research and Education Center, Kobe University, Hyogo 675-2103, Japan

[3] Faculty of Agriculture and Life Science, Hirosaki University, Aomori 036-8561, Japan

[4] Department of Metabolic Biology, John Innes Centre, Norwich NR4 7UH, UK

ABSTRACT

Flowering peach *Prunus persica* cv. Genpei bears pink and variegated flowers on a single tree. The structural genes involved in anthocyanin biosynthesis were expressed strongly in pink petals but only very weakly or not at all in variegated petals. A cDNA clone encoding a *MYB*-like gene, isolated from pink petals was strongly expressed only in pink petals. Introduction of this gene, via biolistics gave magenta spots in the white areas of variegated petals, therefore this gene was named as *Peace* (*pe*achanthocyanin *c*olour *e*nhancement). Differences in *Peace* expression determine the pattern of flower colouration in flowering peach. The R2R3 DNA-binding domain of Peace is similar to those of other plant MYBs regulating anthocyanin biosynthesis. Key amino acids for tertiary structure and the motif for interaction with bHLH proteins were conserved in Peace. Phylogenetic analysis indicates that Peace is closely related to AtMYB123 (TT2), which regulates proanthocyanidin biosynthesis in *Arabidopsis*, and to anthocyanin regulators in monocots rather than to regulators in dicots. This is the first report that a TT2-like R2R3 MYB has been shown to regulate anthocyanin biosynthesis.

INTRODUCTION

Pigmentation is an important trait for ornamental plants especially flowers. Anthocyanins, one of the major pigments in the flower kingdom, are widely distributed among higher plants. Six basic types of anthocyanidin are found in plants (pelargonidin, cyanidin, delphinidin, peonidin, petunidin, and malvidin) which provide different colours in flowers of different species depending on the degree of side-chain decoration, (glycosylation and acylation) (Glover and Martin, 2012). Horticulturalists, breeders, and geneticists have been eager to produce unique cultivars with attractive flower colour traits (Forkmann, 1993; Martin and Gerats, 1993; Holton and Cornish, 1995).

There are two categories of genes involved in the biosynthesis of anthocyanins (Martin *et al.*, 1991; Jackson *et al.*, 1992). The first category includes structural genes encoding enzymes catalysing each step of the biosynthetic pathway such as chalcone synthase (CHS), chalcone isomerase (CHI), flavanone 3-hydroxylase (F3H), dihydroflavonol 4-reductase (DFR), anthocyanidin synthase (ANS), and UDP glucose-flavonoid 3-*O*-glucosyltransferase (UFGT). A second category includes regulatory genes encoding MYB-like transcription factors (Sablowski *et al.*, 1994; Martin and Paz-Ares, 1997; Kranz *et al.*, 1998; Stracke *et al.*, 2001), basic helix-loop-helix (bHLH) transcription factors (Bailey *et al.*, 2003; Heim *et al.*, 2003;Toledo-Ortiz *et al.*, 2003), and WD repeat proteins (de Vetten *et al.*, 1997;Walker *et al.*, 1999) which regulate the expression of the structural genes co-ordinately (Quattrocchio *et al.*, 1993; Baudry *et al.*, 2004; Zimmermann*et al.*, 2004).

Flowering peach, *Prunus persica* (L.) Batsch. cv. Genpei is one of the ornamental peach cultivars which bears fully pigmented pink flowers and variegated flowers on a single tree. The cultivar name 'Genpei' has its origin in Japanese history, therefore it is iconic in Japanese culture with a colour image of red and white. Pink flowers and variegated flowers of Genpei are believed to result from mutation as opposed to a graft chimera. Branches with pink flowers reproducibly bear only pink flowers every year, but branches with variegated flowers occasionally produce fully pigmented flowers. This phenomenon is similar to the unstable variegation caused by insertion and excision of class II transposable elements as observed in snapdragon (Harrison and Carpenter, 1979; Bonas *et al.*, 1984; Martin *et al.*, 1985; Sommer *et al.*, 1985; Upadhyaya *et al.*, 1985; Martin *et al.*, 1988), petunia (Doodeman *et al.*, 1984; Gerats *et al.*, 1990; Snowden and Napoli, 1998; van Houwelingen *et al.*,

1998), morning glory (Epperson and Clegg, 1987; Inagaki *et al.*, 1994; Habu *et al.*, 1998), and carnation (Itoh *et al.*, 2002; Nishizaki *et al.*, 2011). Chaparro *et al.* (1995) suggested the involvement of an active transposable element as a cause of flower colour variegation in flowering peach cv. Pillar. However, no transposon has yet been detected in Pillar.

Bud mutation has commonly been used as a convenient tool to obtain new cultivars from fruit trees, even though the mechanisms of mutation remain unclear (Imai, 1935; Shamel and Pomeroy, 1936; Whitham and Slobodchikoff, 1981). Flowering peach with variegated flowers could be a good system to reveal the mechanisms of bud mutation in woody species. Generally, fruit trees are genetically heterozygous and have long life cycles, so genetic analysis is problematic. Flowering peach Genpei, has two phenotypes, pink and variegated flowers, within one tree, i.e. two genotypes, wild type and mutant, co-exist within the same tree. These two genotypes probably possess the same genetic background except for the difference in flower colour trait and so are equivalent to isogenic lines.

In some fruit species such as apple, pear, peach, grape, and citrus, many cultivars have been derived from a few cultivars via bud mutation (Shamel and Pomeroy, 1936; Butelli *et al.*, 2012). In Japan, more than two-thirds of peach cultivars are thought to have arisen as chance seedlings or sports by mutation from 'Hakuto' or its relatives (Yamamoto *et al.*, 2003a, b). A high frequency of bud mutations in certain cultivars might suggest involvement of a transposable element in their creation. If transposable elements could be found in unstable flowering peach, it might be possible to use them to create new mutations and as tags to isolate the genes responsible for useful traits.

The isolation of the *Peace* gene encoding a Myb transcription factor that regulates petal pigmentation in flowering peach is reported here. *Peace* is expressed strongly only in pink petals and can complement the magenta colour in the white areas of variegated petals when introduced by particle bombardment. Comparison of the deduced amino acid sequence of Peace with other Myb transcription factors, suggests that *Peace* plays a role in inducing the anthocyanin biosynthetic pathway in peach petals.

MATERIALS AND METHODS

Plant Material

A weeping type tree of flowering peach (*Prunus persica* cv. Genpei, Plant ID No. 4F0199) bearing double flowers maintained at the Botanical Gardens of Osaka City University, Japan, was analysed in this study (Fig. 1A). Fully pigmented pink flowers and variegated flowers were borne simultaneously within one tree. One branch bearing pink flowers (P2 branch) and another branch bearing variegated flowers (V2) were marked to collect flower material.

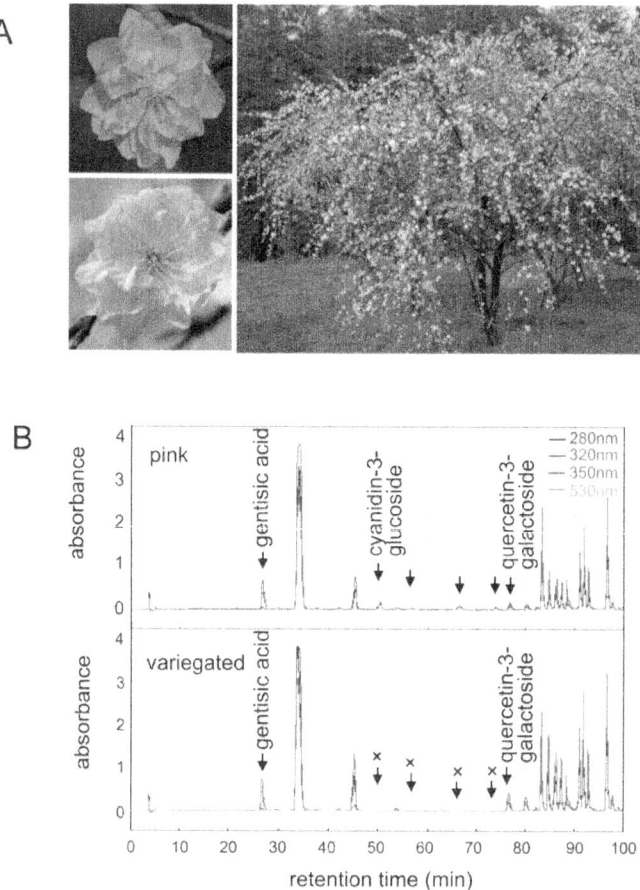

Figure 1. Peach flowers used in this study and pigment analysis. (A) Peach tree 'Genpei' bearing fully pink coloured flowers and variegated flowers within a single tree. (B) HPLC profiles obtained from pink petals and variegated petals, respectively. Absorbance spectrum at the wave length of 280, 320, 350, and 530nm are given.

HPLC Assay for Pigment Analysis

For the pigment analysis using HPLC, flower petals collected from newly opened flowers were frozen in liquid nitrogen and stored at −80 °C until use. Frozen petals from one flower were placed in the pre-cooled mortar with liquid nitrogen and ground thoroughly using a chilled pestle. Extraction buffer containing 300 µl of 15% acetic acid in methanol and 100 µl of 0.1% gentisic acid was added to the ground material and kept on ice for 1h. The extraction mixture was centrifuged at 10 000rpm for 10min at 4 °C. The supernatant was filtered using a Millipore filter of 0.45 µm and then applied for HPLC analysis using NANOSPACE SI-1 (SHISEIDO, Japan) equipped with Photo Diode Array Detector 2017 (SHISEIDO, Japan) and CAPCELL PAK C18 UG 120 column (4.6 mm in diameter × 150 mm in length). Solvent A (1.5% phosphate solution) and solvent B (1.5% phosphate, 20% formic acid, 25% acetonitrile solution) were used with the following gradients: A; 90–70% (0–30min), 70–65% (30–40min), 65–55% (40–70min), 55–0% (70–90min), and B; 100% after 90min. The flow rate was 100 µl min^{-1} and the detection wavelengths were 280, 320, 350, and 530nm.

In order to identify peaks obtained from HPLC profiles, chlorogenic acid,*p*-coumaric acid, eriodictyol, naringenin, quercetin dihydrate, quercetin-3-galactoside, catechin, epicatechin, cyanidin chloride, and cyanidin-3-*O*-glucoside were used as standards for HPLC analysis, each prepared as a 0.1% solution with 15% acetic acid in methanol. For the quantification of identified peaks, 0.1% gentisic acid prepared with 15% acetic acid in methanol was used as an internal standard.

DNA extraction and blotting to the membrane

Total DNA was extracted from frozen petals collected separately from pink flowers and variegated flowers just after blooming using a CTAB method (Katayama and Uematsu, 2003). DNA (10 µg) was digested with *Bam*HI,*Dra*I, *Eco*RI or *Hin*dIII and fractionated by 0.8% agarose gel electrophoresis then blotted onto the Hybond-N$^+$ (GE Healthcare, UK) nylon membranes.

Probe Preparation and Signal Detection

Partial genomic clones of the *CHS*, *CHI*, *F3H*, *DFR*, and *AS* genes were obtained from pink flowers by PCR cloning using degenerate primers as follows:

CHS/F2 (5′-CACCCACYTKGATWSYYTIG-3′),

CHS/R1 (5′-GCWATCCAGAABADVGAGTT-3′),

CHI/F1 (5′-SYYAARTGGAARGGYAARAC-3′),

CHI/R3 (5′-DGAHBCRTCTTTGGARAAGC-3′),

F3H/F2 (5′-CTACAACGACTTCAGCAACG-3′),

F3H/R1 (5'-CARTCTTGCACWGCTTCTCC-3'),

DFR/F8 (5'-GCITCIGGITTYATWGGITCATGGC-3'),

DFR/R8 (5'-CGTAGCATSGTGIGAIGARCAAATG-3'),

AS/F2 (5'-CUATCCCIAAAGAGTACATC-3') and

AS/R1 (5'-GAAAACAGCCCATGAAATCC-3').

Degenerate primers contained the following nucleotides; B=T+C+G, D=A+T+G, H=A+T+C, K=T+G, R=A+G, S=C+G, V=A+C+G, W=A+T, Y=C+T, and I which indicates deoxyinosine. Primers of Malubi/F (5'-ATGATCGAGGTGGTGCTGAA-3') and Malubi/R (5'-AGTCCTTGAGGGTGATGTTGG-3') both designed based on the *Malus*ubiquitin sequence (GenBank DT00167) were used to amplify the ubiquitin gene from flowering peach. Amplified fragments were cloned using a TA cloning kit (Life Technologies, USA). Clones were subjected to sequencing to confirm they had the appropriate insert. Clones of the *CHS, CHI, F3H,DFR, ANS*, and *Ubiquitin* genes were used as homologous probes. cDNA clones of *PAL* and *UFGT* obtained from *Antirrhinum majus* (*A. majus*) were also used as a probes. Hybridization was performed with ECL non-radioactive DNA labeling and detection kits (GE Healthcare, UK).

RNA Extraction and First Strand cDNA Synthesis

Two hundred milligrams of petals were collected separately from pink and variegated flower buds ranging from 6–9mm in diameter. Petals removed from flower buds were immediately frozen in liquid nitrogen and subjected to total RNA extraction using an RNeasy Plant Mini kit (Life Technologies, USA). First strand cDNA was prepared using a cDNA synthesis kit (GE Healthcare, UK) according to the manufacturer's instructions. On occasion, an adapter primer having dT_{17} (B26) was used (Frohman *et al.*, 1988; Schwinn *et al.*, 2006). Six micrograms of total RNA was used for one reaction and cDNA was diluted into 1ml with sterilized water. 10 μl of first-strand cDNA solution was used for the RT-PCR and 3' RACE PCR.

Partial Cloning of a Peach *MYB*-like Gene by 3 RACE PCR

The 3' end of the peach *MYB* gene was obtained by 3' RACE amplification using 10 μl of first-strand cDNA solution as a template with the G1709 primer (forward, 5'-AAAAGCTGCAGACTTAGG   TGGTTGAATTATCTAAAGCC-3') designed for *Ros*1 amplification in snapdragon (Schwinn *et al.*, 2006) and B25 adapter primer (reverse, 5'-GACTCGAGTCGACATCG-3') (Frohman *et al.*,

1988). PCR amplification was performed using Gene Amp® PCR System 9700 (Life Technologies, USA) with the following thermal conditions; 94 °C for 5min then a repeat of 30 or 40 cycles of 94 °C 100%/94 °C for 48 s, 50 °C 50%/50 °C for 75 s, 72 °C 80%/72 °C for 3min and, finally, 72 °C for 10min for extension. Amplified fragments were cloned using pGEM-T Easy Vector System (Promega, USA).

Gene Expression Confirmed by RT-PCR Southern Hybridization

Expression of the structural genes was detected by RT-PCR amplification using the first-strand cDNA solution as a template with primers used for probe preparation and the B25 adapter primer described above. For amplification of *MYB*-like genes, the G1709 forward primer was used (Schwinn *et al.*, 2006). PCR amplification was performed using the same conditions as used for cloning the peach *MYB*-like genes. Amplified fragments were fractionated using 1.4% agarose gel electrophoresis then blotted onto a Hybond-N⁺ membrane. Signals were generated according to the methods described for probe preparation and signal detection.

5′ RACE PCR

The 5′ end of the peach *MYB* gene was determined by 5′ RACE System (Life Technologies, USA) according to the manufacturer's instructions using the PGS240 primer (5′-GCTCTTGTTTCTCTTC TTCTTCGTGTTGTCGTTAA-3′) as a GSP1 (Gene-Specific Primer 1) and PGS42 primer (5′-CTCTTCCTCCTGAGTTA TATTTCCCCT TTTGATAT-3′) as a GSP2 (Gene-Specific Primer 2). Amplified product was cloned using the pGEM-T Easy Vector System (Promega, USA) and sequenced.

Preparation of an Expression Vector Harbouring the Peach MYB Gene

The PMintF primer (5′-TTGCCAATTCGC CGAAGTTTGGA-3′) and the PMintR primer (5′-AGGTTATGCTTATGGTCAAC ATTAT-3′) were designed based on the sequence information from 3′ RACE and 5′ RACE of the peach MYB gene. First-strand cDNA was amplified using this primer set. PCR product was diluted 100-fold and used for GATEWAY cloning (Life Technologies, USA) according to the manufacturer's instructions. PCR amplification was performed with primers in which 25bp *att*B1 and 25bp*att*B2 sequences plus four termial Gs were added to the intermediate F and R primers, respectively. The Entry clone was made by recombination between *att*B-PCR product of the peach MYB gene and pDONR 207 using the BP reaction. The peach MYB gene (*Peace* cDNA sequence, GenBank/EMBL/DDBJ accession AB897865) was then transferred to the expression vector of pJAM1500 (a destination vector derived from pJIT60 with a Gateway destination cassette inserted in the *Sma*I site of the polylinker between the double CaMV 35S promoter and the CaMV terminator sequence) using the LR reaction.

Complementation Analysis by Particle Bombardment

Variegated flower buds (just before opening) were bombarded using a particle inflow helium gun (Vain *et al.*, 1993; Schwinn *et al.*, 2006) with genes encoding the transcription factors: Rosea1, Rosea2, Venos, and Mutabilis from *Antirrhinum majus* or peach *MYB* cDNA cloned into the expression vector. About 2mg of gold particles (size 0.95) were coated with 10 µg of plasmid DNA of each transcription factor and 5 µg of *YFP*gene DNA prepared in 20 µl by adding 100 µl each of the Xho buffer (30 µl of 5M NaCl, 5 µl of 2M TRIS pH 8.0, 965 µl of SDW), 0.1M spermidine, 25% PEG (mol. wt. 1300–1600), and 2.5M $CaCl_2$. This mixture was incubated for a few minutes and then, after pulse centrifugation, the supernatant was removed and the pellet was washed with absolute ethanol. Gold particles were resuspended in 2ml of absolute ethanol. Horizontally placed tubing was filled with this gold particle suspension using a syringe and then left for a few minutes to settle the gold particles at the bottom of tubing. A piece of paper was placed at the end of the tubing to drain the ethanol. The tubing was dried completely with nitrogen gas. This dried tubing was then cut and inserted into the cartridge of the gene gun for shooting. Each flower bud was shot three times at 300 psi pressure. After shooting, branches were kept at room temp (about 20 °C). After 24h, petals were removed from the flowers and placed on glass slides. Magenta spots, generated by complementation, were counted under bright field of a dissection fluorescent microscope (LEICA MZFLIII, Germany) and, in the same field, yellow spots generated by expression of the *YFP* gene were counted under dark field.

Cloning of the *Peace* gene using a Genomic Library

Cloning of the *Peace* gene (*Peace* genomic sequence, GenBank/EMBL/DDBJ accession AB897866) from genomic DNA was carried out following the directions for the Stratagene phage cloning system (Stratagene, USA). Total DNA extracted from V2 petals was partially digested with *Sau3A*I (TAKARA, Japan). After a partial fill-in reaction, total DNA fragments were ligated with lambda FIX/*Xho*I partial fill-in treated DNA (Stratagene, USA), and a packaging reaction was carried out using GigaPack Gold (Stratagene, USA). Recombinant phages were selected by plaque hybridization with*Peace* cDNA (clone cPpP14) as a probe. Plaque hybridization was carried out using an ECL non-radioactive DNA labelling and detection kit (GE Healthcare, UK).

Cloning of the PprMYB10 Homologue from Flowering Peach

A partial genomic sequence homologous to the *PprMYB10* (*Prunus persica*MYB10, Lin-Wang *et al.*, 2010) gene was obtained from DNAs of pink petals by PCR cloning followed by sequencing. Three primer sets:

A2F (5′-ATGGAGGGTTATGACTTGAGTGTGAGA-3′) and A2R (5′-TACTTCATCCTCTGCAAACTCTCCTTTC-3′);

B2F (5′-GTGCAGGAAGAGCTGTAGACTAAGG-3′) and B2R (5′-CTCCCACCAATCACGTTGAGTATGG-3′); and

C2F (5′-AAAGACCATAATAAGGCAACAACCAAG-3′) and C2R (5′-GGTCCACGCTAAAAGAGAAATCACC-3′)

were used for amplification. The A2F primer started from the start codon of *PprMYB10*. Three primer sets were designed to overlap each other and cover almost the entire region of the *PprMYB10* gene.

Phylogenetic Analysis

Amino acid sequences of 21 MYB transcription factors including Peace were aligned using CLUSTALW [version 2.1]. The phylogenetic tree was generated using Njplot [version 2.3].

RESULTS

Pigment Responsible for the Pink Colour in Petals

HPLC analysis revealed that cyanidin-3-glucoside was the major pigment in pink peach flowers (Fig. 1B). Pink flowers contained 62.3 µg of cyanidin-3-glucoside per flower. By contrast, variegated flowers contained only 0.6 µg of cyanidin-3-glucoside per flower, so that the level of anthocyanin in variegated flowers was only 1% of those in pink flowers. Variegated flowers contained higher levels of quercetin-3-galactoside (108.2 µg per flower), which was twice the level of that in pink flowers (56.5 µg). These results suggested that the latter part of the anthocyanin biosynthetic pathway catalysed by DFR, ANS, and UFGT is suppressed in variegated flowers.

Genomic Southern Hybridization

DNA extracted from pink flowers and variegated ones was digested with restriction enzymes, separated by gel electrophoresis, subjected to Southern hybridization and probed with genomic or cDNA clones encoding genes involved in the anthocyanin biosynthetic pathway (Fig. 2). The only polymorphism observed was in the *Eco*RI digestion, for the *CHS* gene. A very weak hybridization signal to a 2.7kb fragment was seen only in DNA from pink petals, but not in DNA from variegated petals. No other polymorphic patterns were observed for other restriction enzymes such as *Bam*HI, *Dra*I, and *Hin*dIII, so it was concluded that it was unlikely that this weak signal was responsible for the difference in flower colour phenotype. Other probes such as *PAL*, *CHI*, *F3H*, *DFR*, *ANS*, and *UFGT* did not show any polymorphisms between pink and variegated petals. No other differences were detected at the DNA level that could explain the different flower colour phenotypes.

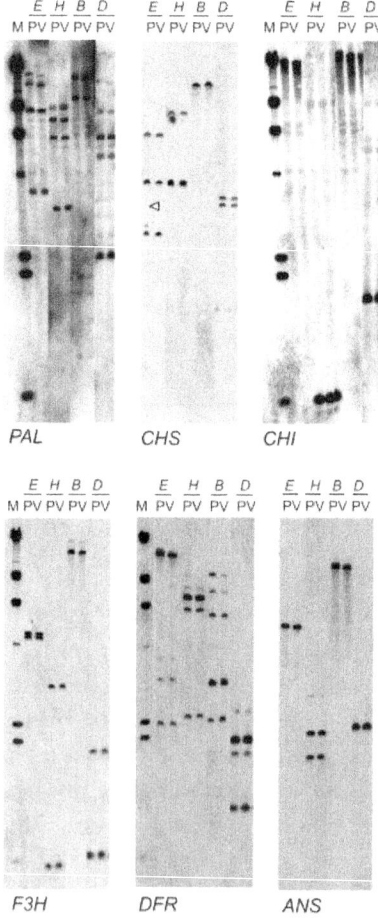

Figure 2. Genomic Southern hybridization probed with *PAL*, *CHS*, *CHI*, *F3H*, *DFR*, and *ANS* genes, respectively. Total DNAs of P, pink petal, and V, variegated petal, were

digested with *E, Eco*RI; *H,Hind*III; *B, Bam*HI; and *D,Dra*I. Only a slight difference indicated by a triangle was seen.

Gene Expression in the Flower Bud

Expression of structural genes, which encode enzymes in the anthocyanin biosynthetic pathway, was analysed by RT-PCR Southern hybridization using RNAs extracted from young petals of pink and variegated flower buds. All genes were very strongly expressed in pink petals, whereas in variegated petals *CHI*, *F3H*, and *ANS* showed very weak expression compared with pink petals (Fig. 3). *PAL*, *CHS*, and *DFR* gene expression in variegated petals was not detected even after 40 cycles of amplification. This result suggested that the white sectors in variegated flowers involved loss of function of a gene regulating the expression of structural genes of the anthocyanin biosynthetic pathway.

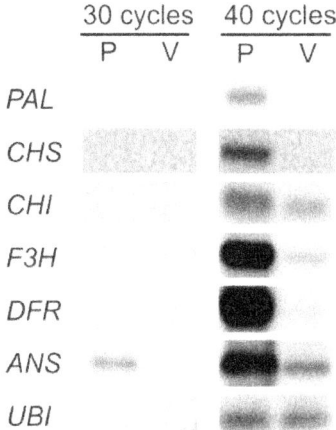

Figure 3. Gene expression in the petals of young flower buds examined by RT-PCR Southern hybridization. P, pink petal; V, variegated petal.

Complementation of Pigment Biosynthesis

To assess whether heterologous transcription factors could regulate the anthocyanin biosynthetic pathway in flowering peach, four genes encoding transcription factors from *Antirrhinum majus*, i.e. *Rosea1*, *Rosea2*, *Venosa*, and *Delila* (Schwinn *et al.*, 2006) were bombarded into the white area of variegated petals together with the *YFP* gene. All introduced genes were cloned into the pJIT60 vector under the control of the double CaMV 35S promoter. Twenty-four hours later, bombarded petals were removed from the flower and observed under the dissection microscope equipped with a fluorescent light source. Magenta spots were observed when an introduced gene for the transcription factor could complement anthocyanin biosynthesis (Table 1). The same area of petal tissue was observed under fluorescent light to count the number of yellow spots generated by expression of the *YFP* gene. The complementation ratio was calculated from the number of magenta spots divided by the number of yellow spots (Table 1). *Rosea1*, which encodes one of the MYB transcription factors controlling anthocyanin biosynthesis in *A. majus*, showed the highest complementation ratio of 25.3%, whereas *Rosea2* and *Venosa* (which also encode MYB transcription factors; Schwinn *et al.*, 2006) and *Delila* (which encodes a bHLH protein controlling anthocyanin production; Goodrich *et al.*, 1992), showed lower complementation ratios. Therefore, expression of a MYB-like transcription factor, functionally homologous to Rosea1, might be disrupted in the white parts of variegated petals of flowering peach, because cyanidin-3-glucoside was not produced there.

View this table:

Table 1. Magenta spot ratio observed after particle bombardment

Transcription factors	Magenta spot/yellow spot (%)
Negative control	0.0
YFP control	0.0
Rosea1	25.3
Rosea2	9.1
Venosa	4.0
Delila	7.7

Isolation of a *MYB*-like Gene from Pink Peach Flower Buds

To confirm the involvement of a MYB-like transcription factor in controlling flower pigmentation, expression of *MYB*-like genes was examined in pink and variegated petals by RT-PCR Southern hybridization. The G1709 primer, based on the petunia *AN2* and maize *C1* gene sequences (Schwinn *et al.*, 2006) and the B25 primer for 3' RACE amplification (Frohman *et al.*, 1988) were used. The *Rosea1* gene was used as a probe. cDNA from both pink and variegated petals gave strong signals for a 1.3kb transcript (Fig. 4A). However, there was a difference in cDNAs from pink and variegated petals for bands of 1.0kb and 1.1kb. For both these fragments, pink petals showed stronger expression than variegated petals. Therefore these 1.0kb and 1.1kb transcripts could be candidates for genes regulating the difference in anthocyanin biosynthesis in flowering peach.

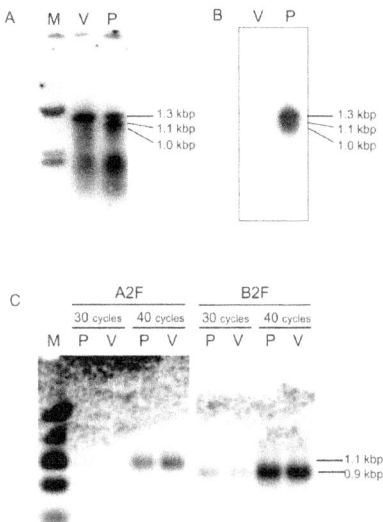

Figure 4. *MYB*-like gene expression in young petals of variegated (V) and pink (P) flower buds. (A) Amplified using the G1709 primer and probed with the *Rosea1* gene of *Antirrhinum majus*. (B) Amplified using the G1709 primer and probed with clone cPpP14 derived from pink flowers of flowering peach. (C) Amplified using the A2F or B2F primer for *PprMYB10* and probed with the 3' region of the *PprMYB10* gene. Number of cycles indicates the amplification cycle of PCR performed.

To obtain genes encoding MYB-like transcription factors regulating the phenotypic difference between pink and variegated petals, 3' rapid amplification of cDNA ends (RACE) PCR was performed for cDNA from pink petals using the G1709 and B25 primers (Fig. 5). Amplified fragments were cloned into the pGEM-T Easy cloning vector. cDNA clones derived from pink petals had various lengths of insert ranging from 860bp to 1020bp, but sequence similarity was very high between these clones. The amino acid sequence of the N-terminal region deduced from the pink clone cPpP14 showed high similarity to other plant MYB transcription factors. Among 21 R2-R3 MYB transcription factors, 33 out of 66 amino acids corresponding to the latter part of R2 and whole R3 repeat were identical (Fig. 8). Furthermore, 50 amino acids in the same region of cPpP14 were identical to *TT2*, *ZmC1*, *ZmPl*, and *OsC1*, whereas 37 were identical to *Rosea1*, *Venosa*, *An2*, *VvMYBA1*, and *PprMYB10*. However, the C-terminal region was different from other MYB proteins, which may reflect the specificity of this gene.

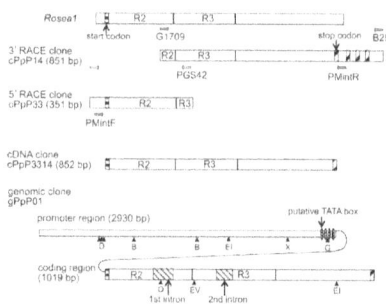

Figure 5. Schematic strategy for the cloning of the peach MYB-like gene based on the sequence information of*Roseal* and the peach MYB-like gene structure obtained from cDNA and genomic DNA. Restriction enzyme sites used in Southern blots were indicated by a triangle with B for *Bam*HI, G for *Bgl*II, D for *Dra*I, EI for *Eco*RI, EV for *Eco*RV, and X for *Xba*I.

When this cPpP14 clone was used as a probe for hybridization to RT-PCR products on Southern blots to confirm the expression of this *MYB*-like gene in flower buds, the 1.0, 1.1, and 1.3kb transcripts were detected only in cDNA from pink flower buds (Fig. 4B).

Another *MYB*-like gene obtained from genomic DNAs of pink petals was almost identical to *PprMYB10*. The sequence of this gene, 1818bp in length, was confirmed by connecting overlapped clones. The *PprMYB10*gene is 1849bp in length. The deduced amino acid sequence of this gene started with the same first codon as *PprMYB10* and the sequence of this gene, including the 1st and 2nd introns, was identical to *PprMYB10*. Based on these results the *MYB10* gene from flowering peach was named as*FPMYB10*. Expression of *FPMYB10* in pink and variegated flower buds was investigated by RT-PCR Southern blots using two forward primers, A2F and B2F, in combination with the B25 reverse primer. The 3' region of*FPMYB10*, amplified using C2F and C2R primers, was used as a probe. Pink and variegated flower buds showed strong and almost identical levels of expression of the *FPMYB10* cDNA amplified using either A2F and B2F primers (Fig. 4C). Therefore at least in petals, differences in expression of the *FPMYB10* gene are not the cause of the differences in colour between pink and variegated flowers.

Function of the *MYB* gene Obtained from Pink Petal cDNA

5' rapid amplification of cDNA ends (RACE) PCR was performed to obtain the 5' region of the cPpP14 cDNA clone using the gene-specific primer PGS42 based on the cPpP14 sequence (Fig. 5). The amplified fragment was cloned into the pGEM-T Easy vector and sequenced. The entire coding sequence of the flowering peach *MYB* gene was reconstructed from the sequence information of the 5' RACE clone, cPpP33, carrying a 351bp fragment as an insert and 3' RACE clone, cPpP14, having a 858bp insert. The full-length gene was then amplified from cDNA from pink petals with gene-specific primers; namely the PMintF primer designed just upstream of the start codon and the PMintR primer just 3' to the first stop codon. The amplified product was diluted and used as a template for *att*B-PCR to add the *att*-B sequences, required for preparation of an entry vector for the GATEWAY cloning system (Life Technologies, USA). Amplified *att*-B product was inserted into the pDONR207 entry vector by BP recombination. The cDNA was transferred by LR recombination to the pJAM1500 vector where expression is driven by the double CaMV 35S promoter. The Peach *MYB* gene in this expression cassette was bombarded into variegated flower buds together with the *YFP* gene. Twenty-four hours later, bombarded petals were observed under the dissection microscope. Many magenta spots were observed surrounding aggregated gold particles (Fig. 6). The complementation ratio of 104.1% (Table 2) was much higher than the ratio achieved by bombardment with *Rosea1* (Table 1). It was concluded that this peach *MYB* gene was controlling the pigmentation of petals of flowering peach. Therefore, this gene was named *Peace* (*pe*achanthocyanin *c*olour *e*nhancement). In variegated flower petals, *Peace*expression was not detected by RT-PCR Southern hybridization when probed with clone cPpP14 (Fig. 4B). Consequently, the lack of *Peace*expression probably resulted in white areas in variegated petals.

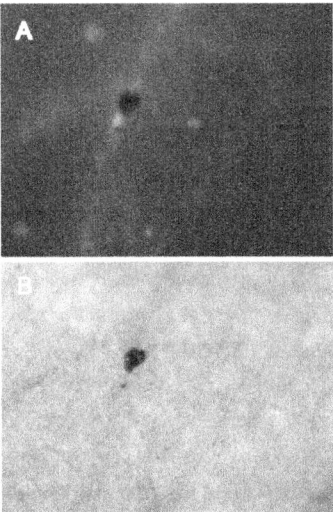

Figure 6. Complementation of pigment synthesis confirmed by particle bombardment

into the white area of variegated petals with the *Peace* gene. (A) Yellow spots were generated by the *YFP* gene. (B) Magenta spots were generated by the *Peace* gene. Both observed in the same field of vision. View this table:

Table 2. Magenta spot ratio acquired following introduction of the peach *MYB*-like gene (*Peace*) by particle bombardment

Transcription factors	Magenta spot/yellow spot (%)
Negative control	0.0
YFP control	0.0
Peach *MYB*-like gene (*Peace*)	104.1

Comparison of *Peace* Genomic and cDNA Sequences between Pink and Variegated Petals

The gene structure of *Peace* was confirmed by comparison of cDNA and genomic sequences. The *Peace* gene was composed of three exons disrupted by two introns, the first intron was 87bp in length located in the R2 domain at 133 from ATG and the second one was 80bp in length located in R3 domain at 261 from ATG in cDNA numbering (Fig. 5). These were identical positions to the introns in other plant *MYB* genes (Martin and Paz-Ares, 1997). Four candidate TATA box sequences were found within 100bp of the region upstream from the ATG, based on the TFBIND (Tsunoda and Takagi, 1999). One of these candidates, the TATA box situated at 53bp upstream of the initiating ATG codon, showed the highest score of 0.846 in TFBIND.

To investigate the reasons why the *Peace* gene was strongly expressed in pink petals but not in variegated petals, a genomic library was constructed from V2 tissue and coding and 5'-UTR regions were compared between pink and variegated tissues. The genomic sequence of the coding region and *ca.* 3kb of promoter region of the *Peace* gene was determined by screening the V2 genomic library. Based on this sequence information, sequences of coding and promoter regions of the P2 genome was determined by PCR sequencing. The sequences of coding and promoter regions were completely identical in DNA from pink petals and from variegated petals. No polymorphisms were detected within coding and promoter regions. To confirm whether there was a difference between pink and variegated genomes, genomic Southern hybridization was performed under high stringency conditions using both the full-length cDNA clone of *Peace* (cPpP3314) and the cPpP14 clone containing just the 3' region of *Peace* cDNA as probes. Both pink and variegated genomes carried a single copy of the *Peace* gene because hybridization patterns showed a single band for each digestion (Fig. 7A). Probes made from the cPpP14 clone were likely to be highly specific for the *Peace* gene because this part of the cDNA lacked most of the sequences encoding the MYB domain which is highly repeated in plants due to their large number of R2R3 MYB genes. Additional weaker signals were observed when the full-length cDNA clone (cPpP3314) was used as a probe (Fig. 7B). These weaker signals are probably due to hybridization to other genes encoding the conserved R2R3 MYB domain. However, no differences were observed between pink and variegated genomes which could explain the phenotypic differences in flower pigmentation.

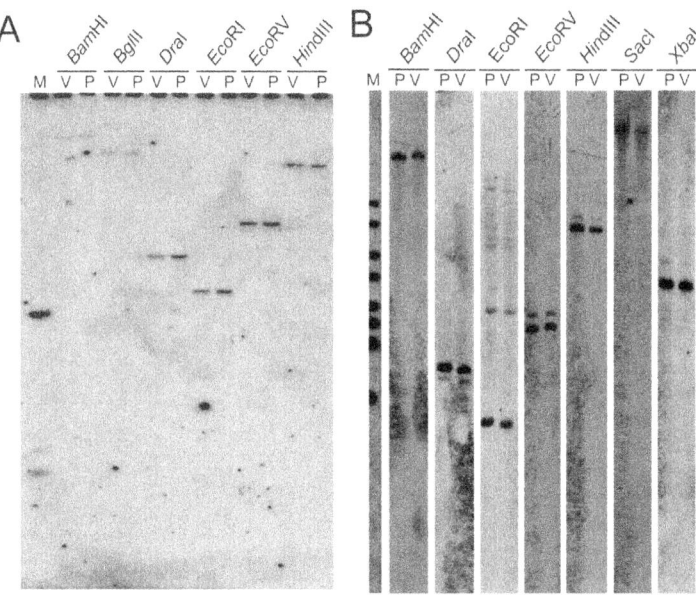

Figure 7. DNA gel blot of genomic DNA obtained from pink (P) and variegated (V) petals. (A) Probed with clone cPpP14, containing the 3' end of the *Peace* gene. (B) Probed with clone cPpP3314, the full-length clone of the *Peace* gene.

To investigate another possible mechanism determining the difference in *Peace* expression between pink and variegated flowers, sites for DNA methylation were searched in the promoter region. Thirty-five CpG dinucleotides were found to be scattered in the promoter region, but there were no CpG islands in which CpG dinucleotides appeared at high density. Therefore, it seems unlikely that differential DNA methylation controls *Peace* gene expression in flowering peach.

Relationship Between PEACE and other Plant MYB Transcription Factors

The amino acid sequence deduced from the full-length *Peace* cDNA revealed the protein to contain well-conserved R2 and R3 MYB domains. Peace aligned well with amino acid sequences of 20 MYB proteins belonging to subgroups 4, 5, 6, 7, and 15, chosen for their functions in regulating flavonoid biosynthesis or by close phylogenetic relationships to regulators of flavonoid metabolism from other plants (Fig. 8) (Stracke *et al.*, 2001; Bailey *et al.*, 2008; Dubos *et al.*, 2010). Among 22 MYB transcription factors, 45 deduced amino acids out of 104 in R2 and R3 repeats were identical.

AtMYB5	---PFGNIIPI SQP---LQMD DCKDGIVGAS --SSSLGHD- --------- --------- --- 249
AtMYB32	RFGFGNGKEC SCNNVKCQTE DSSSSSYSST DISSSIGYDF LGLNNTRVLD FSTLEMK--- --- 274
AtMYBOGL1	LWVHDDDFEL SSLV-MMNFA SGDVEYCL-- --------- --------- --------- --- 228
AtMYB23	FWVLEDDFEL SSLT-MMDFT NG---YCL-- --------- --------- --------- --- 219
AtMYB66Wer	LWVHEDEFEL STLTNMMDFI DG---HCF-- --------- --------- --------- --- 203
AtMYB123TT2	LDCGNVTSLV SSNEIEGELV PAQQNLDLN- ---RPFISCH HR-----GDD EDWLRDFTC- --- 258
Peace	LIMSGINNPD SEFWKICHQL DDNNEGSKN- ---YKWASSH HAPLWFSEDY KDWMGDDDCF HH- 283
ZmC1	GEAGSSDDCS SAASVSLRVG -SHDEPCFSG DGDGDWMDDV RALASFLESD EDWLRCQTAG QLA 273
ZmPl	-EARSSDDCS SAASVSPLVG SSQHDPCFSG DGDGDWMDDV RALASFLESD EEWLRCHTAE QLV 267
OsC1	CDYYCSGSSS AATTTSSSSL PAVVFPCFS- AGD-DWMDDV RALASFLDTD DAWNLCA--- --- 272
AtMYB75PAP1	----EATTTE KGDTLAFDVD QLWSLFDGET VKFD----- --------- --------- --- 248
AtMYB90PAP2	----EATTAE HGATLAFDVE QLWSLFDGET VELD----- --------- --------- --- 249
AtMYB113	----EATETA KGVTLPLDFE QIWARFDEET LFLN----- --------- --------- --- 246
Rosea1	---LFEEAQQ IGN------ --------- --------- --------- --------- --- 220
Rosea2	---LFEDVQQ TGKMSEW--- --------- --------- --------- --------- --- 224
Venosa	--------- --------- --------- --------- --------- --------- --- 194
PhAn2	PTLLHEETAP SVNAESSLTQ GGGSGLSDFS VDIDDIWDLV S--------- --------- --- 255
MdMYB10	----FPEGQS RS--EFSFST DLWNHSKEE- --------- --------- --------- --- 243
PprMYB10	---FSEGLI RG--DFSFSV DPWNHSKEE- --------- --------- --------- --- 274
VvMYBA1	PNDGMIEQIQ GGEGDFPFDV GFWDTPNTQV N---HLI--- --------- --------- --- 250
VvMYBA2	AIKPHPHKFS KALPRFELKT TAVDTFDTQV STSSKLIHVT TTE------ --------- --- 265

Figure 8. Alignment of the deduced amino acid sequences of 21 R2-R3 MYB transcription factors belonging to subgroup 4, 5, 6, 7, and 15. Peace is marked using pink. Conserved tryptophans are marked using green. The amino acid signature motifs of $\{[DE]Lx_2[RK]x_3Lx_6Lx_3R\}$, important for interaction with bHLH proteins, are indicated by a yellow square, the $[A/S/G]N[D/A/N]V$ motif and the $KPXPR[S/T]F$ motif, known to be conserved in anthocyanin regulators, are indicated by a red square and a blue square, respectively. The conserved amino acid sequence [DNEI] is shown by a black square. The horizontal line indicates the amino acid sequence of the R2-R3 region derived from the cPpP14 clone.

The region of highest conservation was the latter part of the R2 domain and middle and C-terminal part of the R3 domain. Five tryptophans, considered to be important to forming the two repeats of the helix-helix-turn-helix structure in the R2 and R3 domains, were completely conserved in Peace. Furthermore, Peace has the conserved amino acid signature motif $[DE]Lx_2[RK]x_3Lx_6Lx_3R$ which is important for interaction with bHLH proteins from sub-group IIIf (Heim *et al.*, 2003; Zimmermann *et al.*, 2004; Lin-Wang *et al.*, 2010). These results suggest *Peace* can act to regulate anthocyanin biosynthesis in petals of flowering peach in combination with bHLH proteins. However, neither the $[A/S/G]N[D/A/N]V$ motif (Lin-Wang *et al.*, 2010) nor the $KPXPR[S/T]F$ motif, both known to be conserved in R2R3MYB proteins promoting anthocyanin biosynthesis, were present in Peace. The C-terminal region of Peace was quite divergent compared with those of other plant MYB proteins.

To compare the amino acid sequences of Peace, phylogenetic analysis was undertaken for it and 20 additional MYB transcription factors (Fig. 9). The phylogenetic tree indicated that Peace was most closely related to AtMYB123 (TT2) of *Arabidopsis* which regulates proanthocyanidin biosynthesis (Nesi *et al.*, 2001; Baudry *et al.*, 2004). Monocotyledonous MYB proteins, C1 and Pl of *Zea mays* and OsC1 of *Oryza sativa*, involved in the anthocyanin biosynthesis (Paz-Ares *et al.*, 1987; Cone *et al.*, 1993) were also closely related to Peace. However, PprMYB10 of *Prunus persica*(peach) was not closely related to Peace. It was clustered together with MdMYB10, VvMYBA1, and VvMYBA2 (Kobayashi *et al.*, 2004).

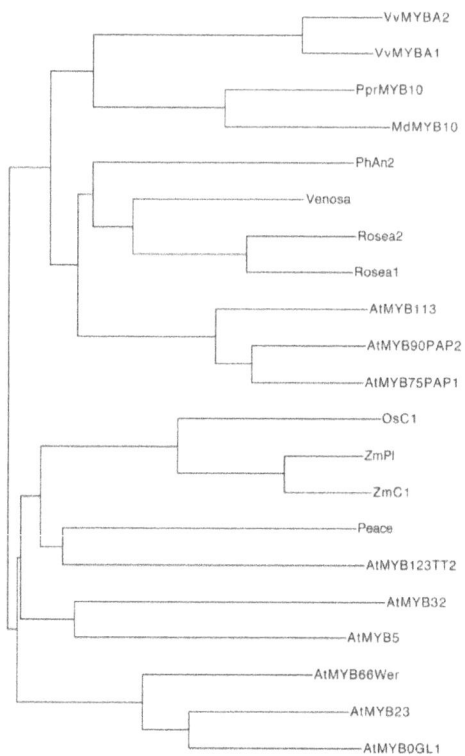

Figure 9. Phylogenetic tree of R2R3 MYB transcription factors showing the relationship between Peace and another 20 accessions belonging to subgroups 4, 5, 6, 7, and 15; the same as shown inFig. 8.

DISCUSSION

The *Peace* gene encoding an R2R3 MYB-like transcription factor that is expressed in pink petals of flowering peach, *Prunus persica* cv. Genpei which bears pink flowers and variegated flowers on a single tree has been identified. This gene is strongly expressed in fully pigmented pink petals but very weakly expressed in variegated petals. Introduction of the *Peace*gene into the white areas of variegated petals by means of particle bombardment complemented pigment biosynthesis and magenta spots appeared. It was, therefore, concluded that the *Peace* gene is likely to be responsible for colouration in the petals of flowering peach.

All structural genes involved in the anthocyanin biosynthetic pathway were strongly expressed in young petals of pink flower buds, whereas *CHI, F3H,* and *ANS* genes showed only weak expression and no expression of *PAL,CHS,* and *DFR* genes were observed in variegated flower buds. These results agreed well with the result of pigment analysis which showed that pink petals contained >100-fold more cyanidin-3-glucoside than variegated petals. It seems likely that a change affecting expression of all the anthocyanin biosynthetic genes has occurred in a gene encoding a transcription factor regulating their expression. In this context, the expression pattern of *Peace*, which is strongly expressed in pink petals but not expressed in variegated petals, supports the view that Peace regulates the anthocyanin biosynthetic pathway and that it is variation in its expression that gives variegated petal pigmentation in flowering peach petals.

When Peace was compared with 20 other MYB transcription factors from plants, high similarity was seen between the 104 amino acids of the R2R3 domain, including the identity in some key amino acid positions which are important for interaction with bHLH proteins. This conservation also suggests that *Peace* encodes a transcription factor controlling anthocyanin biosynthesis in the petals of flowering peach, although a gene encoding a bHLH protein partner has not yet been identified.

Peace does not have the [A/S/G]N[D/A/N]V motif, which appears near the end of the R3 repeat and can be an identifier for anthocyanin regulators (Lin-Wang *et al.*, 2010). However, at the same position, Peace shares the amino acid sequence DNEI with monocotyledonous anthocyanin regulators; ZmC1, ZmPl, and OsC1, and with AtMYB5, AtMYB32, and AtMYB123 (TT2). This suggests that Peace is functionally more similar to AtMYB123 and AtMYB5 which are known to regulate proanthocyanidin production partially redundantly and to monocotyledonous anthocyanin regulators such as ZmC1, ZmPl or OsC1 (Gonzalez *et al.*, 2009; Dubos *et al.*, 2010). VvMYB5a, the protein from grapevine most similar to AtMYB5 has been shown to activate anthocyanin biosynthesis when expressed ectopically in tobacco (Deluc *et al.*, 2006).

Phylogenetic analysis showed that Peace clustered with AtMYB123 and anthocyanin regulators of monocotyledonous plants, such as ZmC1, ZmPl, and OsC1. Stracke *et al.* (2001) showed a correlation between functional conservation and sequence similarity in MYB transcription factors, for example, AtMYB123 and ZmC1 both regulate pigmentation. Therefore it seems possible that Peace has retained the capacity to regulate anthocyanin biosynthesis by a similar manner as the regulators of anthocyanin biosynthesis from monocots or it has arisen *de novo* from a proanthocyanidin regulatory ancestor. It may be that the regulatory activity of Peace reflects the ability of other regulators of proanthocyanidin biosynthesis from dicots to regulate anthocyanidin biosynthesis, a function not previously recognized because of their concomitant activation of ANR and proanthocyanidin production. Recently, Ravaglia *et al.* (2013)suggested that MYB10 and MYBPA1 of nectarine (*Prunus persica*) exclusively regulate anthocyanin biosynthesis and proanthocyanidin biosynthesis, respectively, in the fruit skin and flesh. However, genes similar to Peace may also contribute to regulating anthocyanin biosynthesis and proanthocyanidin biosynthesis in peach.

The mechanisms regulating the difference in the expression of the *Peace*gene between pink and variegated petals remains unclear. *FPMYB10*, a*PprMYB10* homologue, was a possible candidate to control the difference in flower colour or the difference in *Peace* gene expression. However, both in pink and variegated flower buds, *FPMYB10* is expressed at same level (Fig. 4C) and therefore it was difficult to consider this gene to be the cause of the difference in flower colour or *Peace* expression between pink and variegated flowers.

Another possible mechanism regulating the difference in Peace expression, could involve miRNA or siRNA. These are known to regulate expression of transcription factors including MYB proteins (Rhoades *et al.*, 2002; Allen *et al.*, 2007). Furthermore SERRATE in *Arabidopsis* could affect miRNA function to regulate gene expression (Lobbes *et al.*, 2006). Recently, miR156-targeted SPL (SQUAMOSA PROMOTER BINDING PROTEIN-LIKE) gene was revealed negatively regulating anthocyanin biosynthesis in*Arabidopsis* by disrupting the MYB-bHLH-WD40 complex (Gou *et al.*, 2011). In order to understand the mechanism regulating the difference in Peace expression, further investigations are required.

Both pink and variegated genomes possess a single copy of the *Peace*gene, according to the results of genomic Southern hybridization. Sequence analysis, in addition to genomic Southern blots, suggests the*Peace* gene to be identical in pink and variegated flowers, refuting previous suggestions that variegation results from insertion of a transposable element (Chaparro *et al.*, 1995). It is possible that the variegated phenotype in flowering peach cv. Genpei results from a periclinal chimera, where the L1 has been derived from a colourless genotype and the L2/L3 from a pink genotpoye. Pink sectors on flowers and branches with pink flowers could then result from L2 layer invasion, as observed in some grape varieties (Pelsy, 2010). However, no evidence for a chimera was observed in analysis of *Peace* by Southern blot of DNA from pink and variegated flowers. In addition, variegated plants of weeping flowering peach breed true to give progeny with variegated flowers, suggesting that variegation is not due to variegated flowering peach being a periclinal chimera (H Katayama, unpublished observations). Although the *Peace* gene, whose expression controls the pigmentation of flowers in flowering peach, has been identified, the molecular basis for the variegation in flower colouration remains a mystery.

Acknowledgements

This work was supported by the Japan Society for the Promotion of Science(KAKENHI Grant number 07660011), by the Sumitomo Foundation for Grants for Basic Science Research Projects (Grant number 030678), and by The Daiwa Anglo-Japanese Foundation (Daiwa Foundation Small Grants to CU). Special thanks are due to co-author IM, who died suddenly from a brain tumour in 2011 at the age of 32. This report stands as a memorial to her sincere work and warm personality.

REFERENCES

1. Allen RS, Li J, Stahle MI, Dubroue A, Gubler F, Millar AA. 2007. Genetic analysis reveals functional redundancy and the major target genes of the Arabidopsis miR159 family. Proceedings of the National Academy of Sciences, USA 104, 16371—16376.

2. Bailey PC, Dicks J, Wang TL, Martin C. 2008. IT3F: a web-based tool for functional analysis of transcription factors in plants. Phytochemistry 69, 2417—2425.

3. Bailey PC, Martin C, Toledo-Ortiz G, Quail PH, Huq E, Heim MA, Jakoby M, Werber M, Weisshaar B. 2003. Update on the basic helix-loop-helix transcription factor gene family in Arabidopsis thaliana . The Plant Cell 15, 2497—2502.

4. Baudry A, Heim MA, Dubreucq B, Caboche M, Weisshaar B, Lepiniec L. 2004. TT2, TT8, and TTG1 synergistically specify the expression of BANYULS and proanthocyanidin biosynthesis in Arabidopsis thaliana . The Plant Journal. 39, 366—380.

5. Bonas U, Sommer H, Harrison BJ, Saedler H. 1984. The transposable element Tam1 of Antirrhinum majus is 17kb long. Molecular and General

Genetics 194, 138—143.

6. Butelli E, Licciardello C, Zhang Y, Liu J, Mackay S, Bailey P, Reforgiato-Recupero G, Martin C. 2012. Retrotransposons control fruit-specific, cold-dependent accumulation of anthocyanins in blood oranges. The Plant Cell 24, 1242—1255.

7. Chaparro JX, Werner DJ, Whetten RW, O'Malley DM. 1995. Characterization of an unstable anthocyanin phenotype and estimation of somatic mutation rates in Peach. Journal of Heredity 86, 186—193.

8. Cone KC, Cocciolone SM, Burr FA, Burr B. 1993. Maize anthocyanin regulatory gene pl is a duplicate of c1 that functions in the plant. The Plant Cell 5, 1795—1805.

9. de Vetten N, Quattrocchio F, Mol J, Koes R. 1997. The an11 locus controlling flower pigmentation in petunia encodes a novel WD-repeat protein conserved in yeast, plants, and animals. Genes and Development 11, 1422—1434.

10. Deluc L, Barrieu F, Marchive C, Lauvergeat V, Decendit A, Richard T, Carde JP, Merillon JM, Hamdi S. 2006. Characterization of a grapevine R2R3-MYB transcription factor that regulates the phenylpropanoid pathway. Plant Physiology 140, 499—511.

11. Doodeman M, Boersma EA, Koomen W, Bianchi F. 1984. Genetic analysis of instability in Petunia hybrida . Theoretical and Applied Genetics 67, 345—355.

12. Dubos C, Stracke R, Grotewold E, Weisshaar B, Martin C, Lepiniec L. 2010. MYB transcription factors in Arabidopsis . Trends in Plant Science 15, 573—581.

13. Epperson BK, Clegg MT. 1987. Instability at a flower color locus in the morning glory. Journal of Heredity 78, 346—352.

14. Forkmann G. 1993. Genetics of flavonoids. In: Harborne JB , ed. The flavonoids: advances in research since 1986. London: Chapmann & Hall, 537—564.

15. Frohman MA, Dush MK, Martin GR. 1988. Rapid production of full-length cDNAs from rare transcripts: amplification using a single gene-specific oligonucleotide primer. Proceedings of the National Academy of Sciences, USA 85, 8998—9002.

16. Gerats AG, Huits H, Vrijlandt E, Maraña C, Souer E, Beld M. 1990. Molecular characterization of a nonautonomous transposable element (dTph1) of petunia. The Plant Cell 2, 1121—1128.

17. Glover BJ, Martin C. 2012. Anthocyanins. Current Biology 22, R147—

R150.

18. Gonzalez A, Mendenhall J, Huo Y, Lloyd A. 2009. TTG1 complex MYBs, MYB5 and TT2, control outer seed coat differentiation. Developmental Biology 325, 412—421.

19. Goodrich J, Carpenter R, Coen ES. 1992. A common gene regulates pigmentation pattern in diverse plant species. Cell 5, 955—964.

20. Gou JY, Felippes FF, Liu CJ, Weigel D, Wang JW. 2011. Negative regulation of anthocyanin biosynthesis in Arabidopsis by a miR156-targeted SPL transcription factor. The Plant Cell 23, 1512—1522.

21. Habu Y, Hisatomi Y, Iida S. 1998. Molecular characterization of the mutable flaked allele for flower variegation in the common morning glory. The Plant Journal 16, 371—376.

22. Harrison BJ, Carpenter R. 1979. Resurgence of genetic instability in Antirrhinum majus . Mutation Research/Fundamental and Molecular Mechanisms of Mutagenesis 63, 47—66.

23. Heim MA, Jakoby M, Werber M, Martin C, Weisshaar B, Bailey PC. 2003. The basic helix–loop–helix transcription factor family in plants: a genome-wide study of protein structure and functional diversity. Molecular Biology and Evolution 20, 735—747.

24. Holton TA, Cornish EC. 1995. Genetics and biochemistry of anthocyanin biosynthesis. The Plant Cell 7, 1071—1083.

25. Imai Y. 1935. The mechanism of bud variation. The American Naturalist 69, 587—595.

26. Inagaki Y, Hisatomi Y, Suzuki T, Kasahara K, Iida S. 1994. Isolation of a Suppressor- mutator/Enhancer-like transposable element, Tpn1, from Japanese morning glory bearing variegated flowers. The Plant Cell 6, 375—383.

27. Itoh Y, Higeta D, Suzuki A, Yoshida H, Ozeki Y. 2002. Excision of transposable elements from the chalcone isomerase and dihydroflavonol 4-reductase genes may contribute to the variegation of the yellow-flowered carnation (Dianthus caryophyllus). Plant and Cell Physiology 43, 578—585.

28. Jackson D, Roberts K, Martin C. 1992. Temporal and spatial control of expression of anthocyanin biosynthetic genes in developing flowers of Antirrhinum majus . The Plant Journal 2, 425—434.

29. Katayama H, Uematsu C. 2003. Comparative analysis of chloroplast DNA in Pyrus species: physical map and gene localization. Theoretical and Applied Genetics 106, 303—310.

30. Kobayashi S, Goto-Yamamoto N, Hirochika H. 2004. Retrotransposon-induced mutations in grape skin color. Science 304, 982.

31. Kranz HD, Denekamp M, Greco R, et al. 1998. Towards functional characterisation of the members of the R2R3-MYB gene family from Arabidopsis thaliana . The Plant Journal 16, 263—276.

32. Lin-Wang K, Bolitho K, Grafton K, Kortstee A, Karunairetnam S, McGhie T, Espley R, Hellens R, Allan A. 2010. An R2R3 MYB transcription factor associated with regulation of the anthocyanin biosynthetic pathway in Rosaceae. BMC Plant Biology 10, 50.

33. Lobbes D, Rallapalli G, Schmidt DD, Martin C, Clarke J. 2006. SERRATE: a new player on the plant microRNA scene. EMBO Rreports 7, 052—058.

34. Martin C, Carpenter R, Sommer H, Saedler H, Coen ES. 1985. Molecular analysis of instability in flower pigmentation of Antirrhinum majus, following isolation of the pallida locus by transposon tagging. EMBO Journal 4, 1625—1630.

35. Martin C, Gerats T. 1993. The control of flower coloration. In: Jordan BR , ed. The molecular biology of flowering. Wallingford: CAB International, 219—255.

36. Martin C, Mackay S, Carpenter R. 1988. Large-scale chromosomal restructuring is induced by the transposable element tam3 at the nivea locus of Antirrhinum majus . Genetics 119, 171—184.

37. Martin C, Paz-Ares J. 1997. MYB transcription factors in plants. Trends in Genetics 13, 67—73.

38. Martin C, Prescott A, Mackay S, Bartlett J, Vrijlandt E. 1991. Control of anthocyanin biosynthesis in flowers of Antirrhinum majus . The Plant Journal 1, 37—49.

39. Nesi N, Jond C, Debeaujon I, Caboche M, Lepiniec L. 2001. The arabidopsis TT2 gene encodes an R2R3 MYB domain protein that acts as a key determinant for proanthocyanidin accumulation in developing seed. The Plant Cell 13, 2099—2114.

40. Nishizaki Y, Matsuba Y, Okamoto E, Okamura M, Ozeki Y, Sasaki N. 2011. Structure of the acyl-glucose-dependent anthocyanin 5-O-glucosyltransferase gene in carnations and its disruption by transposable elements in some varieties. Molecular Genetics and Genomics 286, 383—394.

41. Paz-Ares J, Ghosal D, Wienand U, Peterson PA, Saedler H. 1987. The regulatory c1 locus of Zea mays encodes a protein with homology to myb

proto-oncogene products and with structural similarities to transcriptional activators. EMBO Journal 6, 3553—3558.

42. Pelsy F. 2010. Molecular and cellular mechanisms of diversity within grapevine varieties. Heredity 104, 331—340.

43. Quattrocchio F, Wing JF, Leppen H, Mol J, Koes RE. 1993. Regulatory genes controlling anthocyanin pigmentation are functionally conserved among plant species and have distinct sets of target genes. The Plant Cell 5, 1497—1512.

44. Ravaglia D, Espley RV, Henry-Kirk RA, Andreotti C, Ziosi V, Hellens RP, Costa G, Allan AC. 2013. Transcriptional regulation of flavonoid biosynthesis in nectarine (Prunus persica) by a set of R2R3 MYB transcription factors. BMC Plant Biology 13, 68.

45. Rhoades MW, Reinhart BJ, Lim LP, Burge CB, Bartel B, Bartel DP. 2002. Prediction of plant microRNA targets. Cell 110, 513—520.

46. Sablowski RW, Moyano E, Culianez-Macia FA, Schuch W, Martin C, Bevan M. 1994. A flower-specific Myb protein activates transcription of phenylpropanoid biosynthetic genes. EMBO Journal 13, 128—137.

47. MedlineWeb of ScienceGoogle Scholar

48. Schwinn K, Venail J, Shang Y, Mackay S, Alm V, Butelli E, Oyama R, Bailey P, Davies K, Martin C. 2006. A small family of MYB-regulatory genes controls floral pigmentation intensity and patterning in the genus Antirrhinum . The Plant Cell 18, 831—851.

49. Shamel AD, Pomeroy CS. 1936. Bud mutations in horticultural crops. Journal of Heredity 27, 487—494.

50. Snowden KC, Napoli CA. 1998. Psl: a novel Spm-like transposable element from Petunia hybrida . The Plant Journal 14, 43—54.

51. Sommer H, Carpenter R, Harrison BJ, Saedler H. 1985. The transposable element Tam3 of Antirrhinum majus generates a novel type of sequence alterations upon excision. Molecular and General Genetics 199, 225—231.

52. Stracke R, Werber M, Weisshaar B. 2001. The R2R3-MYB gene family in Arabidopsis thaliana . Current Opinion in Plant Biology 4, 447—456.

53. Toledo-Ortiz G, Huq E, Quail PH. 2003. The Arabidopsis basic/helix-loop-helix transcription factor family. The Plant Cell 15, 1749—1770.

54. Tsunoda T, Takagi T. 1999. Estimating transcription factor bindability on DNA. Bioinformatics 15, 622—630.

55. Upadhyaya KC, Sommer H, Krebbers E, Saedler H. 1985. The paramutagenic line niv-44 has a 5kb insert, Tam 2, in the chalcone

synthase gene of Antirrhinum majus . Molecular and General Genetics 199, 201—207.

56. Vain P, Keen N, Murillo J, Rathus C, Nemes C, Finer J. 1993. Development of the particle inflow gun. Plant Cell, Tissue and Organ Culture 33, 237––246.

57. van Houwelingen A, Souer E, Spelt K, Kloos D, Mol J, Koes R. 1998. Analysis of flower pigmentation mutants generated by random transposon mutagenesis in Petunia hybrida . The Plant Journal 13, 39—50.

58. CrossRefMedlineWeb of ScienceGoogle Scholar

59. Walker AR, Davison PA, Bolognesi-Winfield AC, James CM, Srinivasan N, Blundell TL, Esch JJ, Marks MD, Gray JC. 1999. The TRANSPARENT TESTA GLABRA1 locus, which regulates trichome differentiation and anthocyanin biosynthesis in Arabidopsis, encodes a WD40 repeat protein. The Plant Cell 11, 1337—1350.

60. Whitham TG, Slobodchikoff CN. 1981. Evolution by individuals, plant–herbivore interactions, and mosaics of genetic variability: The adaptive significance of somatic mutations in plants. Oecologia 49, 287—292.

61. Yamamoto T, Mochida K, Hayashi T. 2003a. Shanhai Suimitsuto, one of the origins of Japanese peach cultivars. Journal of the Japanese Society for Horticultural Science 72, 116—121.

62. Yamamoto T, Mochida K, Imai T, Haji T, Yaegaki H, Yamaguchi M, Matsuta N, Ogiwara I, Hayashi T. 2003b. Parentage analysis in Japanese peaches using SSR markers. Breeding Science 53, 35—40.

63. Zimmermann IM, Heim MA, Weisshaar B, Uhrig JF. 2004. Comprehensive identification of Arabidopsis thaliana MYB transcription factors interacting with R/B-like BHLH proteins. The Plant Journal 40, 22—34.

Chapter 6

METHYLATION MEDIATED BY AN ANTHOCYANIN, O-METHYLTRANSFERASE, IS INVOLVED IN PURPLE FLOWER COLORATION IN PAEONIA

Hui Du[1], Jie Wu[1,2], Kui-Xian Ji[1], Qing-Yin Zeng[3], Mohammad-Wadud Bhuiya[4], Shang Su[1,2], Qing-Yan Shu[1], Hong-Xu Ren[1], Zheng-An Liu[1] and Liang-Sheng Wang[1]

[1] Key Laboratory of Plant Resources/ Beijing Botanical Garden, Institute of Botany, Chinese Academy of Sciences, Beijing 100093, PR China

[2] University of Chinese Academy of Sciences, Beijing 100049, PR China

[3] State Key Laboratory of Systematic and Evolutionary Botany, Institute of Botany, Chinese Academy of Sciences, Beijing 100093, PR China

[4] MOgene Green Chemicals, Saint Louis, MO 63132, USA

ABSTRACT

Anthocyanins are major pigments in plants. Methylation plays a role in the diversity and stability of anthocyanins. However, the contribution of anthocyanin methylation to flower coloration is still unclear. We identified two homologous anthocyanin O-methyltransferase (*AOMT*) genes from purple-flowered (*PsAOMT*) and red-flowered (*PtAOMT*) *Paeonia* plants, and we performed functional analyses of the two genes *in vitro* and *in vivo*. The critical amino acids for AOMT catalytic activity were studied by site-directed mutagenesis. We showed that the recombinant proteins, PsAOMT and PtAOMT, had identical substrate preferences towards anthocyanins. The methylation activity of PsAOMT was 60 times higher than that of PtAOMT *in vitro*. Interestingly, this vast difference in catalytic activity appeared to result from a single amino acid residue substitution at position 87 (arginine to leucine). There were significant differences between the 35S::*PsAOMT* transgenic tobacco and control flowers in relation to their chromatic parameters, which further confirmed the function of *PsAOMT in vivo*. The expression levels of the two homologous*AOMT* genes were consistent with anthocyanin accumulation in petals. We conclude that AOMTs are responsible for the methylation of cyanidin glycosides in *Paeonia* plants and play an important role in purple coloration in *Paeonia* spp.

INTRODUCTION

The colours of flowers, fruits, and leaves are critically important traits in plants for UV protection and attracting pollinators and seed-dispersing organisms (Grotewold, 2006). The major classes of pigments that cause plant coloration are flavonoids, carotenoids, and betalains. The flavonoids are the best-studied class of pigments and are known to be widely distributed among the angiosperms (Tanaka *et al.*, 2008). Flavonoids play important roles in various ecological and physiological processes in plants, including pigmentation, UV absorption, antioxidation, defence responses, and signal transduction, among others (Tahara, 2007).

Anthocyanins are the largest group of flavonoids and the greatest contributors to floral coloration, providing the basis for orange, pink, red, magenta, scarlet, purple, blue, and blue/black flower colours (Tanaka *et al.*, 2009).The basic biosynthetic pathways of three major anthocyanidins, namely pelargonidin, cyanidin, and delphinidin, have been well characterized (Fig. 1A). Further modifications to anthocyanidins including methylation, glycosylation, and acylation in these three aglycones produce a wide variety of anthocyanin compounds that strengthen the flower colour phenotypes (Rausher, 2008; Wessinger and Rausher, 2012). Previous studies that investigated the methylation of flavonoids have focused primarily on flavones and isoflavones (Ibrahim *et al.*, 1998; Noel*et al.*, 2003; Lam *et al.*, 2007; Wang, 2011). The methylation of anthocyanins was first reported in petunia at the 3′ and 5′ positions of aglycones (Wiering and Devlaming, 1977; Jonsson *et al.*, 1983; Brugliera *et al.*, 2003; Provenzano *et al.*, 2014). Anthocyanin *O*-methyltransferase (AOMTs) were isolated from different grape cultivars, named VvAOMT and FAOMT, and characterized *in vitro* and *in vivo*, and they were found to be anthocyanin 3′- and 3′,5′-*O*-methyltransferase (Hugueney *et al.*, 2009;Lucker *et al.*, 2010). Akita *et al.* (2011) revealed a gene (*CkmOMT2*) from purple-flowered fragrant cyclamen, and an enzyme assay with heterologously expressed CkmOMT2 *in vitro* demonstrated that CkmOMT2 exhibited methylation activity with anthocyanins. A red/purple flower colour in a fragrant cyclamen mutant was bred by ion-beam irradiation, which was caused by the loss of the *CkmOMT2* region (Kondo *et al.*, 2009). Previous studies on the chromatic properties of anthocyanins from red grapes

(Heredia *et al.*, 1998) and tree peony petals (Sakata *et al.*, 1995) demonstrated the effects of methylation on the hue; specifically, the higher the number of methoxyl groups, the more pronounced the shift towards purple. In addition, it has been proposed that the methylation of B-ring hydroxyl groups causes a small shift towards red (Tanaka *et al.*, 2008). Although there has been some progress on anthocyanin methylation, a large portion of methyltransferases of plant origin needs to be further characterized, as several questions remain to be answered, such as the molecular mechanism of anthocyanin methylation, the way in which methyltransferase takes part in the methylation of anthocyanins *in vivo*, and the relationship between anthocyanin methylation and floral coloration.

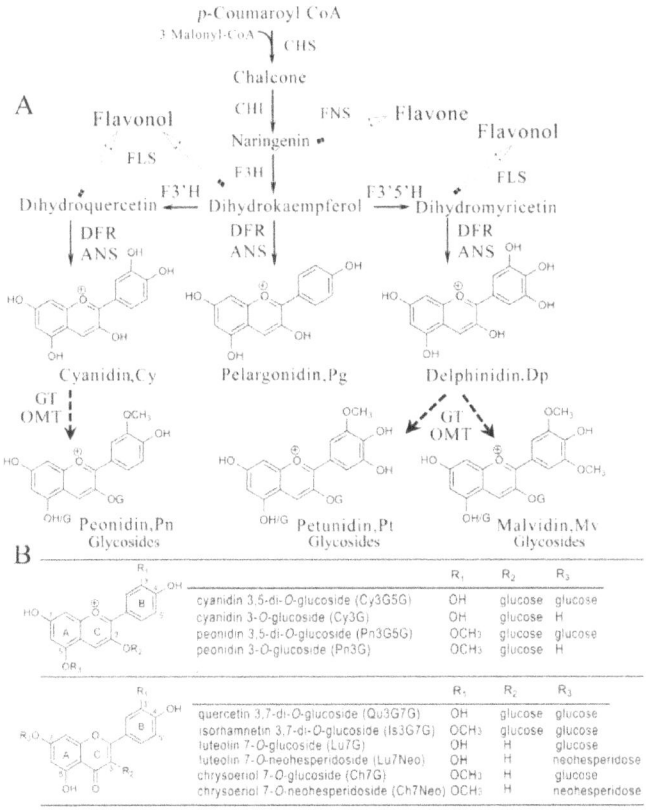

Figure 1. Biochemical pathway and chemical structural information regarding selected flavonoid compounds in plants. (A) Schematic diagram of the biosynthetic pathways of the major flavonoids (Rausher *et al.*, 2008; Wessinger and Rausher, 2012). Dashed arrows represent the unclear steps. The names and structures of three anthocyanidin compounds are indicated. The enzyme names in black boxes are as follows:

CHS, chalcone synthase; CHI, chalcone isomerase; F3H, flavanone 3-hydroxylase; F3′H, flavonoid 3′-hydroxylase; F3′5′H, flavonoid 3′,5′-hydroxylase; DFR, dihydro-flavonol 4-reductase; ANS, anthocyanidin synthase; FNS, flavones synthase; FLS, flavonol synthase; GT, glycosyltransferase; OMT, O-methyltransferase. (B) Chemical structures of anthocyanins, flavones, and flavonols in *Paeonia*flower petals (Wang *et al.*, 2001; Li *et al.*, 2009).

Methylation is an alkylation reaction that transfers an activated methyl group from S-adenosylmethionine (SAM) to the N-, C-, O-, or S-nucleophiles of acceptor molecules (Klimasauskas and Weinhold, 2007). Most methyltransferases methylate hydroxyl and carboxyl moieties, and are referred to as O-methyltransferases (OMTs). A large family of OMTs in plants contributes to the vast structural and functional diversity of plant natural products (Wang, 2011), which can be classified into two types according to their sequence homology and substrate variance (Joshi and Chiang, 1998; Noel *et al.*, 2003; Lam *et al.*, 2007). Type I OMTs include a group of homodimeric OMTs with molecular weights of 38–43kDa and are cation independent, which methylate both flavonoids and isoflavonoids. Type II OMTs have lower molecular weights (23–27kDa) and are cation dependent. Most type II OMTs have been shown to be specific for caffeoyl coenzyme A esters of phenylpropanoids (CCoAOMTs), which are thought to be key enzymes in the biosynthesis of lignin (Ye *et al.*, 1994). However,Ibdah *et al.* (2003) reported that a novel Mg^{2+}-dependent PFOMT from*Mesembryanthemum crystallinum* (ice plant) with a molecular weight of 26.6kDa and high similarity to type II OMTs was specific for the methylation of flavonols and caffeoyl-CoA. Thus, there is a novel subclass within type II OMTs with diverse substrates that are not restricted to lignin synthesis. The identification of cation-dependent OMTs such as ROMT15/17 from *Oryza sativa* (Lee *et al.*, 2008), VvAOMT from *Vitis vinifera* (Hugueney *et al.*, 2009), FAOMT from *V. vinifera* (Lucker *et al.*, 2010), and ObF8OMT (Berim and Gang, 2013) with preferences for flavonoid substrates confirmed the appropriateness of the definition of the new subclass of type II OMTs. This progress highlights the occurrence of novel flavonoid methylation enzymes that differ from type I OMTs and underscores the need to further investigate their catalytic mechanisms.

Plants in the genus *Paeonia* are important ornamentals throughout the world. These beautiful plants have large flowers in a variety of colours and shapes. China has a long history of cultivating and breeding *Paeonia*cultivars and has rich collections of germplasm resources (Ji *et al.*, 2012). However, a large proportion of the available cultivars have a purple or 'purplish' flower, but rarely a red flower. Our previous studies on flavonoids from the flowers of various *Paeonia* spp. demonstrated that peonidin derivatives were the major anthocyanidins that accumulated in most cultivars (Wang *et al.*, 2001; Jia *et al.*,

2008; Li, 2010), indicating that methylation modifications of anthocyanins are prevalent in *Paeonia* spp. As such, these plants provide a good model system for the investigation of methylation mechanisms and their influence on flower coloration.

In this study, we characterized an AOMT (PsAOMT) from a purple-flowered plant from the genus *Paeonia* and characterized its homologue PtAOMT from another plant in the genus *Paeonia* with a vivid red flower using both *in vitro* and *in vivo* methods. The catalytic activity PtAOMT was 60-fold less than that of PsAOMT. By using site-directed mutagenesis, we demonstrated that the vast difference in catalytic activities between these two enzymes was caused by the substitution of one key amino acid. This work characterized the subclass of type II OMTs by integrating biochemical, molecular, and phytochemical analysis, which will support an understanding of the anthocyanin methylating mechanism and shed light on its influence on flower coloration. The efficient enzyme PsAOMT and its key amino acid are responsible for effective activity and could be applied to the specifically targeted molecular breeding of ornamental and crop plants or the development of healthy and beneficial products.

MATERIALS AND METHODS

Chemical Sources

Cyanidin, delphinidin, peonidin, pelargonidin 3-*O*-glucoside (Pg3G), and delphinidin 3-*O*-glucoside (Dp3G) were purchased from Extrasynthese (Genay, France). Quercetin, quercetin 3-*O*-rutinoside (Qu3R), and caffeic acid were obtained from the National Institute for the Control of Pharmaceutical and Biological Products (Beijing, China). Cyanidin 3,5-di-*O*-glucoside (Cy3G5G), cyanidin 3-*O*-glucoside (Cy3G), kaempferol 3-*O*-glucoside (Km3G), kaempferol, luteolin, apigenin, naringenin, and epicatechin were purchased from Sigma-Aldrich (Shanghai, China). Analytical-grade methanol and acetonitrile were obtained from Promptar (Elk Grove, CA, USA).

Plant Materials and Culture Conditions

A tree peony cultivar (*Paeonia suffruticosa* cv. 'Gunpohden') and an herbaceous peony *Paeonia tenuifolia* were used. The plants were grown at the Beijing Botanical Garden. The tobacco and strawberry plants were cultivated in a greenhouse under a 14h light/10h dark photoperiod. The temperature was maintained at 25 °C during the light period and 18 °C during the dark period.

Cloning Candidate *AOMT* cDNA and Phylogenetic Analysis

An open reading frame (ORF) of a segment of expressed sequence tag (FE529149) from a cDNA library of the tree peony (Shu *et al.*, 2009) is highly homologous to genes for the flavonoid OMT (PFOMT) from *M. crystallinum* (Ibdah *et al.*, 2003) and anthocyanin OMT (VvAOMT) from grapevine (Hugueney *et al.*, 2009) and was used for a reference sequence for cloning *AOMT*s from *Paeonia* plants. Total RNA was isolated from the two *Paeonia* petals with an RNAprep pure kit (Tiangen, Beijing, China). One microgram of total RNA was used as the template for cDNA synthesis with Moloney murine leukemia virus reverse transcriptase (Promega, WI, USA). The ORFs of *PsAOMT* and *PtAOMT* were cloned with high-fidelity PrimerSTAR HS polymerase (TaKaRa, Ohtsu, Japan) by using the AOMT forward/reverse primers (Supplementary Table S1, available at *JXB* online) from *P. suffruticosa* cv. 'Gunpohden' and *P. tenuifolia*. The amino acid sequences of PsAOMT and PtAOMT were aligned with CLUSTAL X (Thompson *et al.*, 1997) and refined manually. MEGA 5.1 software was used to reconstruct a phylogenetic tree by using the maximum-likelihood test method (Tamura *et al.*, 2011), with 1000 bootstrap replicates.

Heterologous Expression of AOMTs and Site-Directed Mutagenesis

The sequenced cDNA of *AOMT* was inserted into the pMAL-c5X expression vector (NEB, MA, USA), which contains a maltose-binding protein tag. Recombinant AOMTs were purified with an amylose resin column (NEB). Site-directed mutagenesis was performed by using a Fast Mutagenesis System kit (TransGen, Beijing, China). The sequences of the primers used for this protocol are given in Supplementary Table S1.

Characterizing the Recombinant AOMTs

The assay reaction conditions were optimized prior to performing quantitative analyses. The influence of pH on AOMT activity was assessed within a pH range of 4.5–8.5 using MES (pH 4.5–6.5) and Tris/HCl (pH 7.5–8.5) buffers. The effect of divalent cations on the enzyme activity was estimated by adding aqueous solutions of $MgCl_2$, $CaCl_2$, $ZnCl_2$, $MnCl_2$, $CoCl_2$, or EDTA (all at 10mM final concentration) to the reaction mixture. The optimal concentrations of metal ions were assessed by testing different concentrations of $MgCl_2$ (0.1, 0.2, 0.5, 1.0, 5.0, and 10mM). The optimized conditions were as follows: purified recombinant AOMT (2 µg) was assayed in a final volume of 200 µl containing 200 μM SAM, 1.0mM MgCl$_2$, 14mM β-mercaptoethanol, 100mM Tris/HCl (pH 7.5), and 20 μM flavonoid substrates (the chemical structures are shown in Supplementary Fig. S1, available at *JXB* online). Incubation was performed at 35 °C and stopped with 800 μl of methanol containing 2% formic acid, followed by centrifugation at 12 000rpm for 10min. The upper liquid was prepared for high-performance liquid chromatography (HPLC) analysis, and 20 μl of reaction sample was loaded. Anthocyanin and flavonol profiles were recorded at 525 and 350nm, respectively. The substrates and products were identified by comparison with standards and HPLC electrospray ionization mass spectrometry (HPLC-ESI/MS). For the kinetic studies, purified AOMT was incubated under the above optimized conditions with the exception of a range of substrate concentrations from 5 to 200 μM for K_m determination (Hugueney *et al.*, 2009). Reaction products were analysed by HPLC, using the method reported by Yang *et al.* (2009).

Molecular Modelling of the PsAOMT and PtAOMT Active Sites

Three-dimensional models of PtAOMT and PsAOMT were generated by using the I-TASSER Protein Structure and Function Predictions web server (Roy *et al.*, 2010). Homology models were built by using the known three-domensional (3D) structure of caffeoyl coenzyme, a 3-*O*-methyltransferase (CCoAOMT, PDB code 1SUI) from alfalfa (Ferrer *et al.*, 2005). The best models of PsAOMT and PtAOMT were evaluated based on their template modelling score (0.90), root mean square deviation (1.75), and sequence identity (54.0%) in relation to the 3D structure of CCoAOMT. Because recombinant AOMTs form dimers in solution, PsAOMT and PtAOMT were modelled as homodimers by using the COOT program (Krissinel and Henrick, 2004). Substrate-binding sites were predicted by docking Cy3G5G with the PsAOMT dimer 3D model in the SWISDOCK program (Grosdidier *et al.*, 2011).

Transient Expression in Strawberry Fruit

Fragaria×ananassa cv. 'Hongyan' mature plants with fruits that were just turning red were used for these experiments. A single colony of *Agrobacterium* strain EHA105 with *PsAOMT* or *PtAOMT* and a single colony with *Arabidopsis PRODUCTION OF ANTHOCYANIN PIGMENT1* (*PAP1*; GenBank accession no. AF325123) were cultured and diluted to an OD$_{600}$ of 0.1–0.3, and injected into strawberry fruits according to the method of Hoffmann *et al.* (2006). The infiltrated fruits were harvested after 4 d, and the extracts were analysed according to Yang *et al.* (2009).

Tobacco Transformation

PsAOMT and *PtAOMT* were cloned into the *pBI121* binary vector (Clontech). Sequence-confirmed constructs were then introduced into *Agrobacterium*strain EHA105 by electroporation. A single positive colony was co-cultured with leaf sections of sterile *Nicotiana tabacum* cv. Nc89 according to a previously reported protocol (Horsch *et al.*, 1985). Finally, kanamycin-resistant plantlets were transferred to soil mix, acclimatized, and grown in a greenhouse. Positive transgenic lines were selected by PCR, and empty-plasmid transgenic plantlets were used as controls. The transgenic tobacco lines with similar expression levels of *PsAOMT* and *PtAOMT* were used for further study.

Quantitative PCR (qPCR) Analysis of *AOMT* Expression

Flower petals at five different developmental phases were sampled in triplicate. Total RNA was prepared as described above. Quantitative assays of gene expression were performed using an UltraSYBR Mixture (CWBIO, Beijing, China) and analysed with a Stratagene (CA, USA) Mx3000P instrument as described by Lan *et al.* (2009). The relative quantification of mRNA transcripts was performed in triplicate with normalization to *Actin*(GenBank accession no. JN105299) (Zhao *et al.*, 2012). The primers used in the qPCR analysis were flanked by an intron (Supplementary Table S1). The PCR products were sequenced to confirm that the correct gene was amplified. AOMT expression in the roots, stems, leaves, and sepals was also studied.

HPLC Diode-array Detection (HPLC-DAD) and HPLC-MS anAlyses of Flavonoids Extracts

The flavonoids (anthocyanins, flavones, and flavonols) in petals from *P. suffruticosa* cv. 'Gunpohden' and *P. tenuifolia* at five stages, in transgenic tobacco flowers, and in strawberry fruits were extracted using a method described by Yang *et al.* (2009). The extraction protocol was as follows: transgenic tissues (tobacco and strawberry) were extracted with 2% formic acid/methanol (v/v) assisted by sonication at 20 °C for 20min. All extracts were filtered through a 0.22 μm membrane before injection. The solvent and gradient method for the separation of transgenic tobacco flower extracts, strawberry fruit extracts, and enzyme assay extracts was as follows: solvent A, 10% aqueous formic acid; solvent B, 0.1% formic acid in acetonitrile; constant gradient from 5 to 40% B within 25min, maintain 40% B for 5min, and then return to 5% B in 5min. The flow rate was 0.8ml min^{-1}. The column temperature was maintained at 30 °C, and 10 μl of analyte was injected. DAD data were recorded from 200 to 800nm.

An HPLC-ESI-MS^n system was used as described by Zhu *et al.* (2012). The positive-ion mode was adapted for anthocyanins; both the positive-ion and negative-ion modes were employed for colourless flavonoids. MS spectra were recorded over an *m/z* range of 100–1000.

The identification of anthocyanins and colourless flavonoids was according to past research on *Paeonia* plants (Wang *et al.*, 2001; Li *et al.*, 2009). The anthocyanins and colourless flavonoids were quantified using Cy3G and Qu3R as standards, respectively, by linear regression. The anthocyanin quantity was expressed as μg of Cy3G equivalents g^{-1} of dry weight (DW), and by using the calibration curve, the following was obtained: anthocyanin (mAU)=0.474 [Cy3G (μg ml^{-1})] – 1.275 (r^2=0.999). Colourless flavonoids were expressed as μg of Qu3R equivalents g^{-1} of DW, with the following calibration curve: (mAU)=0.437 [Qu3R (μg ml^{-1})] – 2.485 (r^2=0.997). All samples were analysed in triplicate.

Colour Measurements (CIELab system)

Chromatic analyses were performed based on the CIE 1976 ($L*a*b*$) system (Gonnet, 1998). The colours were expressed as $L*$, $a*$, and $b*$values. The value of $L*$ represents lightness, from black (0) to white (100);$a*$ describes red (positive) to green (negative); $b*$ describes yellow (positive) to blue (negative); and $C*$ represents the chroma or saturation of the colour. The colour parameters were measured with an NF333 spectrophotometer according to Zhu *et al.* (2012).

RESULTS

Flavonoid Accumulation During the Floral Development of *Paeonia* Plants

We investigated the flavonoid profiles of purple flower petals from *P. suffruticosa* cv. 'Gunpohden' and the vivid red flower petals of *P. tenuifolia*at five different developmental stages (S1–S5, from colourless to full floral expansion) (Fig. 2). Petal anthocyanins were identified as Cy3G, Cy3G5G, peonidin 3-*O*-glucoside (Pn3G), and peonidin 3,5-di-*O*-glucoside (Pn3G5G) according to Wang *et al.* (2001) (Fig. 1B). The anthocyanins accumulated during development and reached a maximum at the bloom stage (S5). The total concentration of anthocyanins were 10.73±2.29mg g^{-1} of DW and 13.25±0.76mg g^{-1} of DW at S5 for *P. suffruticosa* cv. 'Gunpohden' and *P. tenuifolia*, respectively. Interestingly, peonidin-derived anthocyanins (Pn3G and Pn3G5G, 94%) were the primary anthocyanins of purple flowers (*P.

suffruticosa cv. 'Gunpohden'), and cyanidin-derived anthocyanins (Cy3G and Cy3G5G, 76%) were the primary anthocyanins in vivid red flowers (*P. tenuifolia*) (Fig. 2 and Supplementary Table S2, available at *JXB* online). Other flavonoid compound profiles of *P. suffruticosa* cv. 'Gunpohden' and *P. tenuifolia* flowers were analysed at 350 and 280nm, respectively (Supplementary Fig. S2, available at *JXB* online). Co-pigments (flavonols and flavones) were abundant in the petals of *P. suffruticosa* cv. 'Gunpohden', and they were primarily quercetin 3,7-di-*O*-glucoside (Qu3G7G), isorhamnetin 3,7-di-*O*-glucoside (Is3G7G), luteolin 7-*O*-glucoside (Lu7G), chrysoeriol 7-*O*-glucoside (Ch7G), luteolin 7-*O*-neohesperidose (Lu7Neo) and chrysoeriol 7-*O*-neohesperidose (Ch7Neo) according to Li *et al.*(2009) (Fig. 1B). The amounts at different developmental stages are shown in Supplementary Fig. S3, available at *JXB* online. By contrast, the red flowers of *P. tenuifolia* contained negligible amounts of co-pigments (Supplementary Fig. S2). Given the hypothesis that anthocyanins are responsible for coloration, the anthocyanin methyltransferase might be the reason for the observed colour difference given the anthocyanins that were characterized from these two plants.

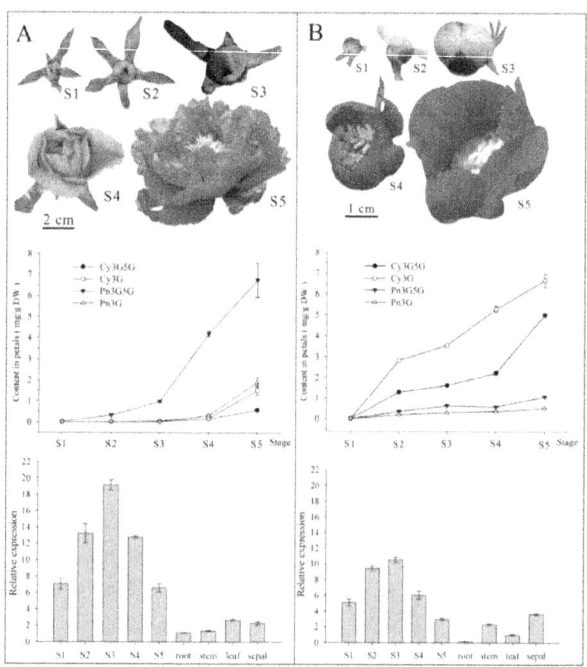

Figure 2. Flowers from five developmental stages and the anthocyanin accumulation at five developmental stages (S1–S5) from *Paeonia suffruticosa* cv. 'Gunpohden' (A) and *P. tenuifolia* (B), as well as expression patterns of *AOMT* transcripts at the corresponding developmental stages and in different tissues, measured by qPCR. The

expression values have been normalized against the *Actin* gene and are expressed as relative abundances.

In silico analysis of candidate *AOMT*s

To confirm our hypothesis, two genes with ORFs of 708bp that encoded for methyltransferase were obtained from *P. suffruticosa* cv. 'Gunpohden' and *P. tenuifolia*. The deduced amino acid sequences of these genes contained a conserved domain similar to those of the SAM or AdoMet_MTases superfamily (cl17173) and the Methyltransf_3: *O*-methyltransferase superfamily (pfam01596), and were designated *PsAOMT* and *PtAOMT*, respectively. The putative peptides of these proteins had 235 aa and a calculated molecular mass of 26.4kDa, and they each contained a conserved structural fold of seven β-sheets (Fig. 3) (Cheng and Roberts, 2001). Surprisingly, there were a total of four amino acids (positions 13, 85, 87, and 205) that differed between PsAOMT and PtAOMT (Fig. 3). A BLAST analysis indicated that PsAOMT and PtAOMT were homologous to several previously characterized type II subclass OMTs. PsAOMT shared 73% identity with VvAOMT from grapevine (Hugueney *et al.*, 2009), 62% with GmAOMT from *Glycine max* (Kovinich *et al.*, 2011), and 58% with PFOMT from *M. crystallinum* (Ibdah *et al.*, 2003). Phylogenetic analysis demonstrated that PsAOMT belongs to the type IIOMTs (with a low molecular weight and Mg^{2+} dependence) and was similar to the typical example of this class of enzyme, CCoAOMT (Fig. 4). A subclass of the type II OMTs specific for flavonoid OMTs has been proposed (Ibdah *et al.*, 2003), and thus PsAOMT is a possibly new member of this subclass together with VvAOMT, McPFOMT, and several other polypeptides described in a patent (Brugliera *et al.*, 2003) (Fig. 3).

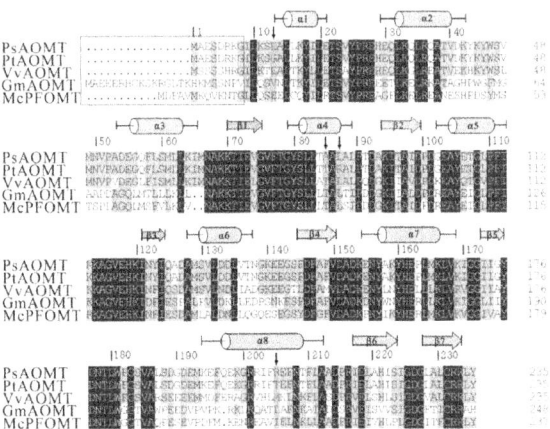

Figure 3. Sequence alignments of PsAOMT and PtAOMT with predicted sec-

ondary structural elements. The reference sequences VvAOMT (grapevine, BQ796057), GmAOMT (black soybean, ADX43927) and McPFOMT (ice plant, AY145521) are also included. α-Helices and β-strands are represented as cylinders and arrows, respectively. The residues conserved in all OMTs are shaded. (This figure is available in colour at *JXB* online.)

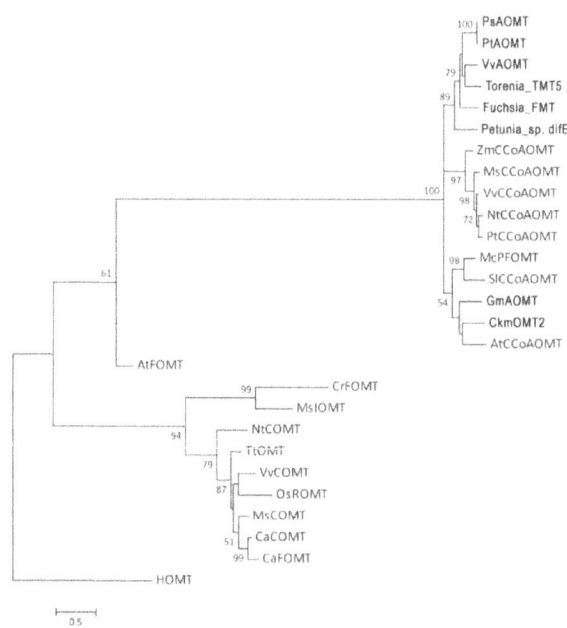

Figure 4. Phylogenetic tree of selected OMT peptides. OMTs in bold are known to act in the methylation of anthocyanins. OMT names and GenBank accession numbers are as follows: *Fuchsia*, FMT, HB975539; *Vitis vinifera*, VvAOMT, BQ796057;*Torenia*, TMT5, HB975529;*Petunia* difE, FMT, HB975519; *Cyclamen persicum×Cyclamen purpurascens*, CkmOMT2, BAK74804; *Mesembryanthemum crystallinum*, McP-FOMT, AY145521; *Stellaria longipes*, SlCCoAOMT, L22203; *Glycine max*, GmAOMT, ADX43927;*Arabidopsis thaliana*, AtCCoAOMT, AAM64800; *Zea mays*, ZmCCoAOMT, AJ242980; *Medicago sativa*, MsCCoAOMT, AAC28973; *Vitis vinifera*, VvCCoAOMT, Z54233;*Nicotiana tabacum*, NtCCoAOMT, U38612;*Populus balsamifera* subsp., PtCCoAOMT, AJ224896; *Homo sapiens*, HOMT, A38459;*Medicago sativa*, MsIOMT, AAC49927; *Catharanthus roseus*, CrAOMT, AY127568; *Oryza sativa*, OsROMT, DQ288259;*Thalictrum tuberosum*, TtOMT, AF064693; *Nicotiana tabacum*, NtCOMT, AF484252; *Vitis vinifera*, VvCOMT, AF239740;*Medicago sativa*, MsCOMT, M63853; *Chrysosplenium americanum*, CaCOMT, AAA86982 and CaAOMT, U16794. The number beside the branches represents the bootstrap values based on 1000 replicates using MEGA 5. Bar, nucleotide substitutions per site.

Biochemical Characterization of Recombinant AOMTs

The observed molecular weight of both recombinant PsAOMT and PtAOMT was approximately 66kDa (comprising 26kDa of target protein plus 40kDa of maltose-binding protein tag) (Supplementary Fig. S4, available at *JXB*online). The reaction conditions for purifying recombinant PsAOMT was optimized, and the results demonstrated that PsAOMT activity was completely absent when EDTA was present (Fig. S5, available at *JXB*online), indicating that it was cation dependent. We also tested the influence of five divalent cations $(Mg^{2+}, Ca^{2+}, Mn^{2+}, Co^{2+},$ and $Zn^{2+})$ on PsAOMT activity with the substrate Qu3R. In comparison with the other four divalent cations, PsAOMT showed the highest activity in the presence of Mg^{2+}, and the optimal Mg^{2+} concentration was 1.0mM (Supplementary Fig. S5). Recombinant PsAOMT was active within a pH range of 6.5–8.5, with an optimum of approximately 7.5 with the substrate Qu3R (Supplementary Fig. S5).

The activities of purified recombinant PsAOMT and PtAOMT were analysed by using a number of potential substrates including anthocyanidins, anthocyanins, flavonols, flavones, flavan-3-ols, and phenolic acid based on optimized conditions (Table 1, Supplementary Figs S1 and Fig. S6, available at *JXB* online). PsAOMT could use anthocyanins as methoxyl acceptors, and they acted to methylate the 3′-hydroxyl group of the B-ring with high affinity and efficiency. Pg3G was the only tested anthocyanin compound that was not a substrate for PsAOMT. With Dp3G, a sequential methylation occurred at the 3′- and 5′-hydroxyl group on the B-ring (Table 1 and Supplementary Table S3, available at *JXB* online). In comparison with the substrates Dp3G, Qu3R, and quercetin, PsAOMT had a higher affinity for Cy3G and Cy3G5G (Table 1), suggesting that cyanidin-derived anthocyanins are high-affinity substrates for PsAOMT. Unlike anthocyanins, anthocyanidin cyanidin could be used as substrate by PsAOMT with a low activity. In addition, we could not detect the methylated products of the anthocyanidin delphinidin (Supplementary Table S3), which suggested that delphinidin might be unstable in the reaction system, or might be due to the low enzymatic activity of PsAOMT. PsAOMT could also methylate Lu7G and Lu with weak affinity. The following substrates could not be methylated by PsAOMT: Km3G, kaempferol, luteolin, apigenin, naringenin, epicatechin, and caffeic acid (Table 1). Surprisingly, although PtAOMT possesses identical substrate specificity and biochemical properties, the catalytic activity of PtAOMT was much lower than that of PsAOMT (approximately 60-fold lower) (Table 1).

View this table:

Table 1. Overview of PsAOMT and PtAOMT activities with various substrates under optimized assay conditions

Substrate	PsAOMT					PtAOMT
	K_m (µM)	V_{max} (nM^{-1})	K_{cat} x 10^{-3} (s^{-1})	K_{cat}/K_m (M^{-1}s^{-1})	Specific activity (pkat mg^{-1})	Specific activity (pkat mg^{-1})
Pelargonidin 3-O-glucoside	–	–	–	–	–	–
Cyanidin 3,5-di-O-glucoside	1.06 (0.01)	17.88 (0.26)	127.70 (1.05)	120886 (563)	1788 (15)	29.60 (3.47)
Cyanidin 3-O-glucoside	1.76 (0.02)	11.57 (0.06)	161.99 (0.73)	91829 (421)	2314 (11)	37.98 (1.05)
Delphinidin 3-O-glucoside	4.11 (0.35)	17.05 (0.27)	94.72 (1.49)	23293 (1538)	1364 (21)	21.50 (0.10)
Quercetin 3-O-rutinoside	12.32 (0.32)	27.44 (1.63)	192.11 (4.41)	15599 (110)	2744 (135)	30.20 (0.24)
Luteolin 7-O-glucoside	146.26 (10.07)	31.61 (1.64)	0.22 (0.00)	1486 (145)	2017 (109)	2.24 (0.18)
Kaempferol 3-O-glucoside	–	–	–	–	–	–
Cyanidin	–	–	–	–	0.40 (0.05)	–
Delphinidin	–	–	–	–	–	–
Quercetin	4.33 (0.07)	1.77 (0.02)	9.83 (0.12)	2271 (24)	142 (2)	17.58 (0.57)
Luteolin	–	–	–	–	0.043 (0.002)	2.08 (0.27)
Kaempferol	–	–	–	–	–	–
Apigenin	–	–	–	–	–	–
Naringenin	–	–	–	–	–	–
Epicatechin	–	–	–	–	–	–
Caffeic acid	–	–	–	–	–	–

A single Amino acid Substitution can Dramatically Alter the Catalytic Activity of AOMTs

The high amino acid sequence identities of PsAOMT and PtAOMT indicated that their divergent enzymatic activity might result from differences in amino acid residues at four positions. Four separate site-directed mutations were constructed by using the low-activity enzyme PtAOMT as a template, namely PtAOMT-G13E, PtAOMT-T85A, PtAOMT-R87L, and PtAOMT-T205R. When assayed for 5min under the optimized reaction conditions with Cy3G5G as a substrate, only Pn3G5G was detected from the PtAOMT-R87L reaction product, and the observed *in vitro* catalytic efficiency of PtAOMT-R87L was equal to that of PsAOMT (Table 2). To further characterize these mutations, the reaction durations were extended from 5 to 30min, and the recombinant G13E, T85A, and T205R enzymes displayed 1.4, 4.1, and 9.1%, respectively, of the PsAOMT catalytic activity (100%). The PtAOMT activity was 1.5% of that of PsAOMT (100%), and the increased duration of the assays did not lead to a detectable increase in PtAOMT activity (Table 2). However, the single R87L mutation of PtAOMT led to an equally high activity (104.5%) as that of PsAOMT (100%). This result indicated that a leucine residue at position 87 plays an important role in the catalytic activity of these AOMTs. To validate that a leucine at position 87 was indeed a key amino acid residue, two mutations (L87R and L87A) of PsAOMT were constructed. Both these mutant enzymes exhibited decreased catalytic activity (similar to the activity of recombinant PtAOMT) (Table 2). A kinetic study of PtAOMT-R87L showed that its substrate specificity was identical to that of PsAOMT and its specific activity was similar to that of PsAOMT (Supplementary Table S4, available at *JXB* online). It is worth noting that the K_m of PtAOMT-R87L for Cy3G was slightly lower than its K_m for Cy3G5G; the opposite was true for PsAOMT. These results demonstrated that Leu-87 is a critically important residue for the catalytic activity of these AOMT enzymes and that a single amino acid mutation can cause a dramatic decrease in activity without a loss of function.

Table 2. Comparison of catalytic activities of the recombinant AOMTs generated using site-directed mutagenesis, using Cy3G5G as the substrate

Polypeptide	Incubation for 5 min [% (±SD)]	Incubation for 30 min [% (±SD)]
PsAOMT	100(0)	100(0)
PtAOMT-G13E	0(0)	1.4(0.1)
PtAOMT-T85A	0(0)	4.1(0.3)
PtAOMT-R87L	100.7(1.1)	104.5(0.7)
PtAOMT-T205R	0(0)	9.4(0.5)
PsAOMT-L87R	0(0)	4.0(0.2)
PsAOMT-L87A	0(0)	4.8(0.4)
PtAOMT	0(0)	1.5(0.3)

Activity (%): recombinant protein activity as a percentage of the 'full' activity measured for PsAOMT; SD, standard deviation with three repetitions.

Structural Basis for Substrate Discrimination in AOMTs

The 3D models of PsAOMT and PtAOMT were computed and analysed to investigate the structural basis for their activity differences (Supplementary Fig. S7, available at *JXB* online). The SAM-binding residues (Met-49, Glu-73, Ser-81, Asp-99, Asp-151, and Asp-153) were conserved in both OMTs and were very similar to those of CCoAOMT (Fig. 5).

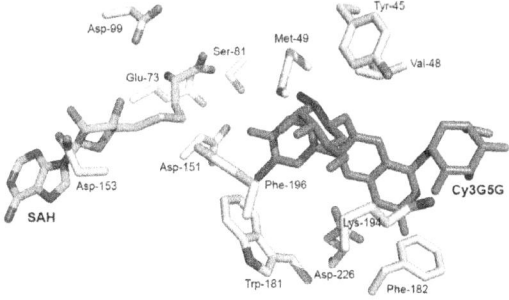

Figure 5. Molecular model of the PsAOMT active site. The SAH (*S*-adenosyl-L-homocysteine), Cy3G5G, and substrate-binding residues are represented as sticks, and labelled in cyan, grey, and white, respectively. The oxygen, nitrogen, and sulfur atoms are labelled in red, blue, and yellow, respectively.

After careful analyses of the PsAOMT model and the substrate-binding site of CCoAOMT, the residues Tyr-45, Val-48, Trp-181, Phe-182, Lys-194, Phe-196, and Asp-226 were predicted to be the putative substrate-binding pocket of Cy3G5G (Fig. 5). To our surprise, the single amino acid residue (Leu-87) that contributes the substrate specificity of PsAOMT is far from the putative substrate-binding pocket (Supplementary Fig. S7).

In vivo Characterization of AOMTs

We first investigated the catalytic activities of the *Paeonia* AOMTs *in vivo* by using transient expression in strawberry fruits. The strawberry fruit mostly contained the pelargonidin type of anthocyanins (da Silva *et al.*, 2007), which are known for not being AOMT substrates.

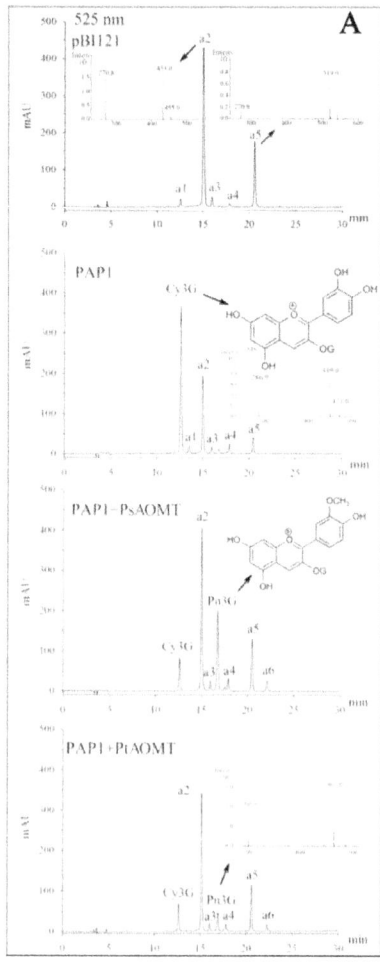

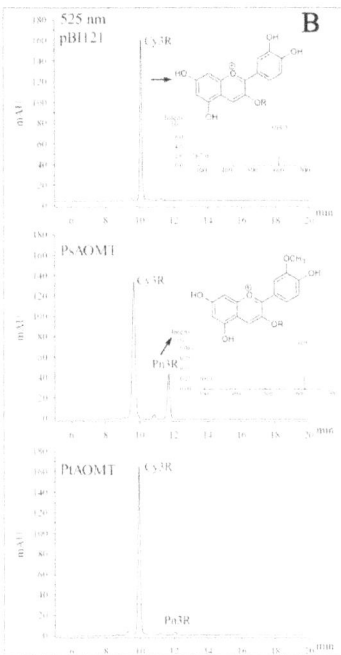

Figure 6. Characterization of AOMT activity *in vivo* using transient expression in strawberry and the stable transformation of tobacco. (A) Anthocyanin profiles of strawberry fruits with transient *PBI121* or *PAP1* expression, and the co-expression of *PAP1/PsAOMT*, and *PAP1/PtAOMT* (4 d of expression). Other anthocyanins are shown in Supplementary Table S5, available at *JXB* online. (B) Anthocyanins from transgenic tobacco flowers (35S::*PsAOMT*, 35S::*PtAOMT* and empty vector control, respectively) were analysed by HPLC-MS.

To provide appropriate substrates for the enzymatic reactions, the anthocyanin biosynthesis-related R2R3 MYB transcription factor *PAP1* from *Arabidopsis* Borevitz *et al.*, 2000) was introduced along with the two *Paeonia AOMT* sequences by agroinfiltration (Hoffmann *et al.*, 2006). The simultaneous expression of *PAP1* and *AOMT* transcripts resulted in the synthesis of Cy3G (*m/z* 449 [M]$^+$) and Pn3G (*m/z* 463 [M]$^+$) in strawberry fruits (Fig. 6A). It should be noted that there are several anthocyanin compounds that occur naturally in strawberries (a1– a6 shown in Supplementary Table S5). Pn3G accumulated to higher levels than those of Cy3G in the *PAP1* and *PsAOMT* transformants. Conversely, the Cy3G levels were higher than the Pn3G levels in the *PAP1* and *PtAOMT* transformants (Fig. 6A).

Next, we used *Agrobacterium*-mediated transformation of tobacco for *in vivo* studies of AOMT function. Tobacco has pink flowers that are known to contain cyanidin-3-*O*-rutinoside (Cy3R). Five and four transgenic tobacco lines were found to accumulate similar mRNA levels of *PsAOMT* and *PtAOMT* (Supplementary Fig. S8, available at *JXB* online), respectively. The anthocyanin profiles of these transgenic and control lines were analysed by HPLC-MS (Fig. 6B). The results demonstrated that five 35S::*PsAOMT* lines accumulated peonidin-3-*O*-rutinoside (Pn3R), with a peak of m/z 609 [M]$^+$, accounting for 21.7±1.8% of total anthocyanin, and Pn3R was also detected in four 35S::*PtAOMT* transgenic lines, accounting for a lower 1.9±0.2% total anthocyanin content. This finding indicated that the proteins encoding *PsAOMT* and *PtAOMT* had a contrasting catalytic activity *in vivo*, which was consistent with the characterization of recombinant proteins *in vitro*.

To evaluate the influence of the introduced *PsAOMT* and *PtAOMT* on flower coloration, chromatic analyses of transgenic tobacco lines and controls were performed. One-way analysis of variance was performed for the colour parameters L^*, a^*, b^*, and C^*, and significant differences were observed between 35S::*PsAOMT* and control flowers (Fig. 7). The 35S::*PsAOMT* transgenic lines had higher values of a^* towards red, and lower values of b^* towards blue, leading to an h change towards a purple hue compared with that of the control, which demonstrated that methylated anthocyanins would resulted in purplish flower colours. The colour parameters of the 35S::*PtAOMT* lines fell in between those of the 35S::*PsAOMT* lines and the control and showed no significant differences among them (Fig. 7).

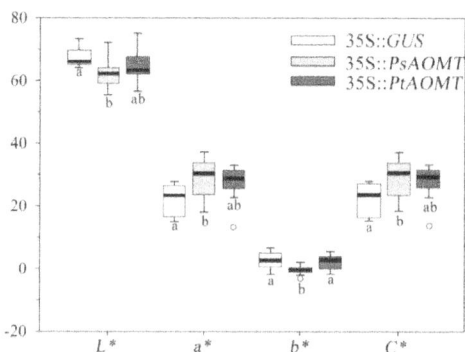

Figure 7. CIELab colour space analysis of transgenic tobacco flower colour. L^* represents lightness, from black (0) to white (100); a^* represents red (positive) to green (negative); b^* represents yellow (positive) to blue (negative); and C^* represents the chroma or saturation of the colour. Boxes of each parameter with no common letter in the figure body indicate a significant difference at $P<0.05$ ($n≥9$).

This result further validated the OMT activity of AOMTs of *Paeonia* and underscored the effects that dramatically different enzymatic efficiencies can cause, leading to differences in the biosynthesis and accumulation of these compounds.

AOMT Expression Patterns are Consistent with Anthocyanin Accumulation

To determine whether there was any correlation between the expression levels of *AOMT* transcripts and flavonoid accumulation during floral development in *Paeonia* spp., the corresponding flower petals from S1–S5 were harvested and assayed for the expression of *AOMT* transcripts. The results indicated that the *PsAOMT* expression levels in *P. suffruticosa* cv. 'Gunpohden' were consistent with the anthocyanin accumulation levels in petals. As shown in Fig. 2, anthocyanins were barely detected at the beginning of the colourless stage (S1), and *PsAOMT* transcripts were also detected at low levels (Fig. 2). From the flower coloration stage onward, the *PsAOMT* expression level increased and reached a maximum when the flower bud became loose and ready to bloom (S3) (Fig. 2). The methylated anthocyanin Pn3G5G accumulated to its maximum amount at S5 because of the continuous expression of *PsAOMT*. The expression levels of *PsAOMT* showed no obvious relationship with the accumulation of flavones and flavonols because the methylated colourless flavonoids (Is/Ch glycosides) already existed before the onset of S1 (Supplementary Fig. S3) when *PsAOMT* transcripts were hardly detected. The concentrations of these compounds did not change with the increasing expression levels of *PsAOMT* transcripts (Fig. 2 and Supplementary Fig. S3). These results indicated that *PsAOMT* was related to and specific for methylated anthocyanin accumulation in *Paeonia*. The expression pattern of *PtAOMT* in *P. tenuifolia* was identical to that of *PsAOMT* in *P. suffruticosa* cv. 'Gunpohden' (Fig. 2). However, there was no noticeable correlation between the *PtAOMT* expression levels and the accumulation of methylated anthocyanin compounds (Fig. 2 and Supplementary Fig. S3). We inferred that this finding was caused by the relatively weak enzyme activity of PtAOMT in comparison with that of PsAOMT.

In addition, *AOMT* expression was evaluated in various vegetative organs (Fig. 2). *PsAOMT* transcripts could be detected at very low levels in roots and stems, and they were detected at slightly higher levels in leaf and sepal tissues, which are known to accumulate low anthocyanin concentrations. *PtAOMT* transcripts were detected in the roots and leaves at very low levels and were noticeable in the stems and sepals.

AOMT Catalyses the Biosynthesis of Methylated Anthocyanins and Leads to Purple Coloration in *PAEONIA* Flowers

We sequenced *AOMT* cDNA from five tree peony cultivars ('Fengdan', 'Hongling', 'Liuguangyicai', 'Chaoyi', and 'Qinglongwomochi'), two herbaceous peony cultivar ('Dafugui' and 'Zixiayingri'), and one intersectional hybrid ('Hexie') between tree peony and herbaceous peony, which all have a red/purple or purple flower colour (Supplementary Table S2, Supplementary Fig. S9, available at *JXB* online). The *AOMT* genes obtained from these plants were nearly identical, and all of them possessed the critically important leucine residue at position 87 (Supplementary Table S2, Supplementary Fig. S10, available at *JXB* online). A chemical analysis of anthocyanins from the five cultivars by HPLC techniques showed that methylated anthocyanins (peonidin glycosides) accounted for 87–95% of the total anthocyanins (Supplementary Table S2,Supplementary Fig. S11, available at *JXB* online) and cyanidin glycosides were detected as minor compounds, indicating that all the cultivars possessed efficient methyltransferases to catalyse the methylation of cyanidin glycosides to peonidin glycosides. As a result, peonidin glycosides accumulated in the petals, leading to the purple coloration of*Paeonia* flowers. These findings reinforced the notion that AOMTs with leucine residues at position 87 contribute to purple flower coloration in*Paeonia* plants.

DISCUSSION

Functional Characterization of AOMTs From *Paeonia* and Their Active Switch

A phytochemical analysis in *Paeonia* plant petals on methylated anthocyanin compounds accumulation suggested that strong methylation activity occurs in the petals (Wang *et al.*, 2001; Zhang *et al.*, 2007; Jia *et al.*, 2008; Li *et al.*, 2009). Therefore, we identified and characterized genes in this study for anthocyanin methyltransferases that belonged to the new type II OMTs. The characterization of recombinant PsAOMT and PtAOMT with a wide range of flavonoid compounds and caffeic acids as substrates demonstrated that they both methylated flavonoids with a vicinal dihydroxy on the B-ring and a 3-hydroxyl on the C-ring. The kinetic parameters of PsAOMT showed that the specific activities of anthocyanins and flavonols were similar to their activity in VvAOMT, and the K_m values for anthocyanins were smaller than those in VvAOMT (Hugueney *et al.*, 2009), displaying a better affinity with the optimum substrates. In addition, PsAOMT showed much lower catalytic activities with anthocyanidins, quercetin, and luteolin, indicating that it was able to methylate glycosylated flavonoids rather than aglycones. In peony flowers, the major anthocyanins are both methylated and glycosylated, and the order of aglycone modifications *in vivo* remain unknown. The present work suggested that methylation may occur after glycosylation during anthocyanin biosynthesis, which is consistent with reports in grape berries (Ford *et al.*, 1998; Hugueney *et al.*, 2009). To clarify the order of methylation and glycosylation *in vivo*, the further characterization of the flavonoid glycosyltransferase in *Paeonia* will be meaningful.

The dimer of type II OMTs lacking a domain for dimerization in the N-terminal region could be formed by hydrophobic interactions. Each monomer could interact with substrate(s) and cofactor(s), unlike type I OMTs, which require a homodimeric structure to perform methylation (Noel *et al.*, 2003). Surprisingly, we found that the leucine at position 87 of PsAOMT was vital for the high activity level, and a mutant with an arginine residue at the same position in PsAOMT led to a remarkable decrease in enzymatic activity.

OMT-catalysed methylation reactions require a methyl donor (SAM), a methyl accepter (substrate), and an appropriately conformed reaction centre. Typically, the SAM-binding residues comprise a highly glycine-rich sequence of E/DXGXGXG known as motif I, which is highly conserved in the N-terminal region of OMTs (Martin and McMillan, 2002). The N terminus and a variable insertion loop near the C terminus played important roles in the substrate specificity (Kopycki *et al.*, 2008). The key amino acid at position 87 was located in the α4 helix proximal to motif I (Fig. 3), suggesting that a mutation in this amino acid may influence the binding of SAM or the methyl transfer. Although the PtAOMT and PsAOMT activities differ greatly in terms of catalytic efficiency, these enzymes shared the same substrate specificities, indicating that the substrate-binding regions of both proteins were probably conserved. The SAM-binding site might be altered by the single amino acid substitution at position 87 in the two proteins, and this mutation may be responsible for the observed differences in the enzyme turnover rates but may not affect substrate regioselectivity. The substitution may also affect the methylation reaction. The currently recognized catalytic mechanisms of OMTs are S_N2-like reactions, which are metal dependent, acid/base residues, or mediation by proximity and desolvation (Liscombe *et al.*, 2012). The arginine residue is polar, unlike leucine (a neutral amino acid), which may change the hydrophobic environment around SAM or the catalytic centre. However, Leu-87 of PsAOMT is far from the putative substrate-binding pocket; however, mutations at distant residues can alter the properties of the active site region and change the activity as reported for glutathione*S*-transferase (Ketterman *et al.*, 2001), alcohol dehydrogenase (Nagel *et al.*, 2013), and regulator proteins

(Freeman *et al.*, 2011). Mutations at distant residues might cause dynamic motion at the active site that is involved in the conformational change of the substrate-binding region and then alter the catalytic efficiency and substrate specificity. To elucidate the mechanism precisely, X-crystal structures of AOMTs must be further investigated.

AOMTs are Primarily Responsible for Anthocyanin Methylation in Peonies

There were four major anthocyanins and nine major colourless flavonoids in peony flowers in this study. The anthocyanins consisted of two anthocyanidins, namely, non-methylated cyanidin and monomethylated peonidin, with mono- or diglucosylation. The colourless flavonoids included glycosides derived from non-methylated quercetin, luteolin, apigenin, kaempferol, monomethylated isorhamnetin and chrysoeriol (Li,*et al.*, 2009). Anthocyanins (Fig. 2 and Supplementary Fig. S11) and the other six colourless flavonoids (Supplementary Fig. S3) as well as *AOMT*expression were investigated during flower development. The results indicated that *PsAOMT* expression was notably consistent with the accumulation of monomethylated peonidin glycosides in *P. suffruticosa* cv. 'Gunpohden' flowers. In contrast, peonidin glycosides were barely accumulated in *P. tenuifolia* flowers, because of weak PtAOMT activity, while the AOMT activity and accumulation of methylated colourless flavonoids seemed less relevant. Although recombinant PsAOMT could employ flavonols and flavones as substrates *in vitro* (Table 1), the methylated colourless flavonoids in tree peony flowers were minor components, such as Is3G7G, Ch7G, and Ch7Neo (Supplementary Fig. S3). Moreover, these methylated colourless flavonoids were detected at early stages before colouration, and the content increased slowly, displaying no obvious correlation with the *PsAOMT* expression. This finding might be explained by the specificity of the substrate structure and anthocyanin competition, or there may be other flavone and flavonol OMTs in peonies.

Methylation of Anthocyanins is Involved in the Purplish Flower Colouration of *PAEONIA* spp.

In ornamental plants, the mechanism of blue flower colouration has been extensively studied, and consists primarily of delphinidin anthocyanins (Yoshida *et al.*, 2009). However, the basis of coloration of purple flowers from cyanidin-based anthocyanins remains unclear at present. Cyanidin-based anthocyanins confer red/purple and purple colours to plants, with some exceptions such as cornflower (a blue colour) (Shiono *et al.*, 2005). The flower colour phenotype is controlled primarily by pigment chemical structures and the concentrations of particular pigments (Nakayama *et al.*, 2012). In the present study, the two plant materials presented purple and vivid red colours, although they contained the same four cyanidin-based derivatives and similar total concentrations of anthocyanins. The vast difference in methylated anthocyanins (peonidin glycosides) was dominant in purple flowers and for cyanins as the major pigment in red flowers. This finding suggested that methylation varied the colour of cyanidin-based anthocyanins from red to purple. To elucidate the influence of methylation on coloration, transgenic tobacco lines with overexpressed *PsAOMT* were generated. The flower colour of transgenic tobacco plants with different methylation profiles corresponding to those in Fig. 6 demonstrated significant changes in chromatic parameters towards a purple hue, compared with that of the control (Fig. 7). A similar phenomenon was reported in fragrant cyclamen (Akita *et al.*, 2011). A purple-flowered fragrant cyclamen lacking an enzyme for anthocyanin OMT (*CkmOMT2*) (ion-beam irradiation deletion) produced red/purple flower phenotypes (Akita *et al.*, 2011).

Sakata *et al.* (1995) studied 38 tree peony cultivars in Japan and concluded that hydroxylation and methylation both contributed considerably to the blueing of flower colours. Chemical and chromatic analyses of hundreds of tree peony cultivars showed that the methylation level of anthocyanins was significantly correlated with the purpleness of the flower colour (Li, 2010). Supplemental information including the flower colour, chromatic parameters, anthocyanin methylation levels and the key amino acid of corresponding AOMTs for 10 typical purple or red-flowered cultivars are summarized in Supplementary Table S2. Taken together, these findings show that there was a noticeable correlation for cultivars that contained a key leucine amino acid and then generated a high level of methylated anthocyanins, resulting in a purple flower.

In comparison with the methylation of anthocyanins in tree peony, the minor differences in glycosylation might play less of a role in flower coloration because it may simply increase the solubility or decrease anthocyanic vacuolar inclusions (Morita *et al.*, 2005). The former study demonstrated that, in tree peony flowers, glycosylation was present only with mono- and diglucosides, and the number of glucose moieties did not contribute to the blueing of the tree peony flower (Sakata *et al.*, 1995; Li, 2010).

Co-pigments (colourless flavonoids) are another difference between the two plant materials. Other flavonoid compound profiles of *P. suffruticosa* cv. 'Gunpohden' and *P. tenuifolia* flowers were analysed at 350 and 280nm (Supplementary Fig. S2). Red flowers contained negligible amounts of co-pigments, and co-pigments were abundant in purple flowers (Supplementary Fig. S2). Co-pigments interacting with anthocyanins at the proper concentrations and ratios can result in a bathochromic shift (Asen*et al.*, 1971). However, a study of Japanese tree peony cultivars showed that there were pink-flowered cultivars with high co-pigment contents and purple-flowered cultivars with low co-pigment contents, suggesting that co-pigmentation might not be a major factor in the blueing of tree peony flowers (Sakata *et al.*, 1995). As the content of flavonols increased, the blue flower of lisianthus changed towards red (Nielsen *et al.*, 2002). Flavone accumulation in transgenic torenia made the flower bluer (Aida *et al.*, 2000). Nevertheless, these studies were based primarily on blue flowers with delphinidin-based anthocyanins. Whether these compounds influence coloration in flowers with cyanidin-based anthocyanins, and the mechanism of any such influence, will require further investigation.

Similar chemical and chromatic analyses of lotus cultivars have also been investigated, and the results suggested that the methylation level of anthocyanins was significantly correlated with the hue towards purpleness (Yang, 2009; Yang *et al.*, 2009). Heredia *et al.* (1998) reported a chromatic characterization of five anthocyanins from red grape skins obtained in a spectroscopic study, and showed that as the degree of methylation increased, a shift towards purple was observed. This finding suggested that anthocyanin methylation might commonly contribute to purple coloration, especially in cyanidin-based pigment systems.

In conclusion, we characterized an AOMT in *Paeonia* plants. Based on *in vitro* and *in vivo* experiments, we grouped this AOMT into a recently defined subclass of type II OMTs and identified the specificity for anthocyanin substrates. In addition, we characterized the biochemical properties of recombinant PsAOMT and found that a mutation in a key amino acid residue could dramatically change the catalytic efficiency. Moreover, the AOMT is likely to be responsible for the methylation of anthocyanins and to contribute to the purple coloration of flowers in*Paeonia* plants. Thus, research on AOMTs in *Paeonia* is an important step for understanding anthocyanin modifications, and their contribution to flower coloration innovation is relevant for ornamental plants.

SUPPLEMENTARY DATA

Supplementary data are available at *JXB* online.

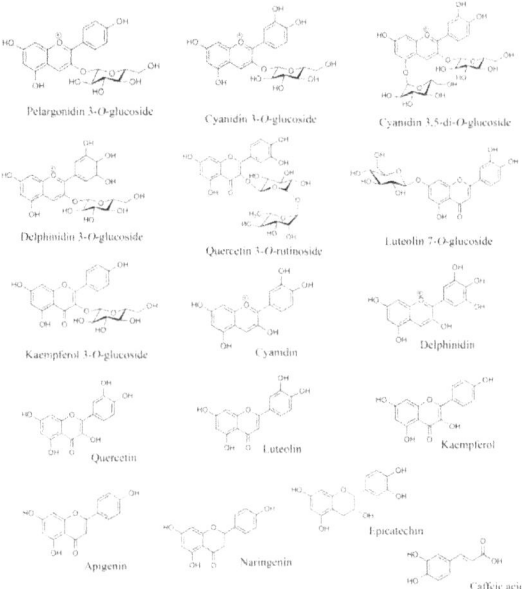

Figure S1. The chmical structure of substrates used for *in vitro* analysis of recombinant PsAOMT and PtAOMT.

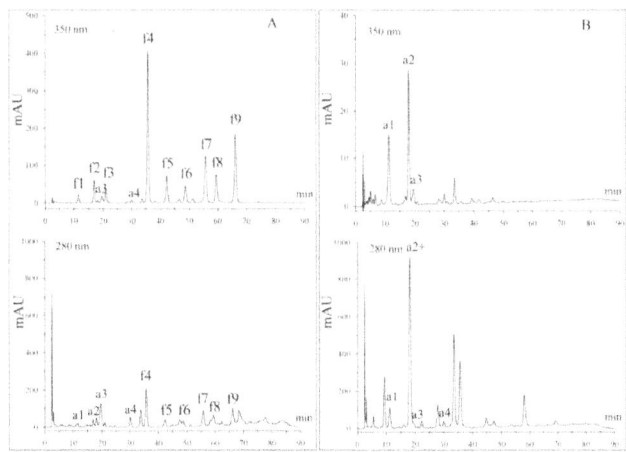

Figure S2. HPLC profiles of flower extracts from *Paeonia suffruticosa* cv. 'Gunpohden' (A) and *Paeonia tenuifolia* (B) at 350 and 280nm.

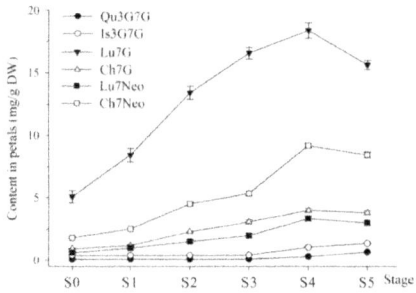

Figure S3. The accumulation of colourless flavonoids in*Paeonia suffruticosa* cv. 'Gunpohden' petals during flower development.

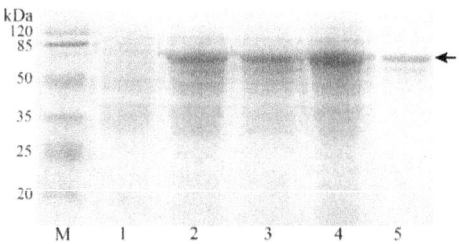

Figure S4. SDS-PAGE of recombinant PsAOMT.

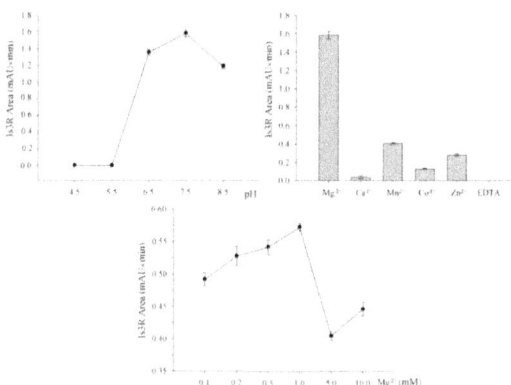

Figure S5. The catalytic activity of PsAOMT with quercetin 3-*O*-rutinoside (Qu3R) as a substrate under different reaction conditions, including a pH gradient of buffer solutions, the presence of different divalent cations, and various Mg^{2+} concentrations.

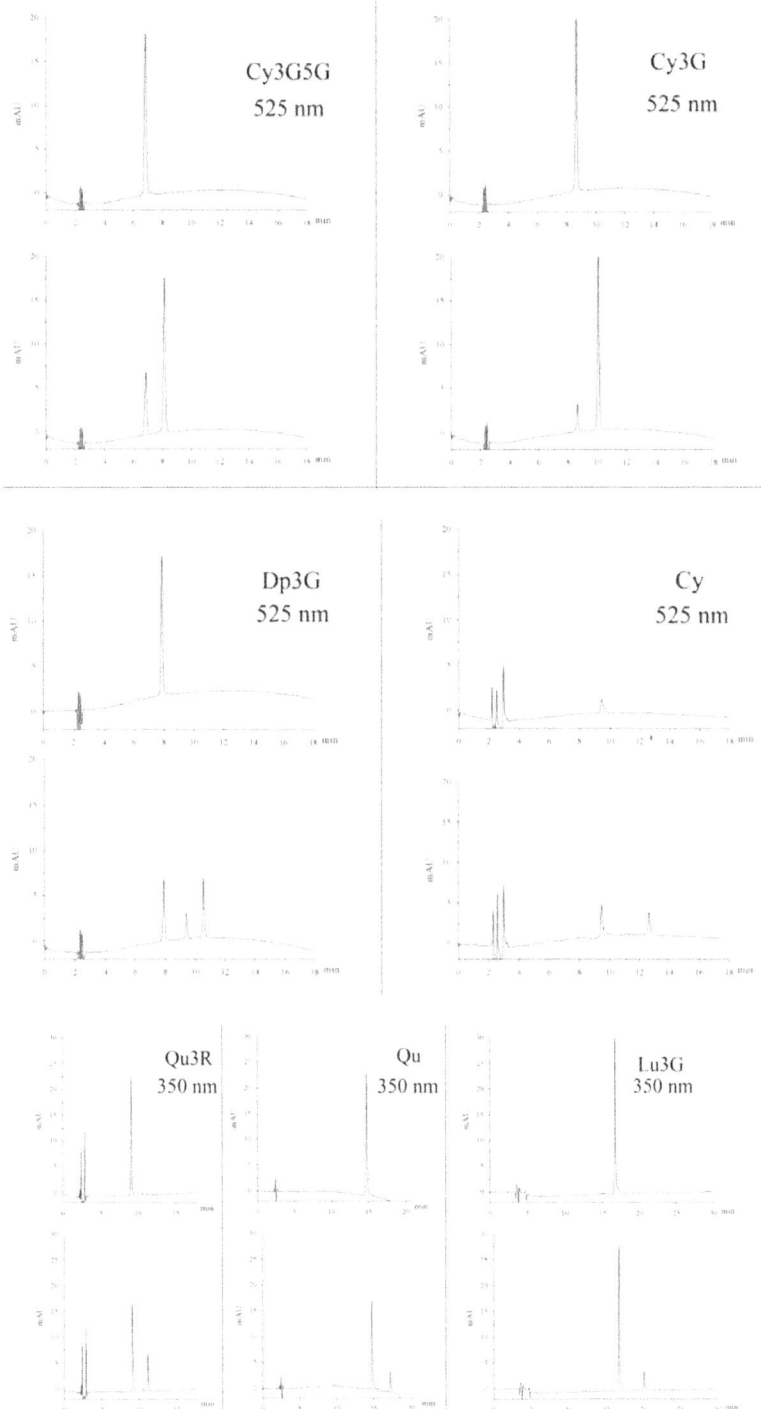

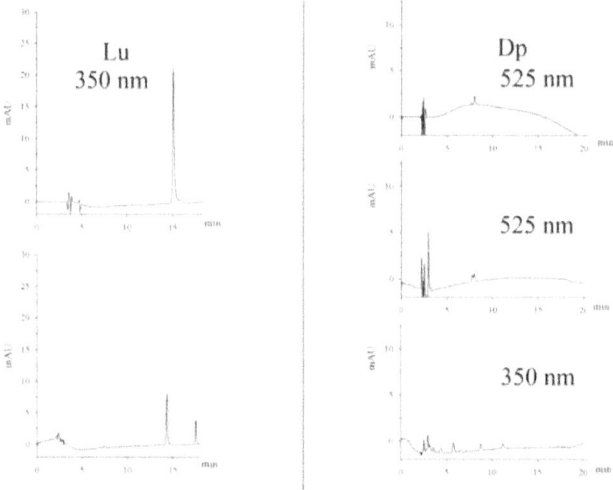

Figure S6. Profiles of products methylated by recombinant PsAOMT using a series of substrates in the *in vitro* system by HPLC analysis.

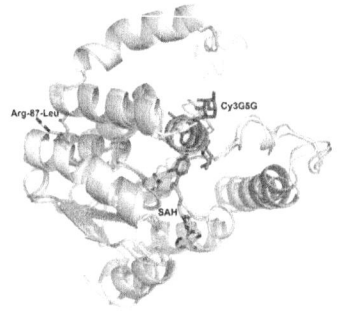

Figure S7. Superimposed 3D models of PsAOMT and PtAOMT represented as cartoons.

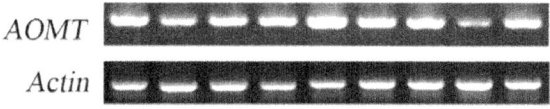

Figure S8. The expression of *AOMT*s in transgenic tobacco lines according to RT-PCR.

P. suffruticosa cv. 'Gunpohden'   P. tenuifolia   P. suffruticosa cv.'Fengdan'

P. suffruticosa cv.'Hongling'   P. suffruticosa cv. 'Liuguangyicai'   P. hybrid 'Hexie'

P. lactiflora cv. 'Dafugui'   P. suffruticosa cv.'Chaoyi'   P. suffruticosa cv. 'Qinglongwomochi'   P. lactiflora cv. 'Zixiayingri'

Figure S9. Flower phenotypes of different *Paeonia* cultivars used in this study.

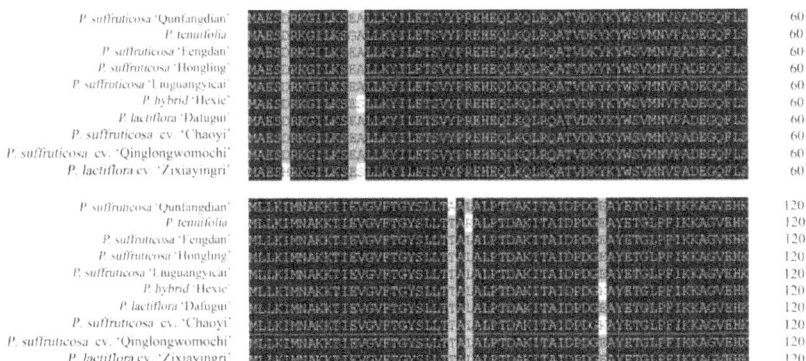

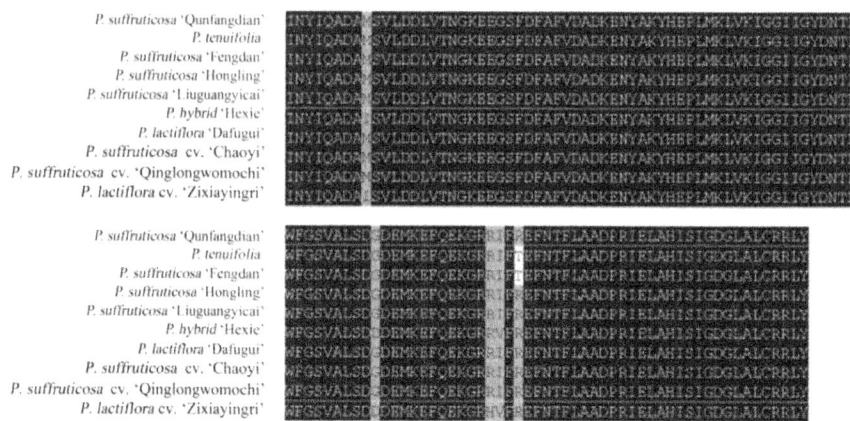

Figure S10. Amino acid sequence alignments of the AOMTs from 10 *Paeonia* plants.

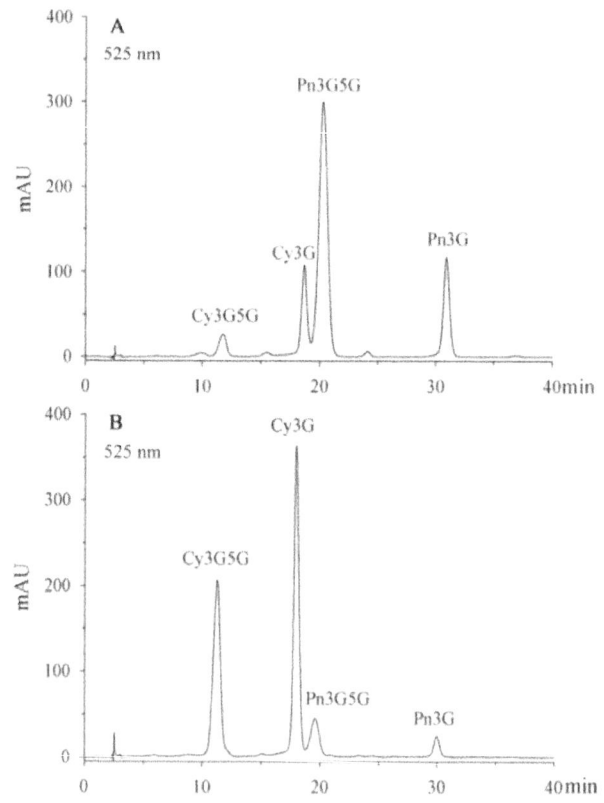

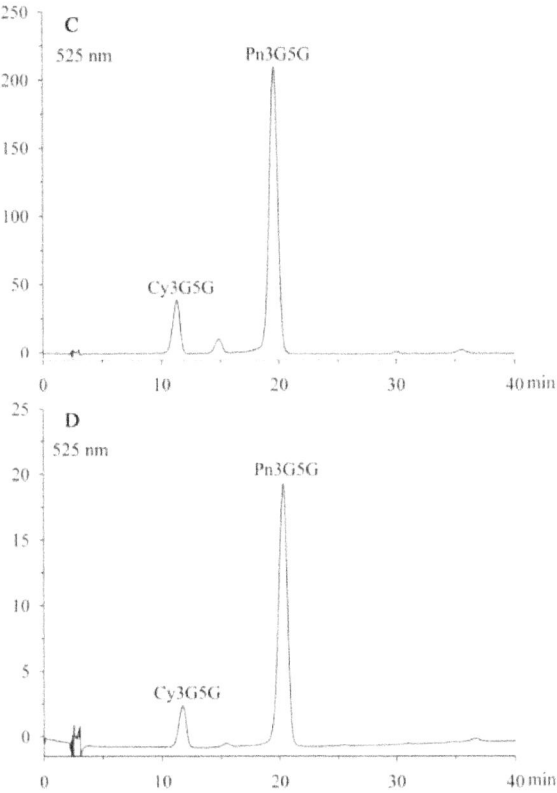

Figure S11. HPLC profiles of anthocyanins in flower petals.

Table S1. The primer sequences used in site-directed mutagenesis and qPCR.

Primer Name	Oligonucleotide sequence
AOMT-forward	ATGGCCGAGTCAGATAGGAAGGGT
AOMT-reverse	TAAAATGATCCTTCTGCCATGCT
PtAOMT–G13E-forward	TATTCTTAAAAGTGAAGCCCTTTTG
PtAOMT–G13E-reverse	TCACTTTTAAGAATACCCTTCCTAT
PtAOMT–T85A-forward	TTATTCTCTTCTCACTGCCGCTCGT
PtAOMT–T85A-reverse	CAGTGAGAAGAGAATAACCAGTAAAG
PtAOMT–R87L-forward	TCACTACCGCTCTTGCTCTGCCTAC
PtAOMT–R87L-reverse	AGAGCGGTAGTGAGAAGAGAATAAC
PtAOMT–T205R-forward	GGCGCATCTTTAGAGAGTTCAACAC
PtAOMT–T205R-reverse	CTAAAGATGCGCCTGCCCTTTTCCTG
PsAOMT–L87R-forward	CTTCTCACTGCCGCTCGTGCTCTGC
PsAOMT–L87R-reverse	CGAGCGGCAGTGAGAAGAGAATAACC
PsAOMT–L87A-forward	TCTCACTGCCGCTGCTGCTCTGCCT
PsAOMT–L87A-reverse	GCAGCGGCAGTGAGAAGAGAATAACC
AOMT-qPCR-forword	TAAGAAGGCTGGAGTGGAGCATAAG
AOMT-qPCR- reverse	GGCATAGTTTTCCTTGTCAGCATCC
Actin-forward (For *Paeonia*)	GCAGTGTTCCCCAGTATT
Actin-reverse (For *Paeonia*)	TCTTTTCCATGTCATCCC
Actin-forward (For transgenic lines)	GCCAACAGAGAGAAAATGACCC
Actin-reverse (For transgenic lines)	TCATGGATGGCTGGAAGAGGACTTC

Table S2. A summary of the 10 specimens, including information regarding the flower colour, chromatic parameters, total anthocyanin content, methylation levels of the anthocyanins, and the key amino acid(s) of the AOMTs.

	Paeonia suffruticosa cv 'Gunpohden'	P. tenuifolia	P. suffruticosa cv 'Fengdan' (flower disc)	P. suffruticosa cv 'Hongling'	P. suffruticosa cv 'Luguangyi cai'	P. hybrid 'Hexie'	P. lactiflora cv 'Dafugui'	P. suffruticosa cv 'Chaoyi'	P. suffruticosa cv 'Qinglongwo mochi'	P. lactiflora 'Zixiayingri'
Flower color	purple-red	vivid red	purple-red	purple-red	Purple-red	Purple-red	Purple-red	purple-red	purple-red	red-purple
RHSCC	77B	61A	—	61C	82C	72B	71D	71B	71A	61A
L^*	35.5	31.5	75.9	51.9	75.8	35.3	46.1	33.94	25.26	22.47
a^*	36.2	41.5	-7.4	40.7	-0.7	47.1	45.2	47.35	33.51	34.37
b^*	-16.0	9.4	5.6	-7.2	5.6	-17.0	-13.1	-10.60	-4.15	0.64
C^*	39.7	42.6	9.3	41.4	5.8	50.2	47.2	48.55	33.85	34.37
Hue	-23.9	-12.9	-10.2	-10.15	-15.05	-20.2	-16.5	-12.59	-6.86	1.07
Total anthocyanins (mg/g DW)	10.73	13.25	2.32	13.05	2.54	7.64	4.86	27.62	45.25	15.93
TF	32.70	0.72	—	42.28	33.1	38.23	23.37	24.18	41.71	23.55
CI	3.05	0.05	—	3.24	13.06	5.28	4.81	0.88	0.92	1.48
Methylation%	94	24	95	91	86	87	89	55	67	68
Amino acid at 87 position	Leucine	Arginine	Leucine	Leucine	Leucine	Leucine	Leucine	Leucine	Leucine	Leucine

The data of chroma parameters (RHSCC, L^*, a^*, b^*, C^* and hue), TA, TF and CI were cited from Li CH (2010, The flavonoid composition in tree peony petals and their effects on the coloration, PhD dissertation) and Jia N (2008, Studies on petal coloration mechanism and chemotaxonomy of herbaceous peony, Master dissertation)

Table S3. Identification of products methylated by recombinant PsAOMT using a series of substrates in the *in vitro* system.

Substrate (μM) and reaction duration (min)	Wavelength nm	Retention time min	m/z positive mode	m/z negative mode	Identification
Cy3G5G 25 μM-10 min	525	4.9	287.1(3.59),449.1(8.97), 611.1(100)		Cyanidin 3,5-di-O-glucoside
		6.2	301.0(3.39), 625.1(100)		Peonidin 3,5-di-O-glucoside
Cy3G 25 μM-10 min	525	6.4	287.0(89.94), 449.1(54.29)		Cyanidin 3-O-glucoside
		7.8	301.1(98.30), 463.1(100)		Peonidin 3-O-glucoside
Dp3G 25 μM-10 min	525	6.1	303.1(100), 465.1(75.44)		Delphinidin 3-O-glucoside
		7.6	317.0(98.05), 479.1(100)		Petunidin 3-O-glucoside
		8.9	331(74.91), 493.1(100)		Malvidin 3-O-glucoside
Qu3R 25 μM-10 min	350	9.8	303.0(100), 465.1(35.01), 611.1(28.62)		Quercetin 3-O-rutinoside
		11.9	317.1(100), 479.0(30.55), 625.1(22.67)		Isorhamnetin 3-rutinoside
Qu 25 μM-10 min	350	14.6	303.1(100)	301.9(100)	Quercetin
		18.1	149.0(100), 317.1(33.64)	311.0(100)	Isorhamnetin
Lu7G 25 μM-10 min	350	10.2	287.0(100), 449.1(49.36)		Luteolin 7-O-glucoside
		12.4	301.1(100), 463.1(46.08)		Chrysoeriol 7-O-glucoside
Lu 25 μM-10 min	350	15.1	287.1(100)		Luteolin
		18.2	301.1(100)		Chrysoeriol
Cy 50 μM-10 min	525	11.4	287.1(100)		Cyanidin
			301.2(31.19), 287.0(20.42), 276.0(72.62), 149.0(100)		Peonidin
Dp 50 μM-10 min	350	8.1	393.0(100), 149.0(79.45), 332.0(75.05), 471.8(23.54)	468.9(100), 409.0(29.44),379.9(22.32),276.8(14.04)	Unknown
		11.4	132.0(100), 276.1(85.36), 395.0(76.54), 471.9(52.11)	362.8(84.36), 468.8(100)	Unknown
		13.8	132.0(100),149.0(51.39), 276.0(37.42), 309.7(39.57), 319.0(42.72)	297.0(100), 326.9(99.49)	Unknown

Table S4. PtAOMT-R87L activities with major substrates under optimized assay conditions.

Substrates	K_m μM	V_{max} nM s^{-1}	PtAOMT-R87L $K_{cat}× 10^{-3}$ s^{-1}	K_{cat}/K_m M^{-1}s^{-1}	Specific Activity pkat mg^{-1}
Cyanidin 3,5-di-O-glucoside	8.51 (1.39)	14.94 (0.73)	106.71 (5.21)	13010 (1428)	1494 (73)
Cyanidin 3-O-glucoside	3.53 (0.31)	17.59 (0.30)	125.66 (2.16)	36058 (3045)	1759 (30)
Delphinidin 3-O-glucoside	27.27 (1.49)	19.06 (0.88)	0.14 (0.01)	4999 (82)	1906 (88)
Quercetin 3-O-rutinoside	8.09 (0.25)	9.31 (0.97)	0.07 (0.01)	8253 (923)	931 (97)
Luteolin 7-O-glucoside	48.88 (2.74)	1.26 (0.24)	0.01 (0.00)	189 (47)	126 (24)

Data are expressed as means (± SE) of triplicate assays.

Table S5. Anthocyanins in strawberry fruits as detected by HPLC-MS (demonstrated in Fig. 6).

Peak No.	Compound	Peak No.	Compound
a1	Pg 3,5-di-*O*-glucoside	a4	Cy 3-*O*-malonylglucoside
a2	Pg 3-*O*-glucoside	a5	Pg 3-*O*-malonylglucoside
a3	Pg 3-*O*-rutinoside	a6	Pn 3-*O*-malonylglucoside

Acknowledgements

We thank Professor Si-Lan Dai (Beijing Forestry University) for tobacco NC89. We appreciated the kind help of Dr Oliver Yu (Conagen Inc., USA) in protein structure analysis. This work was supported by the National Natural Science Foundation of China (grant nos 30771521 and 31071824), and by the National High Technology Research and Development Program of China (863 program, grant no. 2011AA10020702).

REFERENCES

1.   Aida R Yoshida K Kondo T Kishimoto S Shibata M . 2000. Copigmentation gives bluer flowers on transgenic torenia plants with the antisense dihydroflavonol-4-reductase gene. Plant Science 160, 49–56.

2.   Akita Y Kitamura S Hase Y et al. 2011. Isolation and characterization of the fragrant cyclamen O-methyltransferase involved in flower coloration. Planta 234, 1127–1136.

3.   Asen S Stewart RN Norris KH . 1971. Co-pigmentation effect of quercetin glycosides on absorption characteristics of cyanidin glycosides and color of Red Wing azalea. Phytochemistry 10, 171–175.

4.   Berim A Gang DR . 2013. Characterization of two candidate flavone 8-O-methyltransferases suggests the existence of two potential routes to nevadensin in sweet basil. Phytochemistry 92, 33–41.

5.   Borevitz JO Xia YJ Blount J Dixon RA Lamb C . 2000. Activation tagging identifies a conserved MYB regulator of phenylpropanoid biosynthesis. The Plant Cell 12, 2383–2393.

6.   Brugliera F Demelis L Koes R Tanaka Y . 2003. Genetic sequences having methyltransferase activity and uses therefor. WO Patent 2,003,062,428.

7.   Cheng XD Roberts RJ . 2001. AdoMet-dependent methylation, DNA methyltransferases and base flipping. Nucleic Acids Research 29, 3784–3795.

8.   da Silva FL Escribano-Bailon MT Alonso JJP Rivas-Gonzalo JC Santos-

Buelga C . 2007. Anthocyanin pigments in strawberry. LWT—Food Science and Technology 40, 374–382.

9.   Ferrer JL Zubieta C Dixon RA Noel JP . 2005. Crystal structures of alfalfa caffeoyl coenzyme A 3-O-methyltransferase. Plant Physiology 137, 1009–1017.

10.   Ford CM Boss PK Hoj PB . 1998. Cloning and characterization of Vitis vinifera UDP-glucose: flavonoid 3-O-glucosyltransferase, a homologue of the enzyme encoded by the maize Bronze-1 locus that may primarily serve to glucosylate anthocyanidins in vivo. Journal of Biological Chemistry 273, 9224–9233.

11.   Freeman AM Mole BM Silversmith RE Bourret RB . 2011. Action at a distance: amino acid substitutions that affect binding of the phosphorylated CheY response regulator and catalysis of dephosphorylation can be far from the CheZ phosphatase active site. Journal of Bacteriology 193, 4709–4718.

12.   Gonnet JF . 1998. Colour effects of co-pigmentation of anthocyanins revisited—1. A colorimetric definition using the CIELAB scale. Food Chemistry 63, 409–415.

13.   Grosdidier A Zoete V Michielin O . 2011. SwissDock, a protein-small molecule docking web service based on EADock DSS. Nucleic Acids Research 39, W270–W277.

14.   Grotewold E . 2006. The genetics and biochemistry of floral pigments. Annual Review of Plant Biology 57, 761–780.

15.   Heredia FJ Francia-Aricha EM Rivas-Gonzalo JC Vicario IM Santos-Buelga C . 1998. Chromatic characterization of anthocyanins from red grapes—I. pH effect. Food Chemistry 63, 491–498.

16.   Hoffmann T Kalinowski G Schwab W . 2006. RNAi-induced silencing of gene expression in strawberry fruit (Fragaria×ananassa) by agroinfiltration: a rapid assay for gene function analysis. The Plant Journal 48, 818–826.

17.   Horsch RB Fry JE Hoffmann NL Eichholtz D Rogers SG Fraley RT . 1985. A simple and general method for transferring genes into plants. Science 227, 1229–1231.

18.   Hugueney P Provenzano S Verries C Ferrandino A Meudec E Batelli G Merdinoglu D Cheynier V Schubert A Ageorges A . 2009. A novel cation-dependent O-methyltransferase involved in anthocyanin methylation in grapevine. Plant Physiology 150, 2057–2070.

19.   Ibdah M Zhang XH Schmidt J Vogt T . 2003. A novel Mg 2+-dependent

O-methyltransferase in the phenylpropanoid metabolism of Mesembryan themum crystallinum . Journal of Biological Chemistry 278, 43961–43972.

20. Ibrahim RK Bruneau A Bantignies B . 1998. Plant O-methyltransferases: molecular analysis, common signature and classification. Plant Molecular Biology 36, 1–10.

21. Ji LJ Wang Q da Silva JAT Yu XN . 2012. The genetic diversity of Paeonia L. Scientia Horticulturae 143, 62–74.

22. Jia N Shu QY Wang LS Du H Xu YJ Liu ZA . 2008. Analysis of petal anthocyanins to investigate coloration mechanism in herbaceous peony cultivars. Scientia Horticulturae 117, 167–173.

23. Jonsson LMV Devlaming P Wiering H Aarsman MEG Schram AW . 1983. Genetic control of anthocyanin-O-methyltransferase activity in flowers of Petunia hybrida . Theoretical and Applied Genetics 66, 349–355.

24. Joshi CP Chiang VL . 1998. Conserved sequence motifs in plant S-adenosyl-L-methionine-dependent methyltransferases. Plant Molecular Biology 37, 663–674.

25. Ketterman AJ Prommeenate P Boonchauy C Chanama U Leetachewa S Promtet N Prapanthadara L . 2001. Single amino acid changes outside the active site significantly affect activity of glutathione S-transferases. Insect Biochemistry and Molecular Biology 31, 65–74.

26. Klimasauskas S Weinhold E . 2007. A new tool for biotechnology: AdoMet-dependent methyltransferases. Trends in Biotechnology 25, 99–104.

27. Kondo E Nakayama M Kameari N Tanikawa N Morita Y Akita Y Hase Y Tanaka A Ishizaka H . 2009. Red-purple flower due to delphinidin 3,5-diglucoside, a novel pigment for Cyclamen spp., generated by ion-beam irradiation. Plant Biotechnology 26, 565–569.

28. Kopycki JG Rauh D Chumanevich AA Neumann P Vogt T Stubbs MT . 2008. Biochemical and structural analysis of substrate promiscuity in plant Mg 2+-dependent O-methyltransferases. Journal of Molecular Biology 378, 154–164.

29. Kovinich N Saleem A Arnason JT Miki B . 2011. Combined analysis of transcriptome and metabolite data reveals extensive differences between black and brown nearly-isogenic soybean (Glycine max) seed coats enabling the identification of pigment isogenes. BMC Genomics 12, 381–398.

30. Krissinel E Henrick K . 2004. Secondary-structure matching (SSM), a new tool for fast protein structure alignment in three dimensions. Acta Crystallographica Section D—Biological Crystallography 60, 2256–2268.

31. Lam KC Ibrahim RK Behdad B Dayanandan S . 2007. Structure, function, and evolution of plant O-methyltransferases. Genome 50, 1001–1013.

32. Lan T Yang ZL Yang X Liu YJ Wang XR Zeng QY . 2009. Extensive functional diversification of the Populus glutathione S-transferase supergene family. The Plant Cell 21, 3749–3766.

33. Lee YJ Kim BG Chong Y Lim Y Ahn JH . 2008. Cation dependent O-methyltransferases from rice. Planta 227, 641–647.

34. Li CH . 2010. The flavonoid composition in tree peony petals and their effects on the coloration . PhD thesis, University of Chinese Academy of Sciences, China.

35. Li CH Du H Wang LS Shu QY Zheng YR Xu YJ Zhang JJ Yang RZ Ge Y . 2009. Flavonoid composition and antioxidant activity of tree peony (Paeonia section moutan) yellow flowers. Journal of Agricultural and Food Chemistry 57, 8496–8503.

36. Liscombe DK Louie GV Noel JP . 2012. Architectures, mechanisms and molecular evolution of natural product methyltransferases. Natural Product Reports 29, 1238–1250.

37. Lucker J Martens S Lund ST . 2010. Characterization of a Vitis vinifera cv. Cabernet Sauvignon 3′,5′-O-methyltransferase showing strong preference for anthocyanins and glycosylated flavonols. Phytochemistry 71, 1474–1484.

38. Martin JL McMillan FM . 2002. SAM (dependent) I AM: the S-adenosylmethionine-dependent methyltransferase fold. Current Opinion in Structural Biology 12, 783–793.

39. Morita Y Hoshino A Kikuchi Y et al. 2005. Japanese morning glory dusky mutants displaying reddish-brown or purplish-gray flowers are deficient in a novel glycosylation enzyme for anthocyanin biosynthesis, UDP-glucose: anthocyanidin 3-O-glucoside-2″-O-glucosyltransferase, due to 4-bp insertions in the gene. The Plant Journal 42, 353–363.

40. Nagel ZD Cun SJ Klinman JP . 2013. Identification of a long-range protein network that modulates active site dynamics in extremophilic alcohol dehydrogenases. Journal of Biological Chemistry 288, 14087–14097.

41. Nakayama M Tanikawa N Morita Y Ban Y . 2012. Comprehensive

analyses of anthocyanin and related compounds to understand flower color change in ion-beam mutants of cyclamen (Cyclamen spp.) and carnation (Dianthus caryophyllus). Plant Biotechnology 29, 215–221.

42. Nielsen K Deroles SC Markham KR Bradley MJ Podivinsky E Manson D . 2002. Antisense flavonol synthase alters copigmentation and flower color in lisianthus. Molecular Breeding 9, 217–229.

43. Noel JP Dixon RA Pichersky E Zubieta C Ferrer JL . 2003. Structural, functional, and evolutionary basis for methylation of plant small molecules. Integrative Phytochemistry: From Ethnobotany to Molecular Ecology 37, 37–58.

44. Provenzano S Spelt C Hosokawa S et al. 2014. Genetic control and evolution of anthocyanin methylation. Plant Physiology 165, 962–977.

45. Rausher MD . 2008. Evolutionary transitions in floral color. International Journal of Plant Sciences 169, 7–21.

46. Roy A Kucukural A Zhang Y . 2010. I-TASSER: a unified platform for automated protein structure and function prediction. Nature Protocols 5, 725–738.

47. Sakata Y Aoki N Tsunematsu S Nishikouri H Johjima T . 1995. Petal coloration and pigmentation of tree peony bred and selected in Daikon Island (Shimane Prefecture). Journal of the Japanese Society for Horticultural Science 64, 351–357.

48. Shiono M Matsugaki N Takeda K . 2005. Structure of the blue cornflower pigment - Packaging red-rose anthocyanin as part of a 'superpigment' in another flower turns it brilliant blue. Nature 436, 791–791.

49. Shu QY Wischnitzki E Liu ZA Ren HX Han XY Hao Q Gao FF Xu SX Wang LS . 2009. Functional annotation of expressed sequence tags as a tool to understand the molecular mechanism controlling flower bud development in tree peony. Physiologia Plantarum 135, 436–449.

50. Tahara S . 2007. A journey of twenty-five years through the ecological biochemistry of flavonoids. Bioscience Biotechnology and Biochemistry 71, 1387–1404.

51. Tamura K Peterson D Peterson N Stecher G Nei M Kumar S . 2011. MEGA5: Molecular Evolutionary Genetics Analysis using maximum likelihood, evolutionary distance, and maximum parsimony methods. Molecular Biology and Evolution 28, 2731–2739.

52. Tanaka Y Brugliera F Chandler S . 2009. Recent progress of flower colour modification by biotechnology. International Journal of Molecular Sciences 10, 5350–5369.

53. Tanaka Y Sasaki N Ohmiya A . 2008. Biosynthesis of plant pigments: anthocyanins, betalains and carotenoids. The Plant Journal 54, 733–749.

54. Thompson JD Gibson TJ Plewniak F Jeanmougin F Higgins DG . 1997. The CLUSTAL_X windows interface: flexible strategies for multiple sequence alignment aided by quality analysis tools. Nucleic Acids Research 25, 4876–4882.

55. Wang LS Shiraishi A Hashimoto F Aoki N Shimizu K Sakata Y . 2001. Analysis of petal anthocyanins to investigate flower coloration of Zhongyuan (Chinese) and Daikon Island (Japanese) tree peony cultivars. Journal of Plant Research 114, 33–43.

56. Wang XQ . 2011. Structure, function, and engineering of enzymes in isoflavonoid biosynthesis. Functional & Integrative Genomics 11, 13–22.

57. Wessinger CA Rausher MD . 2012. Lessons from flower colour evolution on targets of selection. Journal of Experimental Botany 63, 5741–5749.

58. Wiering H Devlaming P . 1977. Glycosylation and methylation patterns ofanthocyanins in Petunia hybrida. II. Genes Mf 1 and Mf 2 . Plant Breeding 78, 113–123.

59. Yang RZ . 2009. Studies on petal pigment composition and coloration mechanism of lotus cultivars. Master thesis, University of Chinese Academy of Sciences.

60. Yang RZ Wei XL Gao FF Wang LS Zhang HJ Xu YJ Li CH Ge YX Zhang JJ Zhang J . 2009. Simultaneous analysis of anthocyanins and flavonols in petals of lotus (Nelumbo) cultivars by high-performance liquid chromatography-photodiode array detection/electrospray ionization mass spectrometry. Journal of Chromatography A 1216, 106–112.

61. Ye ZH Kneusel RE Matern U Varner JE . 1994. An alternative methylation pathway in lignin biosynthesis in Zinnia . The Plant Cell 6, 1427–1439.

62. Yoshida K Mori M Kondo T . 2009. Blue flower color development by anthocyanins: from chemical structure to cell physiology. Natural Product Reports 26, 884–915.

63. Zhang JJ Wang LS Shu QY Liu ZA Li CH Zhang J Wei XL Tian D . 2007. Comparison of anthocyanins in non-blotches and blotches of the petals of Xibei tree peony. Scientia Horticulturae 114, 104–111.

64. Zhao DQ Tao J Han CX Ge JT . 2012. Flower color diversity revealed by differential expression of flavonoid biosynthetic genes and flavonoid accumulation in herbaceous peony (Paeonia lactiflora Pall.). Molecular Biology Reports 39, 11263–11275.

65. Zhu ML Zheng XC Shu QY Li H Zhong PX Zhang HJ Xu YJ Wang LJ

Wang LS . 2012. Relationship between the composition of flavonoids and flower colors variation in tropical water lily (Nymphaea) cultivars. PLoS One 7, e 34335.

Chapter 7

ISOLATION AND CHARACTERIZATION OF GTMYBP3 AND GTMYBP4, ORTHOLOGUES OF R2R3-MYB TRANSCRIPTION FACTORS THAT REGULATE EARLY FLAVONOID BIOSYNTHESIS, IN GENTIAN FLOWERS

Takashi Nakatsuka, Misa Saito, Eri Yamada, Kohei Fujita, Yuko Kakizaki and Masahiro Nishihara

Iwate Biotechnology Research Center, 22-174-4 Narita, Kitakami, Iwate 024-0003, Japan

ABSTRACT

Flavonoids are one of the major plant pigments for flower colour. Not only coloured anthocyanins, but also co-pigment flavones or flavonols, accumulate in flowers. To study the regulation of early flavonoid biosynthesis, two R2R3-MYB transcription factors, GtMYBP3 and GtMYBP4, were identified from the petals of Japanese gentian (Gentiana triflora). Phylogenetic analysis showed that these two proteins belong to the subgroup 7 clade (flavonol-specific MYB), which includes Arabidopsis AtMYB12, grapevine VvMYBF1, and tomato SlMYB12. Gt MYBP3 and GtMYBP4 transcripts were detected specifically in young petals and correlated with the profiles of flavone accumulation. Transient expression assays showed that GtMYBP3 and GtMYBP4 enhanced the promoter activities of early biosynthetic genes, including flavone synthase II (FNSII) and flavonoid 3′-hydroxylase (F3′H), but not the late biosynthetic gene, flavonoid 3′,5′-hydroxylase (F3′5′H). GtMYBP3 also enhanced the promoter activity of the chalcone synthase (CHS) gene. In transgenic Arabidopsis, overexpression of Gt MYBP3 and Gt MYBP4 activated the expression of endogenous flavonol biosynthesis genes and led to increased flavonol accumulation in seedlings. In transgenic tobacco petals, overexpression of Gt MYBP3 and Gt MYBP4 caused decreased anthocyanin levels, resulting in pale flower colours. Gt MYBP4-expressing transgenic tobacco flowers also showed increased flavonols. As far as is known, this is the first functional characterization of R2R3-MYB transcription factors regulating early flavonoid biosynthesis in petals.

INTRODUCTION

Natural flower pigments generally consist of complex secondary metabolites, such as flavonoids, carotenoids, and betalains, depending on the plant species. Flavonoids are the most characterized secondary metabolites in higher plants and have various biological functions, such as flower pigmentation, pollen fertility, plant–microbe interaction, and protection from UV radiation (reviewed in Mol et al., 1998). Determination of the flavonoid biosynthetic pathway is of great interest to plant scientists and many structural genes involved in flavonoid biosynthesis have been cloned and studied in maize kernels (Zea mays), petunia flowers (Petunia hybrida), snapdragon flowers (Antirrhinum majus), and in Arabidopsisseed coats (Arabidopsis thaliana) (Fig. 1; Holton and Cornish, 1995; Mol et al., 1998; Winkel, 2006).

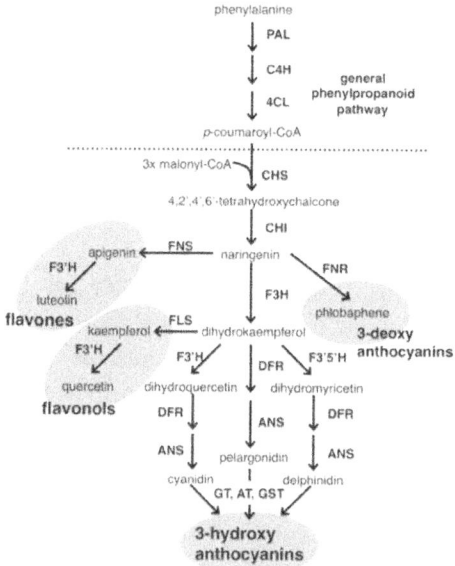

Figure 1. Flavonoid biosynthetic pathway in higher plants. 4CL, 4-coumarate:CoA ligase; ANS, anthocyanidin synthase; AT, anthocyanin acyltransferase; C4H, cinnamate 4-hydroxylase; CHI, chalcone isomerase; CHS, chalcone synthase; DFR, dihydroflavonol 4-reductase; F3'5'H, flavonoid 3',5'-hydroxylase; F3'H, flavonoid 3'-hydroxylase; F3H, flavanone 3-hydroxylase; FLS, flavonol synthase; FNR, flavanone

reductase; FNS, flavone synthase; GST, glutathione S-transferase; GT, anthocyanin O-glucosyltransferase; PAL, phenylalanine ammonia lyase.

Japanese endemic gentian plants (Gentiana triflora) have brilliant blue flowers and are popular floricultural plants in Japan (Nishihara et al., 2008). The petals of Japanese gentian accumulate a polyacylated anthocyanin, termed gentiodelphin (delphinidin 3-O-glucoside-5,3'-O-caffeoylglucoside), and flavone derivatives (Goto et al., 1982; Yoshida et al., 2000; Nakatsuka et al., 2005). The structural genes involved in the gentiodelphin and flavone biosynthetic pathways have been isolated and characterized (Tanaka et al., 1996; Kobayashi et al., 1998; Nakatsuka et al., 2005, 2008a). Thus, the blue pigmentation and its development have been extensively studied in Japanese gentian (Yoshida et al., 2009).

In flavonoid biosynthesis, two clusters of co-regulated structural genes can generally be distinguished: early biosynthetic genes, which are involved in the synthesis of flavones, flavonols, and phlobaphenes; and late biosynthetic genes, which are involved in proanthocyanin and anthocyanin biosynthesis (Fig. 1; Mol et al., 1998; Quattrocchio et al., 2006; Gonzalez et al., 2008). Transcription factors for the anthocyanin and proanthocyanin biosynthetic pathways are mainly members of the R2R3-MYB, basic helix–loop–helix (bHLH), and WD40 repeats (WDR) protein families. Complexes of these transcription factors regulate the expression of structural genes in the late biosynthetic pathway (Broun, 2005; Koes et al., 2005). In maize, C1 (R2R3-MYB) requires interaction with R (bHLH) to activate anthocyanin biosynthesis (Lloyd et al., 1992). In petunia flowers, AN2 (R2R3-MYB) interacts with two distinct bHLH factors, JAF13 or AN1, both of which share a high sequence similarity with the maize R and snapdragon DELILA proteins (Spelt et al., 2000). PAP1/PAP2 are also C1 homologues in Arabidopsis and are involved in proanthocyanin accumulation in seed coats (Borevitz et al., 2000). In gentian, GtMYB3 interacts with GtbHLH1, and the complex of these two proteins activates the expression of flavonoid 3',5'-hydroxylase (F3'5'H) and anthocyanin 5,3'-aromatic acyltransferase (5,3'AT') genes, which belong to the late flavonoid biosynthetic pathway (Nakatsuka et al., 2008b). However, the GtMYB3/GtbHLH1 complex could not activate the expression of the chalcone synthase (CHS) gene, which belongs to the early biosynthetic pathway. Thus, the activation of many R2R3-MYB proteins regulating flavonoid biosynthetic pathways depends on an interaction with bHLH proteins. R2R3-MYB proteins that interact with bHLH proteins share a common motif in the N-terminal R3 repeat (Grotewold et al., 2000). However, the transcription factors regulating early flavonoid biosynthesis in gentian flowers remain unknown. Maize P1 (ZmP1) is an R2R3-MYB that is active without binding a bHLH protein.

ZmP1 controls a subset of genes for 3-deoxyflavonoids and phlobaphene biosyntheses in kernel pericarp and cob tissue (Grotewold et al., 1994). A transient expression assay using a maize cell suspension demonstrated that P1 can activate the transcription of the flavanone reductase gene (A1, FNR) but cannot activate the flavonoid 3-O-glucosyltransferase gene (Bz1, 3GT). Genetic analysis also showed that theZm P1 locus coincides with a major quantitative trait locus determining the levels of maysin, a C-glycosyl flavone (Byrne et al., 1996; Zhang et al., 2003; Cocciolone et al., 2005). In Arabidopsis, AtMYB12/PFG2, AtMYB11/PFG1, and AtMYB111/PFG3, which share amino acid sequence similarity with ZmP1, regulate individual flavonol accumulation in differential organs of developing seedlings (Mehrtens et al., 2005; Strackeet al., 2007). They activate the expression of CHS, chalcone isomerase (CHI), flavanone 3-hydroxylase (F3H), flavonol synthase (FLS), and flavonoid 3'-hydroxylase (F3'H), but do not influence the expression of dihydroflavonol 4-reductase (DFR). In grapevine, the expression of VvMYBF1 correlated with flavonol accumulation and the expression of VvFLS1 (Czemmel et al., 2009). In addition, deficiency of SlMYB12, the tomato orthologue of AtMYB12, resulted in pink fruit lacking the ripening-dependent accumulation of the yellow flavonoid naringenin chalcone in the fruit peel (Ballester et al., 2010). These R2R3-MYBs were classified into subgroup 7 (Dubos et al., 2010) and activated specific gene sets of the early flavonoid biosynthetic pathway (Stracke et al., 2007). However, in spite of the demonstrated accumulation of early flavonoid biosynthetic metabolites, such as flavonol and flavone, no transcription factors regulating the early steps of flavonoid biosynthesis have been identified in the petals of floricultural plants.

In the present study, two P1 orthologues, Gt MYBP3 and Gt MYBP4, were isolated and characterized in Japanese gentian. The expression profiles of both genes correlated with flavone accumulation in petals and activated the expression of CHS, flavone synthase II (FNSII), and F3'H, belonging to early flavonoid biosynthetic pathway. Functional analyses of Gt MYBP3 andGt MYBP4 were performed in transgenic Arabidopsis and tobacco plants. The results strongly suggested that both GtMYBP3 and GtMYBP4 regulate the early steps of the flavonoid biosynthetic pathway. This is the first report to investigate the function of P1 orthologues in floral organs.

MATERIALS AND METHODS

Plant Materials

Japanese gentian (G. triflora cv. Maciry) plants were grown in the fields of the Iwate Agricultural Research Center (Iwate prefecture, Japan). The developmental stages of petal samples were defined as described byNakatsuka et al. (2005).

Isolation of Gentian R2R3-MYB Transcription Factors

Total RNA was isolated from petals at flower developmental stages 1–3. cDNA was synthesized using a Takara RNA PCR kit (AMV) version 3.0 (Takara-bio, Ohtsu, Japan). Degenerate primers were designed from the conserved DNA-binding domain of R2R3-MYBs controlling flavonoid biosynthesis from other plant species (Supplementary Table S1, available at JXB online). Other degenerate primers (pMybU and pMybL; Rabinowiczet al., 1999) were also used. Reaction conditions consisted of pre-heating at 94 °C for 90 s; 40 cycles of 95 °C for 20 s, 50 °C for 40 s, and 72 °C for 1min; and extension at 72 °C for 10min. Amplified fragments of approximately 180bp were subcloned into the pCR4 TA TOPO cloning vector (Invitrogen, Carlsbad, CA, USA) and sequenced using a Big-Dye terminal cycle sequencing kit version 1.1 and an ABI PRISM 3130 DNA sequencer (Applied Biosystems, Foster City, CA, USA). Nucleotide sequences were conceptually translated into amino acid sequences using GENETYX-MAC version 12 (GENETYX, Tokyo, Japan) and compared using the BLAST network service from the National Center for Biotechnology Information (NCBI). A phylogenetic tree was constructed using MEGA version 4 (Tamura et al., 2007).

To obtain full-length cDNAs of Gt MYBP3 and Gt MYBP4, 3′- and 5′-rapid amplification of cDNA ends (RACE) was performed using a GeneRacer kit (Invitrogen). The amplified fragments were subcloned and sequenced as described above.

Northern blot analysis was performed in gentian using Gt MYBP3 and GtMYBP4 probes (Supplementary Table S1), as described by Nakatsuka et al.(2005).

Yeast two-hybrid analysis

The yeast two-hybrid assay was performed using the Matchmaker Yeast Two-Hybrid System 3 (Clontech, Mountain View, CA, USA) as described previously (Nakatsuka et al., 2008b). Briefly, the open reading frame sequences of GtMYBP3 and GtMYBP4 were cloned into the pGAD-T7 and pGBK-T7 vectors. GtMYB3 and GtbHLH1 constructs (Nakatsuka et al., 2008b) were also used.

Isolation of 5′-UTRs of *GtFNSII* and *GtF3′H* in gentian

The 5′-untranscribed regions of gentian Gt FNSII and Gt F3′H were identified by inverse PCR, using primers shown in Supplementary Table S1and as described by Nakatsuka et al. (2008b). Amplified fragments of about 1.8kb for Gt FNSII and 1.6kb for Gt F3′H were subcloned and sequenced as described above. Reporter vectors were constructed to contain the firefly luciferase (LUC) gene under the control of the gentian GtFNSII (accession number AB733018) or Gt F3′H (AB733017) promoters. GtCHSpro-LUC and GtF3′5′Hpro-LUC constructs (Nakatsuka et al., 2008b) were also used.

Transient expression assay using *Arabidopsis* suspension cells

To evaluate whether Gt MYBP3 and Gt MYBP4 are responsible for the regulation of early flavonoid biosynthesis, transient expression assays were performed using protoplasts from Arabidopsis T87 suspension cells (Axelos et al., 1992). The open reading frame sequences of Gt MYBP3 andGt MYBP4 were inserted into the p35Spro expression vector under the control of the CaMV35S promoter and NOS terminator, resulting in p35Spro-GtMYBP3 and p35Spro-GtMYBP4, respectively. pBI221 (Clontech) was used as a negative control vector. p35Spro-RLUC, the Renillaluciferase (RLUC) gene under the control of the CaMV35S promoter, was used as a transformation control. Protoplast isolation and PEG-transfection experiments were performed as described by Hartmann et al.(1998). Firefly and Renilla luciferase activities were measured by the dual-Glo luciferase assay system (Promega, Madison, WI, USA) and Luminescencer JNR II (ATTO, Tokyo, Japan), according to the manufacturers′ instructions. At least five independent transfections were done for each plasmid combination to demonstrate reproducibility.

Vector construct and production of stable *Arabidopsis* and tobacco transformants

p35Spro-GtMYBP3 and p35Spro-GtMYBP4 was inserted into binary vectors, pSkan35SGUS (kanamycin resistance) and pSMABR35S-sGFP (Mishiba et al., 2010), to produce pSkan-35Spro-GtMYBP3 and pSMABR-35Spro-GtMYBP4, respectively. The constructs were then transformed intoAgrobacterium tumefaciens EHA101. A. thaliana ecotype col-1 was transformed by floral dip method, as described by Clough and Bent (1998). Positive transformants were selected on germination medium supplemented with 50mg l^{-1} kanamycin or 6mg l^{-1} bialaphos, and then T_2seeds were obtained following self-pollination.

Tobacco plants (Nicotiana tabacum cv. SR-1), aseptically grown from seeds for about 1 month, were transformed via an A. tumefaciens-mediated leaf disc procedure (Horsch et al., 1985) and selected using 200mg l^{-1} kanamycin or

5mg l^{-1} bialaphos. After rooting and acclimatization, regenerated plants were grown in a greenhouse to set seeds by self-pollination. T$_1$ transgenic plant lines were used for further analyses.

Flavonoid analysis in transgenic *Arabidopsis* and tobacco plants

The flavonol accumulation in T$_2$ transgenic Arabidopsis seedlings was visualized by diphenylboric acid 2-aminoethyl ester (DPBA), as described by Stracke et al. (2007). Five-day-old seedlings grown on germination medium with 3mg l^{-1} norflurazon were stained by 0.25% (w/v) DPBA. Fluorescence images were visualized under UV light on a stereoscopic microscope (Olympus, Tokyo, Japan). The amount of anthocyanin and flavonol pigments in petals of transgenic tobacco plants were measured as described by Nakatsuka et al. (2007).

Quantitative real-time PCR in transgenic *Arabidopsis* and tobacco plants

Total RNA was isolated from the seedlings of transgenic Arabidopsis after germination for 7 days using an RNeasy Plant Mini kit (Qiagen, Venlo, The Netherlands). Total RNA of transgenic tobacco was also isolated from their petals at floral developmental stage 3 using a FastRNA Green kit (Q-Bio gene, Irvine, CA, USA).

Quantitative real-time PCR (qRT-PCR) was performed on a StepOne plus (Applied Biosystems) using SYBR GreenER qPCR Super Mix (Invitrogen). cDNA was synthesized from total RNA after removal of genomic DNA using PrimeScript RT reagent kit with gDNA eraser (Takara-bio), according to the manufacturer's instructions. Reaction mixtures (10 µl) included the following components: 1×master mix, 0.2 µM of each primer (Supplementary Table S2) and 1 µl cDNA. Cycle conditions were 95 °C for 20 s and then 40 cycles of 95 °C for 1 s and 60 °C for 20 s. The expression levels of each gene were calibrated by the expression of actin (At ACT, Arabidopsis) or ubiquitin (Nt UBQ, tobacco) genes.

RESULTS

Cloning of P1 orthologues from gentian flowers

Degenerate PCR was used to isolate R2R3-MYB transcription factor genes from gentian flowers. More than 200 clones were classified into 24 groups based on a similarity search. Among them, the deduced amino acid sequences of two clones exhibited high similarities with ZmP1 and AtMYB12. Full-length cDNAs for these two clones were determined using 5'/3' RACE techniques and designated GtMYBP3 and GtMYBP4.

Phylogenetic analysis showed that GtMYBP3 and GtMYBP4 belonged to P1/subgroup 7 (Fig. 2). The GtMYBP3 cDNA (accession no. AB733016) was 1474bp in length and encoded a protein of 329 amino acid residues, whereas the Gt MYBP4 cDNA (accession no. AB289446) was 1303bp encoded a 376 amino acid protein. The deduced amino acid sequence of GtMYBP3 was 79.8% identical to that of GtMYBP4 in the R2R3-MYB DNA-binding domain, and 43.5% identical overall (Supplementary Fig. S1). GtMYBP3 was 84.6, 89.4, and 83.7% identical to Arabidopsis AtMYB12, grape VvMYBF1, and maize ZmP1, respectively, whereas GtMYBP4 was 78.8, 80.8, and 79.8% identical to those three genes, respectively. Although the sequence similarity between R2R3-MYB proteins is generally restricted to the N-terminus, the SG7 motif (GRTxRSxMK; Stracke et al., 2001) and the SG7-2 motif ([W/x][L/x]LS; Czemmel et al., 2009) are characteristic of R2R3-MYB proteins regulating flavonol biosynthesis. Both SG7 motifs were partially conserved in GtMYBP3 and GtMYBP4. GtMYBP3 and GtMYBP4 shared 56% identity (five out of nine amino acid residues) with the SG7-1 motif and 75% (3/4) and 50% (2/4) identity with the SG7-2 motif, respectively (Supplementary Fig. S1).

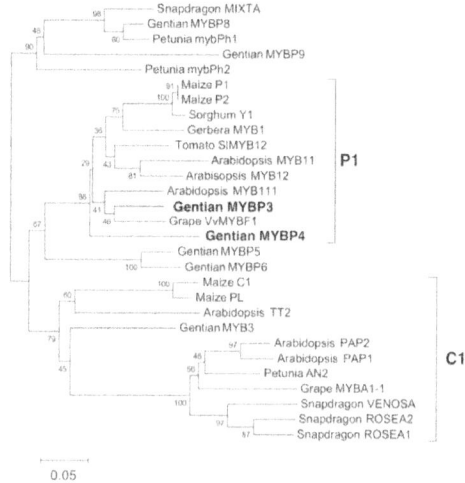

Figure 2. Phylogenetic analysis of deduced amino acid sequences of R2R3-MYB transcription factors in higher plants. The phylogenetic tree of the R2R3-MYB domain was generated using MEGA software (Tamura et al., 2007). Numerals next to branch nodes indicate bootstrap values from 1000 replications. The bar indicates an evolutionary

distance of 0.05%. Accession numbers in the GenBank/EMBL/DDBJ database are as follows: gentian: MYBP3 (AB733016), MYBP4 (AB289446), MYBP5 (AB733616), MYBP6 (AB733617), MYBP8 (AB733618), MYBP9 (AB733619), and MYB3 (AB289445); Arabidopsis: MYB11 (NM_116126), MYB12 (DQ224277), MYB111 (NM_124310), TT2, (AJ299452), PAP1 (AF325123). and PAP2 (AF325124); gerbera: MYB1 (AJ554697); grape: MYBF1 (FJ948477), and MYBA1-1 (AB073010); maize: P1 (M73028), P2 (AF210616), C1 (MZEMYBAA), and PL (AF015268);petunia: mybPh1 (Z13996) and AN2 (AF146702);, snapdragon, MIXTA (X79108), VENO-SA (DQ275531), ROSEA1 (DQ275529), and ROSEA2 (DQ275530); sorghum: Y1 (AY860968); and tomato: SlMYB12 (EU419748).

Expression profiles of Gt*MYBP3* and Gt*MYBP4* transcripts in gentian

To reveal the temporal and spatial expression of Gt MYBP3 and Gt MYBP4in gentian plants, northern blot analysis was performed using total RNA isolated from petals (at four developmental stages), leaves, and stems (Fig. 3). Gt MYBP3 and Gt MYBP4 transcripts were detected at the early flower development stages prior to pigmentation, but disappeared during later flower pigmented stages. Therefore, Gt MYBP3 and Gt MYBP4 would not be expected to regulate anthocyanin accumulation in gentian flowers. They could not be detected in vegetative tissues, such as leaves and stems. These expression profiles coincide with the accumulation profiles of flavones, which are abundantly present in young flower petals. The expression of Gt FNSII and GtF3'H genes, which are involved in flavone biosynthesis, were also detected at early flower developmental stages (Nakatsuka et al., 2005). Therefore, the temporal and spatial profiles of GtMYBP3 and Gt MYBP4 transcripts correlated with those of flavone biosynthesis.

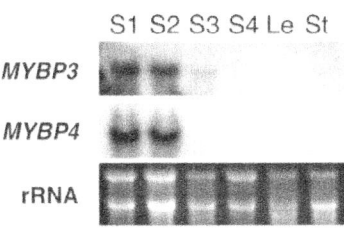

Figure3. Accumulation profiles ofGtMYBP3 and GtMYBP4transcripts in gentian plants. Northern blot analysis was performed using total RNA isolated from petal samples at four different flower developmental stages (S1–S4) and from leaf (Le) and stem (St) samples. RNA was blotted onto nylon membrane and hybridized with DIG-labelled Gt MYBP3 or Gt MYBP4 probes. Ethidium bromide-stained rRNA bands are shown as loading controls.

Protein–protein interactions between *Gt*MYBs and *Gt*bHLH1

To investigate protein–protein interactions, this study employed a yeast two-hybrid system. A preliminary experiment showed that GAL4 DNA-binding domain-fused GtMYBs all led to false-positive results; therefore, the interaction among GtMYB proteins could not be revealed in this study. GtbHLH1, a transcription factor regulating the anthocyanin biosynthetic pathway, interacted with GtMYB3 (Supplementary Fig. S2). However, neither GtMYBP3 nor GtMYBP4 interacted with GtbHLH1 in this study (Supplementary Fig. S2). Therefore, it was assumed that the activities of GtMYBP3 and GtMYBP4 were independent of the bHLH co-activator, as well as ZmP1.

Activation ability of *GtMYBP3* and *GtMYBP4* on promoters of flavonoid-biosynthetic genes by transient expression assay

To gain an insight into the activation ability of GtMYBP3 and GtMYBP4 on flavonoid biosynthesis, a transient expression assay in protoplasts of cultured Arabidopsis T87 suspension cells was used. About 1 kb of the 5'-upstream sequences of Gt CHS (Kobayashi et al., 1998), Gt F3'5'H (Nakatsuka et al., 2008b), Gt FNSII, and Gt F3'H were isolated from the gentian genome and connected to a reporter firefly luciferase (LUC) gene.

GtCHSpro, GtFNSIIpro, and GtF3'Hpro contained the binding sequences of vertebrate MYB protein ($^C/_TAAC^T/_GG$, black box), a P-recognition element ($CC^T/_AACC$, diagonal box; Grotewold et al., 1994), and an ACGT-containing element (CACGT; Hartmann et al., 1998), as shown in Fig. 4A. By contrast, GtF3'5'Hpro contained the vertebrate MYB protein and P-recognition element, but not the ACGT-containing element (Nakatsuka et al., 2008b).

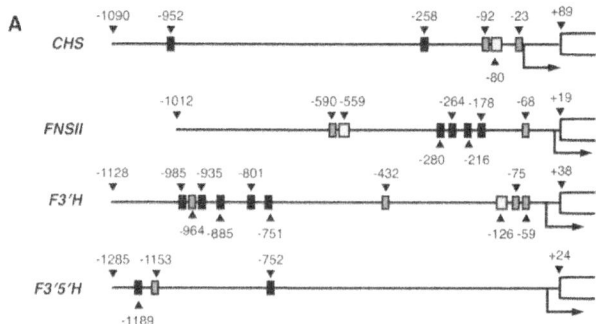

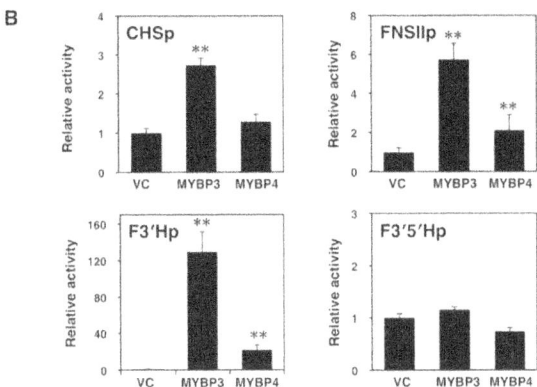

Figure 4. 5'-flanking sequences of flavonoid biosynthetic genes from gentian and transient expression assay. (A) Structures of the promoters of Gt CHS, Gt FNSII, Gt F3'H, and GtF3'5'H. Filled, striped, and dotted boxes indicate MYB recognition elements, vertebrate MYB (Urao et al., 1993), P-recognition element (Grotewold et al. 1994), and an ACGT-containing element (ACE, Hartmann et al., 1998, 2005), respectively. The positions upstream of the transcription initiation site (arrow) are indicated as numerals above the filled triangles. The coding regions are indicated as open boxes. (B) Effect of GtMYBP3 and GtMYBP4 on promoter activities of four flavonoid biosynthetic genes. Transient expression assays were performed by transfecting the reporter and effector plasmid DNA into the protoplasts from Arabidopsis T87 cells. GtCHSpro-LUC, GtFNSIIpro-LUC, GtF3'Hpro-LUC, or GtF3'5'Hpro-LUC were used as reporters, and p35Spro-GtMYBP3, p35Spro-GtMYBP4, or pBI221 (negative control) were used as the effector. p35Spro-RLUC was also used as a transformation control. The promoter activation activities are indicated as relative values compared with that of the negative control. Asterisks indicate statistically significant differences between the means for negative control (pBI221) and tested genes, as judged by Student's t-test (P < 0.01).

GtMYBP3 had about 2-, 5-, and 120-fold activation ability on Gt CHS, GtFNSII and Gt F3'H promoters, respectively, compared with the vector control (Fig. 4B). Similarly, GtMYBP4 also enhanced Gt FNSII and Gt F3'Hpromoter activity, but its ability was weaker than GtMYBP3. Conversely, neither GtMYBP3 nor GtMYBP4 could induce Gt F3'5'H promoter activity. No synergistic effect of GtMYBP3 and GtMYBP4 was observed in the transient expression analysis (Supplementary Fig. S3).

Gt*MYBP3* and Gt*MYBP4* overexpression in transgenic *Arabidopsis* plants

To confirm whether the Gt MYBP3 and Gt MYBP4 genes can regulate early flavonoid biosynthesis in planta, this study produced transgenic

Arabidopsis plants expressing Gt MYBP3 and Gt MYBP4 under the control of a constitutive CaMV35S promoter.

Both constructs were transformed into Arabidopsis plants using the floral dip method, and two independent T_2 lines were produced, respectively. No morphological changes were observed in Gt MYBP3- and Gt MYBP4-expressing Arabidopsis plants compared with vector control plants (data not shown). Flavonoid accumulation of transgenic Arabidopsis seedlings was visualized with DPBA and imaged by epifluorescence microscopy (Fig. 5). In wild-type and vector control plants, cotyledons showed intense orange fluorescence implying abundant accumulation of flavonol derivatives. Their hypocotyls and roots showed yellow and blue fluorescence implying some flavonol accumulation. However, Gt MYBP3- and Gt MYBP4-expressing Arabidopsis showed intense orange fluorescence over the entire plant.

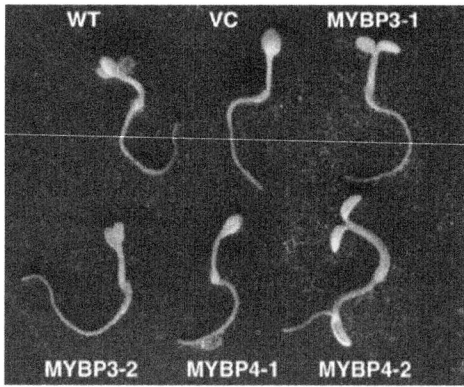

Figure 5. Visualized flavonoid accumulation in transgenicArabidopsis seedlings. Flavonoid staining in wild type (WT), vector control (VC), and transgenic seedlings. Two independent transgenic lines are shown. Norflurazon-bleached seedlings were stained with diphenylboric acid 2-aminoethyl ester and photographed under UV light.

In both Gt MYBP3- and Gt MYBP4-expressing transgenic seedlings, the expression levels of each transgene in clone no. 1 was approximately 2-fold higher than in clone no. 2 (Supplementary Fig. S4A). Among genes involved in phenylpropanoid pathway, the expression levels of At PAL1and At PAL2 were significantly enhanced in transgenic plants. At PAL1 andAt PAL2 transcripts were upregulated by 1.4–1.8-fold in both Gt MYBP3- and GtMYBP4-expressing plants compared with the vector control (Fig. 6). Induction of Arabidopsis 4-coumarate:CoA ligase (At 4CL3) expression, which contributes to flavonoid biosynthesis, was observed during overexpression of Gt MYBP3 (2.8-fold) and Gt MYBP4 (7.3-fold). InGtMYBP3-

expressing transgenic seedlings, the expression levels of several genes, including At PAL4, At 4CL2, and At 4CL5, were reduced in both lines.

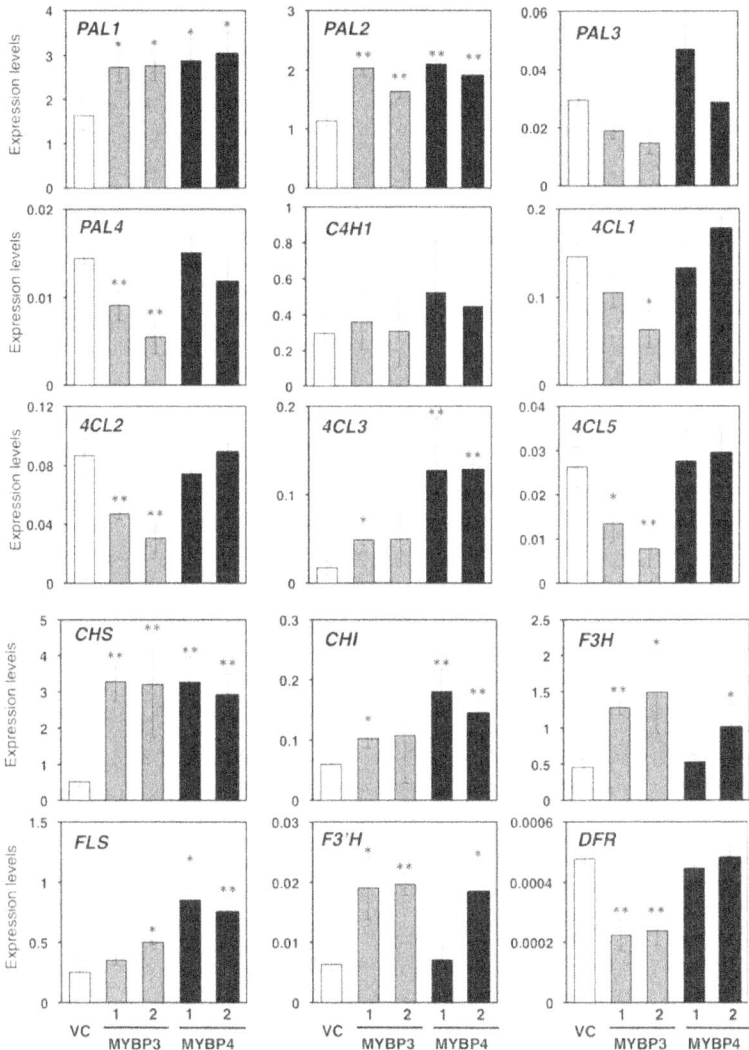

Figure 6. Expression analysis of endogenous flavonoid biosynthetic genes in transgenic Arabidopsis. The effects of Gt MYBP3 and GtMYBP4 overexpression on endogenous phenylpropanoid and flavonoid biosynthetic genes were investigated using qRT-PCR analyses in vector control (VC) and 5-day-old transgenic seedlings. Two independent transgenic lines shown in Fig. 5 were analysed. Asterisks indicate statistically significant differences between the means for vector control and transgenic lines, as judged by Student's t-test (*, P < 0.05; **, P < 0.01).

With regard to flavonoid biosynthetic genes, the expression of At CHS, AtCHI, At F3H and At F3'H were upregulated by 6.3-, 2.7-, 3.2-, and 3.0-fold, respectively, in Gt MYBP3-expressing Arabidopsis plants compared with the vector control. Conversely, the overexpression of GtMYBP4 induced the expression of At CHS, At CHI, and At FLS transcripts by 6.3-, 4.8-, and 2.9-fold. The At F3H and At F3'H genes were also induced in GtMYBP4-expressing Arabidopsis line no. 2. The expression of At DFR, encoding an enzyme of a later step of the anthocyanin biosynthetic pathway, was suppressed in Gt MYBP3-expressing plants, whereas no difference was detected in Gt MYBP4-expressing plants. Expression of endogenous flavonol-specific R2R3-MYB genes, including At MYB12, AtMYB11, and At MYB111, did not markedly change in Gt MYBP3 and GtMYBP4-expressing seedlings (Supplementary Fig. S5). Thus, the overexpression of Gt MYBP3 and Gt MYBP4 in Arabidopsis enhanced or decreased the expression of several genes encoding enzymes of the phenylpropanoid and flavonoid biosynthetic pathways. However, there was a slight difference in the affected gene sets and intensities between GtMYBP3 and Gt MYBP4.

GtMYBP3 and GtMYBP4 overexpression in transgenic tobacco plants

As shown above, transgenic Arabidopsis showed clear changes in flavonoid biosynthesis in vegetative tissues. However, flavonoid compositions, including anthocyanin pigments, were difficult to analyse in the floral tissues. Thus, to investigate the effect of Gt MYBP3 and GtMYBP4 overexpression on floral flavonoid biosynthesis, this study produced transgenic tobacco plants. The petals of tobacco accumulate flavonol and anthocyanin derivatives (Nakatsuka et al., 2007). Tobacco leaf sections were transformed by A. tumefaciens harbouring the same binary vector construct used for Arabidopsis transformation. Over 20 independent transgenic lines were grown in a greenhouse, and the T_1 seeds were collected after self-pollination. Two representative lines for each construct were chosen and subjected to further analysis.

The pigmentation of the petals of Gt MYBP3- and Gt MYBP4-expressing tobacco plants was decreased compared with the wild type (Fig. 7A). Anthocyanin levels in Gt MYBP3-expressing transgenic tobacco petals were 22–46% lower compared with the wild type (Fig. 7B). However, the amounts of flavonol were the same between wild-type and Gt MYBP3-expressing plants (Fig. 7C). Anthocyanin levels in Gt MYBP4-expressing transgenic tobacco petals were 25–28% lower compared with wild type (Fig. 7B). In addition, the amounts of flavonols were increased by 1.5–2.6-fold in the petals of Gt MYBP4-expressing tobacco plants (Fig. 7C) and were inversely correlated with the accumulation levels of anthocyanins. No difference in flavonol and anthocyanin components was observed between any transgenic and wild-type petals by HPLC analysis (data not shown).

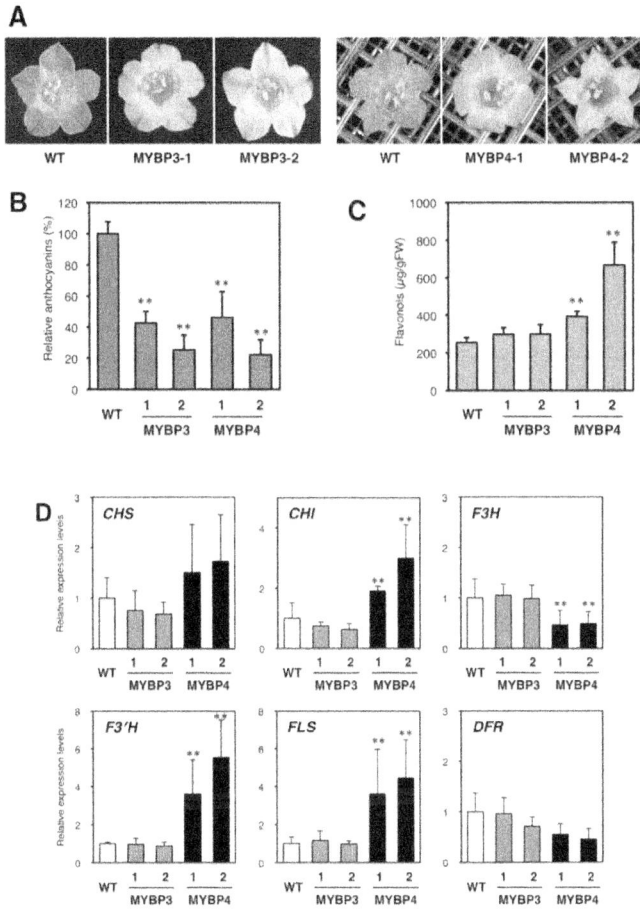

Figure 7. Phenotype and flavonoid analysis in transgenic tobacco flowers. (A) Typical flower phenotypes of wild-type (WT) and Gt MYBP3-and Gt MYBP4-expressing transgenic tobacco plants. Two independent transgenic lines per construct are shown. (B) Anthocyanin concentrations of WT and GtMYBP3- and Gt MYBP4-expressing transgenic petals. (C) Flavonol concentrations of WT and Gt MYBP3- and GtMYBP4-expressing transgenic petals. (D) Expression analyses of endogenous flavonoid biosynthetic genes in transgenic flowers. The effects of Gt MYBP3 and Gt MYBP4 ov erexpression on endogenous flavonoid biosynthetic genes were investigated using qRT-PCR analyses in WT and transgenic petals. Asterisks represent statistically significant differences between the means for wild-type and transgenic lines, as judged by Student's t-test (*, P < 0.05; **, P < 0.01).

The expression of the transgenes were confirmed (Supplementary Fig. S4B). Expression of Nt CHS, Nt CHI, Nt F3'H, and Nt FLS was enhanced 2.1-, 3.5, 6.0-, and 5.1-fold in the petals of Gt MYBP4-expressing transgenic plants (Fig. 7D). The expression of the Nt F3H gene decreased in GtMYBP4-expressing plants. However, no change was detected in the NtDFR gene, which is involved in late flavonoid biosynthesis. On the other hand, no significant change in expression of flavonoid biosynthetic genes was observed in Gt MYBP3-expressing tobacco plants (Fig. 7D).

DISCUSSION

A previous study revealed that two transcription factors, GtMYB3 and GtbHLH1, regulate gentiodelphin biosynthesis in the petals of Japanese gentian (Nakatsuka et al., 2008b). The complex of GtMYB3 and GtbHLH1 proteins activated the expression of genes encoding enzymes involved in the anthocyanin biosynthetic pathway after the Gt F3H gene, but could not induce the transcripts of early flavonoid biosynthetic genes. Recently, some transcription factors controlling early flavonoid biosynthesis, especially flavonol accumulation, have been reported in several plant species, including Arabidopsis seedlings (Mehrtens et al., 2005; Stracke et al., 2007), grapes (Czemmel et al., 2009), and tomatoes (Ballester et al., 2010). However, no transcription factor controlling early flavonoid biosynthesis in the petals has been identified.

In this study, two R2R3-MYB transcription factors from the petals of Japanese gentian were identified. GtMYBP3 and GtMYBP4 were classified into P1/subgroup 7 (Fig. 2), which were reported to regulate early flavonoid biosynthesis (Dubos et al., 2010). Arabidopsis R2R3-MYB subgroup 7 contains AtMYB12/PFG2, AtMYB11/PFG1, and AtMYB111/PFG1, and these are known to control flavonol biosynthesis individually in the different organs (Mehrtens et al., 2005; Stracke et al., 2007; Dubos et al., 2010). The deduced amino acid sequence of GtMYBP3 is highly similar to GtMYBP4 in the R2R3-MYB DNA-binding domain (79.8%), suggesting that they might be functionally redundant genes (Fig. 2 and Supplementary Fig. S1). SG7 and SG7-2 motifs are conserved in the C-termini of R3R3-MYBs belonging to subgroup 7 in Arabidopsis and grapevine (Stracke et al., 2001; Czemmel et al., 2009; Dubos et al., 2010). However, the SG7 or SG7-2 motifs were not conserved in tomato SlMYB12, which has been identified as a flavonol regulator in tomato fruits (Ballester al., 2010). The SG7 and SG7-2 motifs of GtMYBP3 and GtMYBP4 also had amino acid substitutions at some positions (Supplementary Fig. S1). Therefore, conservation of SG7 motifs in GtMYBP3 and GtMYBP4 might not be so important for regulation of early flavonoid biosynthesis in gentian petals.

Gt MYBP3 and Gt MYBP4 had similar temporal and spatial expression profiles, expressing strongly during early developmental stages (stages 1 and 2) of flower petals in gentians (Fig. 3). These expression profiles corresponded well with the accumulation profiles of FNSII and F3'Htranscripts and flavone derivatives (Nakatsuka et al., 2005). In grapevine,Vv MYBF1 transcripts were detected during flowering and in skins of ripening berries and were correlated with the accumulation of flavonol and expression of Vv FLS1 (Czemmel et al., 2009).

Transient expression in Arabidopsis suspension cells showed that both GtMYBP3 and GtMYB4 could enhance the promoter activities of gentian GtFNSII and Gt F3'H, which encode enzymes of the early flavonoid biosynthesis pathway (Fig. 4B). GtMYBP3 also enhanced the promoter activity of gentian Gt CHS. However, the promoter activity of Gt F3'5'H, encoding an enzyme of the late flavonoid biosynthesis pathway, could not be activated by either GtMYBP3 or GtMYBP4. Nakatsuka et al. (2008b) demonstrated that GtMYB3 and GtbHLH1, which are anthocyanin biosynthetic regulators in gentian flowers, induced promoter activity of GtF3'5'H, but not of Gt CHS. These results suggested that GtMYBP3 and GtMYBP4 control the expression of early flavonoid biosynthetic genes, unlike GtMYB3 and GtbHLH1. For the promoters of all early flavonoid biosynthetic genes, the activation intensities of GtMYBP3 were higher than those of GtMYBP4 (Fig. 4B). The Gt FNSII promoter was also activated by GtMYBP3 or GtMYBP4 in a transient assay using gentian mesophyll protoplasts, whereas Gt CHS was activated by GtMYBP3 only (Supplementary Fig. S6), confirming the results of transient expression assays in Arabidopsis suspension cells (Fig. 5B). The reason for the different activation ability between GtMYBP3 and GtMYBP4 is not yet clearly understood, but it might depend on the different DNA-binding activities on each promoter and/or on interactions with other transcription factor(s) to regulate flavonoid biosynthesis. Two cis-elements, a P-recognition element and an ACGT-containing element, were present in close proximity to the transcription initiation site of the promoters of GtCHS, Gt FNSII, and Gt F3'H, but not GtF3'5'H (Fig. 4A). The P-recognition element has been identified as a cis-binding site for the maize P1 protein, controlling 3-deoxyanthocyanin phlobaphene and C-glycosyl flavone maysin biosynthesis (Grotewold et al., 1994). Therefore, GtMYBP3 and GtMYBP4 might bind to the P-recognition element and activate the transcription of downstream genes. This is reasonable, because yeast two-hybrid analysis also showed that the GtMYBP3 or GtMYBP4 proteins did not interact with GtbHLH1 (Supplementary Fig. S2). Moreover, ACGT-containing element potentially binds to bZIP transcription factors, which work together with R2R3-MYB in light-inducible pigmentation (Hartmannet al., 2005). Further analysis is necessary to elucidate the functions of GtMYBP3 and GtMYBP4 in relation to other transcription factors, including bZIP.

Overexpression of Gt MYBP3 and Gt MYBP4 led to flavonol accumulation in transgenic Arabidopsis seedlings (Fig. 5). Overexpression of At MYB12also resulted in increased flavonol amounts in Arabidopsis plants (Mehrtens et al., 2005). qRT-PCR analyses of endogenous phenylpropanoid and flavonol biosynthetic genes showed the activated endogenous gene sets differed somewhat between Gt MYBP3- and GtMYBP4-expressing Arabidopsis (Fig. 6). The expression levels of three endogenous flavonol-specific transcription factor genes, At MYB12, AtMYB11, and At MYB111, were not markedly affected in the transformants and there was no relationship between the expression levels of foreign genes and endogenous genes (Supplementary Figs. S4A and S5); therefore, an indirect effect via activation of endogenous transcription factors was excluded. GtMYBP3 activated the transcripts of At PAL1, At PAL2, At 4CL3,At CHS, At CHI, At F3H, and At F3'H, whereas GtMYBP4 enhanced At PAL1,At PAL2, At 4CL3, At CHS, At CHI, and   At   FLS   transcripts   in   transgenicArabidopsis   (Fig. 6). Arabidopsis genome comprises four PAL genes (AtPAL1 to At PAL4), among them At PAL1 and At PAL2 have a redundant role in flavonoid biosynthesis (Huang et al., 2010). 4CL also comprises a multigene family, At 4CL1, At 4CL2, At 4CL3, and At 4CL5, in Arabidopsis(Ehlting et al., 1999; Costa et al., 2005). It is notable that At 4CL3 is likely to participate in the biosynthetic pathway leading to flavonoids, whereasAt 4CL1 and At 4CL2 are probably involved in lignin formation and in the production of additional phenolic compounds other than flavonoids (Ehlting et al., 1999). Therefore, the observed activation of At PAL1, AtPAL2, and At 4CL3 transcripts in transgenic Arabidopsis plants would be reasonable in light of the functions of GtMYBP3 and GtMYBP4. GtMYBP4 induced stronger expression of almost all phenylpropanoid and flavonol biosynthetic genes, except for At F3H and At F3'H, than GtMYBP3 (Fig. 6).At F3'H was not activated by AtMYB12 or ZmP1 (Mehrtens et al., 2005). The differences in the controlled gene sets between GtMYBP3 and GtMYBP4 might reflect the differences in their C-terminal regions. ZmP1 activates the promoters of At CHS, At F3H, and At FLS, although At FLSinduction was only 18% of that observed for AtMYB12 (Mehrtens et al., 2005). Thus, the intensity of transcriptional activation for individual flavonol biosynthetic genes seems to be different among P1 orthologues. The suppression of several biosynthetic genes in the Gt MYBP3-expressing transgenic Arabidopsis seedlings also suggested that other phenylpropanoid metabolism, such as lignins, organic acids, and proanthocyanidins, were regulated; therefore, detailed analyses are necessary in the future.

To further investigate the functions of GtMYBP3 and GtMYBP4 in flowers, they were expressed in tobacco plants, which are well known for flavonoid biosynthesis in their petals (Nishihara et al., 2005; Nakatsuka et al., 2007, 2008a). The flowers of Gt MYBP3- and Gt MYBP4-expressing transgenic tobacco plants showed decreased colour intensity (Fig. 7A). The phenotype of Gt MYBP4-expressing tobacco flowers arose from increased flavonol and reduced anthocyanin accumulation (Fig. 7B, C). qRT-PCR analysis also confirmed that Gt MYBP4 increased the expression levels of four endogenous tobacco genes, CHI, F3'H, and FLS, by 3–6-fold, but did not activate the expression of flavonoid biosynthesis genes, F3H and DFR (Fig. 7D). These results suggested the increased flavonol accumulation resulted from the activation of genes encoding enzymes catalysing the early steps in flavonoid biosynthesis, but not from the suppression of anthocyanin biosynthetic genes in transgenic tobacco plants. The inverse correlation between anthocyanin and flavonol levels in the flowers probably reflects competition between these two branches for flavonoid metabolites (Davieset al., 2003). The overexpression of At MYB12, a regulator of flavonol biosynthesis, induced the enhanced expression of Nt PAL, Nt CHS, Nt CHI, and Nt FLS and increased flavonol accumulation in tobacco petals (Luo et al., 2008). Therefore, heterologous expression of Gt MYBP4 could control the early flavonoid biosynthesis in tobacco plants. Conversely, Gt MYBP3-expressing tobacco flowers showed reduced anthocyanin accumulation, but showed no increase in flavonol accumulation (Fig. 7B, C). No significantly different expression levels of flavonol and anthocyanin biosynthetic genes were detected between wild-type and Gt MYBP3-expressing plants (Fig. 7D). Arabidopsis At MYB4 and strawberry Fa MYB1, classified into subgroup 4, suppress the expression of cinnamate 4-hydroxylase and anthocyanidin synthase, respectively (Jin et al., 2000; Aharoni et al., 2001). Therefore, GtMYBP3 might affect the expression of genes that were not investigated in this study, such as the phenylpropanoid biosynthetic and flavonoid modification genes, in tobacco petals. Proanthocyanidins (condensed tannins) compounds also derived from the flavonoid biosynthetic pathway, and two key enzymes, anthocyanidin reductase (ANR) and leucoanthocyanidin reductase (LAR), are involved in their biosynthesis (Tanner et al., 2003; Xie et al., 2003). Recently, Han et al. (2012) reported that ectopic expression of apple ANRin tobacco led to downregulation of both CHI and DFR in the flowers, leading to loss of anthocyanin. Similarly, in some cases, metabolic engineering of the flavonoid biosynthetic pathway has been known to induce feedback suppression of endogenous related genes. Therefore, the reduced anthocyanin accumulation in Gt MYBP3-expressing tobacco flowers might result from such unexpected regulation of genes encoding phenylpropanoid and flavonoid biosynthetic enzymes, although further metabolic and gene expression analyses are necessary to explain this observation. As shown above, the endogenous gene sets activated by GtMYBP3 and GtMYBP4 were different between transgenic Arabidopsisseedlings and tobacco flowers (Figs. 6 and 7). However, the targeted gene sets of GtMYBP3 and GtMYBP4 mostly overlapped in the early flavonoid biosynthetic genes; therefore, the functions of these two genes were thought to be complementary in gentian flowers. Moreover, the differences in the targeted gene sets among plant hosts (Arabidopsis and tobacco) suggested that full activation of GtMYBP3 and GtMYBP4 transcription factors probably required cofactors, such as bZIP (Hartmannet al., 2005).

Petunia flowers have been extensively studied as models for flavonoid biosynthesis and are proposed as a model showing that early- and late-biosynthetic genes are controlled by two different regulators (Quattrocchioet al., 1993). However, although it has been revealed that AN1/AN2 regulate the expression of anthocyanin biosynthetic genes, no transcription factor involved in the flavonol biosynthetic pathway has been identified. Uniquely, GhMYB1 was isolated from the petals of Gerbera hybrida, but its function has not been characterized (Elomaa et al., 2003). Therefore, Gt MYBP3 and Gt MYBP4 are the first characterized transcription factors that are involved in the early flavonoid biosynthesis in petal organs. Their identification will advance the understanding of flavonoid biosynthesis in floricultural crops.

SUPPLEMENTARY MATERIAL

Supplementary data are available at JXB online.

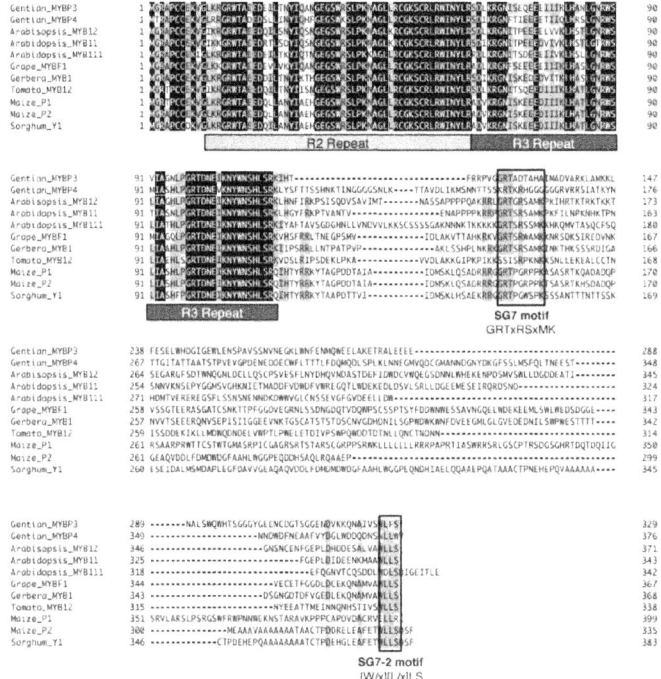

Figure S1. Alignment of P1 orthologue proteins in higher plants.

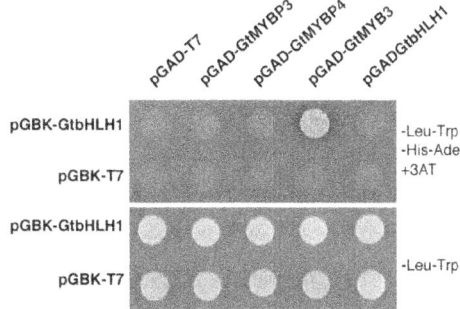

Figure S2. Yeast two-hybrid analysis to examine the protein–protein interaction between GtMYBs and GtbHLH1. The c to the GAL4 binding domain (BD) and assayed for its ability to bind the GtMYBs and GtbHLH1 fused to the GAL4 activation domain (AD). The transformed yeasts were grown on quadruple dropout medium (without leucine, tryptophan, histidine and adenine, upper) supplemented with 15 mM 3AT and on double dropout medium (without leucine and tryptophan, lower) at 30°C for 3 days. Protein-protein interactions are shown by yeast growth on quadruple dropout medium.

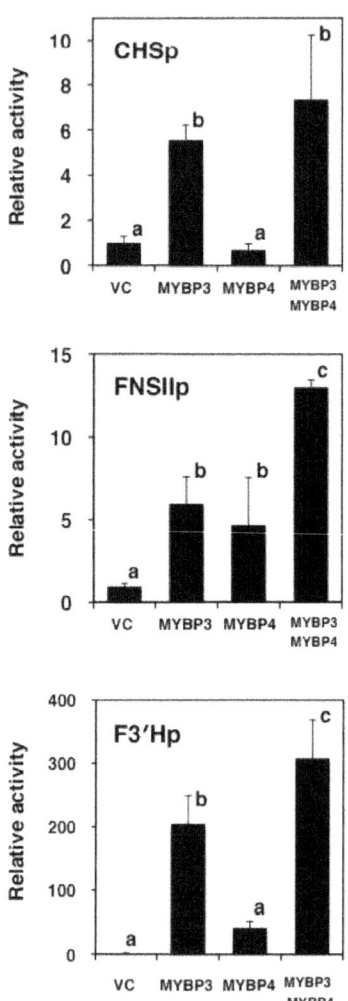

Figure S3. Effect of co-transfection of GtMYBP3 and GtMYBP4 on promoter activities of Gt CHS, Gt FNSII, and GtF3'H.

Transient expression assays were performed using protoplasts from Arabidopsis T87 cells, as described in Fig. 4B. Letters are the results from Tukey's multiple comparisons test where different letters represent a significant difference at $P<0.05$.

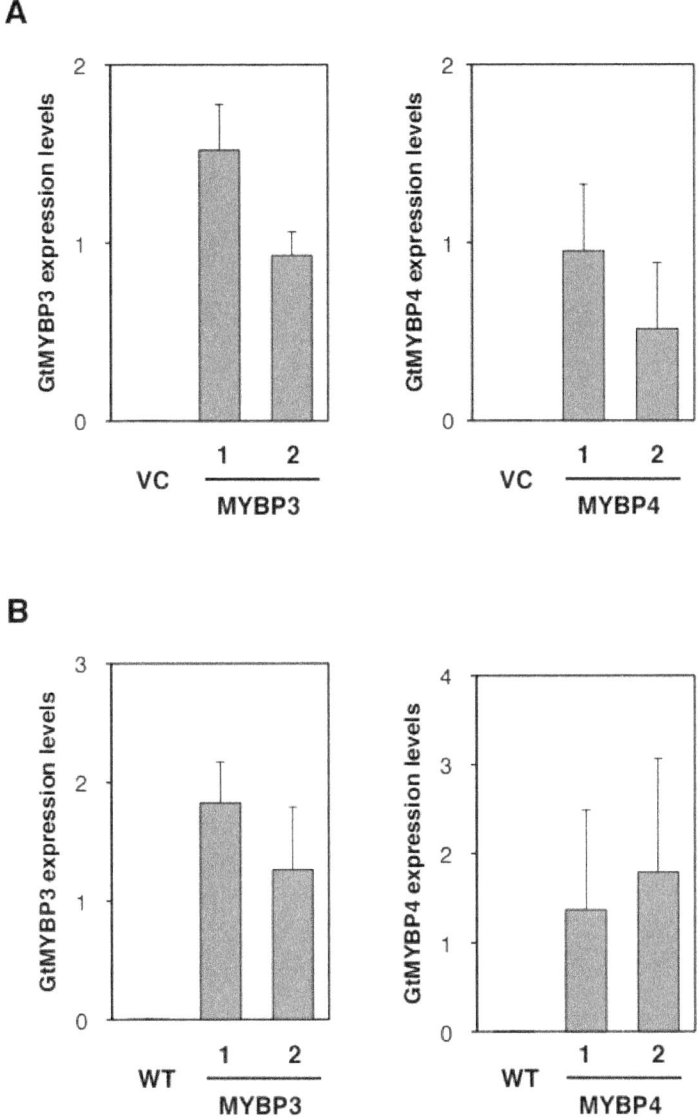

Figure S4. Confirmation of the expression of transgenes in transgenic Arabidopsis and tobacco. The expressions of GtMYBP3 and GtMYBP4 were investigated in transgenic 5-day-old Arabidopsis seedlings (A) and in transgenic tobacco petals (B). VC indicates vector control.

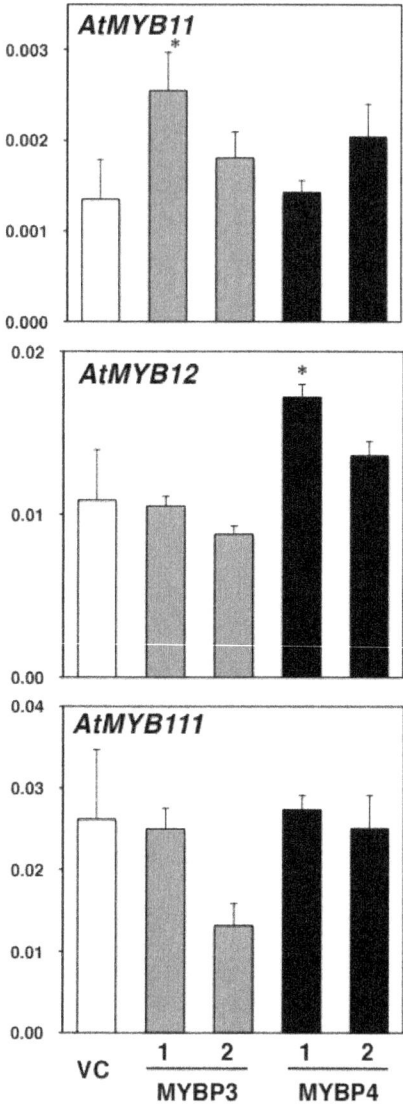

Figure S5. Expression analyses of endogenous flavonol-specific transcription factor genes in transgenic Arabidopsis. The effects of GtMYBP3 and GtMYBP4 overexpression on endogenous flavonol-specific R2R3MYB genes were investigated using qRT-PCR analyses in vector control and 5-dayold transgenic seedlings. The two independent transgenic lines shown in Fig. 6 were analysed. Asterisks (*, $P < 0.05$) represent statistically significant differences between the means for vector control and transgenic lines, as judged by Student's t-test.

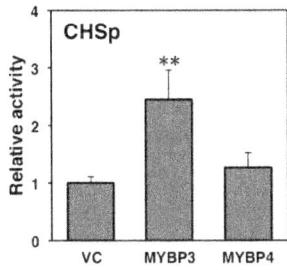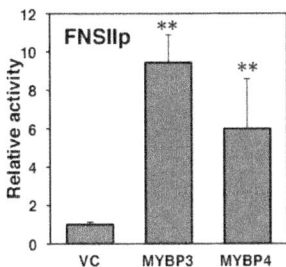

Figure S6. Transient expression assay in gentian mesophyll protoplast. The effect of GtMYBP3 and GtMYBP4 on promoter activities of two early flavonoid biosynthetic genes, GtCHS and GtFNSII, were investigated in mesophyll protoplasts of gentian. Transient expression assays were performed by the dualGlo luciferase assay system, as described in Fig. 4B. Asterisks (**, P < 0.01) represent statistically significant differences between the means for negative control (pBI221) and tested genes, as judged by Student's t-test.

Table S1. Degenerate PCR, probe, and inverse PCR primers used in this study.

		Forward	Reverse
Degenerate PCR	Set 1	GRB TDM GRA ARG GTK CWT GGA	GCW ATH ARD GAC CAY CTR TT
	Set 2	AAR WSI TGY MGI YTI MGI TGG AYN AAY TA	CCA RTA RTT YTT IAM YTC RTT RTC
Probe	MYBP3	ATG CAA ATC TGG GTA ACA GGT GGT	CGT CTC TTT CGC TAA TTC TTC CCA
	MYBP4	TTC ATC CAA GCG GAC CAA GC	CCA TCC CCT CAT TAT TTA GC
Inverse PCR	FNSII pro	CGT TGT ATT GTA GTC TCC ACC CCT GAA CTA GC	GGA GGA TTG AGA GGA CGA AAA TGG AAG CAG AG
	F3'H pro	TCC TTA TAC GAG CAA TAG CGA GTG GCG GTC AT	TGA GGA AGA TGA AGA TGA AGA CGA CGA AAA GG

Table S2. Primers used for quantitative RT-PCR analysis.

	Primer sequence (5' - 3')	
	Forward	Reverse
Arabidopsis		
AtCHS	CTAAGGATCTCGCCGAGAACA	CGGCTGTGATCTCAGAGCAG
AtCHI	GTCACCGGCCTCCTCCA	TGGATATCAAGGCCTCGGAC
AtF3H	TGAAGGAGCGTTTGTCGTCA	TGAACCTCCCATTGCTCAAAA
AtF3'H	CCTCCACCTCCGACTAGGGT	TGCTCGGCCACGGATTTA
AtFLS	CAACATTCCGAGGTCCAACG	TCTTCGTCGGGATCGCTTAG
AtDFR	CGAGATGACGGCAGCTTTG	AGCGGCGACATGGAAGAC
AtMYB11	TCGCCAATACCGTCGAGAAT	CGGATCTGCTGGTTCTTCCA
AtMYB12	CGTAAAACGAAGAAAACGTCTGC	GCTTCTTTATCAGCCCCAGCT
AtMYB111	CAATGTTTCTCACAACCTAAGGAGC	CCAAAGACTCTCCTTCAAAATTACCA
AtACT2	ACCCGATGGGCAAGTCATC	CGAGGGCTGGAACAAGACTTC
Tobacco		
NtPAL	TGCTTAACCACAATGTCACTCCA	CGAGATCACCAGAGGCGGT
NtCHS	GCCGGTGGCACGGTACT	ACTCGAGCGCCCTTGTTGT
NtCHI	TGAAGCAGTGCTGGAATCCA	TTTCGGCGATACTACACTTTGC
NtF3H	TTTTTACCCAAAGTGTCCACAGC	GGGTGATGGTTCCTGGATCA
NtF3'H	TGGATTAACCCATTCATTTGGAT	TTCCAAAAGGCTCAACACTTCTC
NtDFR	TGAGTTTAAAGGCATCGATAAGGA	GAATTGAAACCCCATATCCGTC
NtANS	TCCTCCACAATATGGTGCCTG	GGGTGTCCCCAATATGCATG
NtFLS	GGCCTAAAAATCCTCCCTCCT	TTCTCCACAACTTCTCGCAGC
NtUBQ	AAGATTCAGGACAAGGAAGGCA	AGCTGCTTACCTGCGAAAATCA

Acknowledgements

The authors thank Messrs. K. Fujiwara and T. Nakasato, Iwate Agricultural Research Center, for providing the Japanese gentian materials. They also thank Dr. M. Ohme-Takagi (National Institute of Advanced Industrial Technology and Science, Japan) for providing p35Spro-RLUC. The authors are grateful to Mses. A. Kubota and C. Yoshida, Iwate Biotechnology Research Center, for their technical support. This work was financially supported by Grants-in-Aid for Scientific Research from the Japan Society for the Promotion of Science to T.N. (no. 18789005), M.N. (no. 24380024), and by the Iwate prefectural government.

REFERENCES

1. Aharoni A, De Vos CH, Wein M, Sun Z, Greco R, Kroon A, Mol JN, O'Connell AP. 2001 The strawberry FaMYB1 transcription factor suppresses anthocyanin and flavonol accumulation in transgenic tobacco The Plant Journal 28 319–332

2. Axelos M, Curie C, Mazzolini L, Bardet C, Lescure B. 1992 A protocol for transient gene expression in Arabidopsis thaliana protoplast isolated from cell suspension cultures Plant Physiology and Biochemistry 30 123–128

3. Ballester AR, Molthoff J, de Vos R, et al. 2010 Biochemical and molecular analysis of pink tomatoes: deregulated expression of the gene encoding transcription factor SlMYB12 leads to pink tomato fruit color Plant Physiology 152 71–84

4. Borevitz JO, Xia Y, Blount J, Dixon RA, Lamb C. 2000 Activation tagging identifies a conserved MYB regulator of phenylpropanoid biosynthesis The Plant Cell 12 2383–2394

5. Broun P. 2005 Transcriptional control of flavonoid biosynthesis: a complex network of conserved regulators involved in multiple aspects of differentiation in Arabidopsis Current Opinion in Plant Biology 8 272–279

6. Byrne PF, McMullen MD, Snook ME, Musket TA, Theuri JM, Widstrom NW, Wiseman BR, Coe EH. 1996 Quantitative trait loci and metabolic pathways: genetic control of the concentration of maysin, a corn earworm resistance factor, in maize silks Proceedings of the National Academy of Sciences, USA 93 8820–8825

7. Clough SJ, Bent AF. 1998 Floral dip: a simplified method for Agrobacterium-mediated transformation of Arabidopsis thaliana The

Plant Journal 16 735–743

8. Cocciolone SM, Nettleton D, Snook ME, Peterson T. 2005 Transformation of maize with the p1 transcription factor directs production of silk maysin, a corn earworm resistance factor, in concordance with a hierarchy of floral organ pigmentation Plant Biotechnology Journal 3 225–235

9. Costa MA, Bedgar DL, Moinuddin SG, et al. 2005 Characterization in vitro and in vivo of the putative multigene 4-coumarate:CoA ligase network in Arabidopsis: syringyl lignin and sinapate/sinapyl alcohol derivative formation Phytochemistry 66 2072–2091

10. Czemmel S, Stracke R, Weisshaar B, Cordon N, Harris NN, Walker AR, Robinson SP, Bogs J. 2009 The grapevine R2R3-MYB transcription factor VvMYBF1 regulates flavonol synthesis in developing grape berries Plant Physiology 151 1513–1530

11. Davies KM, Schwinn KE, Deroles SC, Manson DG, Lewis DH, Bloor SJ, Bradley JM. 2003 Enhancing anthocyanin production by altering competition for substrate between flavonol synthase and dihydroflavonol 4-reductase Euphytica 131 259–268

12. Dubos C, Stracke R, Grotewold E, Weisshaar B, Martin C, Lepiniec L. 2010 MYB transcription factors in Arabidopsis Trends in Plant Science 15 573–581

13. Ehlting J, Buttner D, Wang Q, Douglas CJ, Somssich IE, Kombrink E. 1999 Three 4-coumarate:coenzyme A ligases in Arabidopsis thaliana represent two evolutionarily divergent classes in angiosperms The Plant Journal 19 9–20

14. Elomaa P, Uimari A, Mehto M, Albert VA, Laitinen RA, Teeri TH. 2003 Activation of anthocyanin biosynthesis in Gerbera hybrida (Asteraceae) suggests conserved protein–protein and protein–promoter interactions between the anciently diverged monocots and eudicots Plant Physiology 133 1831–1842

15. Gonzalez A, Zhao M, Leavitt JM, Lloyd AM. 2008 Regulation of the anthocyanin biosynthetic pathway by the TTG1/bHLH/Myb transcriptional complex in Arabidopsis seedlings The Plant Journal 53 814–827

16. Goto T, Kondo T, Tamura H, Imagawa H, Iino A, Takeda K. 1982 Structure of gentiodelphin, an acylated anthocyanin isolated from Gentiana makinori that is stable in dilute aqueous solution Tetrahedron Letters 23 3695–3698

17. Grotewold E, Drummond BJ, Bowen B, Peterson T. 1994 The myb-homologous P gene controls phlobaphene pigmentation in maize floral

organs by directly activating a flavonoid biosynthetic gene subset Cell 76 543–553

18. Grotewold E, Sainz MB, Tagliani L, Hernandez JM, Bowen B, Chandler VL. 2000 Identification of the residues in the Myb domain of maize C1 that specify the interaction with the bHLH cofactor R Proceedings of the National Academy of Sciences, USA 97 13579–13584

19. Han Y, Vimolmangkang S, Soria-Guerra RE, Korban SS. 2012 Introduction of apple ANR genes into tobacco inhibits expression of both CHI and DFR genes in flowers, leading to loss of anthocyanin Journal of Experimental Botany 63 2437–2447

20. Hartmann U, Sagasser M, Mehrtens F, Stracke R, Weisshaar B. 2005 Differential combinatorial interactions of cis-acting elements recognized by R2R3-MYB, BZIP, and BHLH factors control light-responsive and tissue-specific activation of phenylpropanoid biosynthesis genes Plant Molecular Biology 57 155–171

21. Hartmann U, Valentine WJ, Christie JM, Hays J, Jenkins GI, Weisshaar B. 1998 Identification of UV/blue light-response elements in the Arabidopsis thaliana chalcone synthase promoter using a homologous protoplast transient expression system Plant Molecular Biology 36 741–754

22. Holton TA, Cornish EC. 1995 Genetics and biochemistry of anthocyanin biosynthesis The Plant Cell 7 1071–1083

23. Horsch RB, Fry JE, Hoffmann NL, Eicholtz D, Rogers SG, Fraley RT. 1985 A simple method for transferring genes into plants Science 227 1229–1231

24. Huang J, Gu M, Lai Z, Fan B, Shi K, Zhou YH, Yu JQ, Chen Z. 2010 Functional analysis of the Arabidopsis PAL gene family in plant growth, development, and response to environmental stress Plant Physiology 153 1526–1538

25. Jin H, Cominelli E, Bailey P, Parr A, Mehrtens F, Jones J, Tonelli C, Weisshaar B, Martin C. 2000 Transcriptional repression by AtMYB4 controls production of UV-protecting sunscreens in Arabidopsis The EMBO Journal 19 6150–6161

26. Kobayashi H, Oikawa Y, Koiwa H, Yamamura S. 1998 Flower-specific expression directed by the promoter of a chalcone synthase gene from Gentiana triflora in Petunia hybrida Plant Science 131 173–180

27. Koes R, Verweij W, Quattrocchio F. 2005 Flavonoids: a colorful model for the regulation and evolution of biochemical pathways Trends in Plant Science 10 236–242

28. Lloyd AM, Walbot V, Davis RW. 1992 Arabidopsis and Nicotiana anthocyanin production activated by maize regulators R and C1 Science 258 1773–1775

29. Luo J, Butelli E, Hill L, Parr A, Niggeweg R, Bailey P, Weisshaar B, Martin C. 2008 AtMYB12 regulates caffeoyl quinic acid and flavonol synthesis in tomato: expression in fruit results in very high levels of both types of polyphenol The Plant Journal 56 316–326

30. Mehrtens F, Kranz H, Bednarek P, Weisshaar B. 2005 The Arabidopsis transcription factor MYB12 is a flavonol-specific regulator of phenylpropanoid biosynthesis Plant Physiology 138 1083–1096

31. Mishiba K, Yamasaki S, Nakatsuka T, Abe Y, Daimon H, Oda M, Nishihara M. 2010 Strict de novo methylation of the 35S enhancer sequence in gentian PLoS One 5–e9670

32. Mol J, Grotewold E, Koes R. 1998 How genes paint flowers and seeds Trends in Plant Science 3 212–217

33. Nakatsuka T, Abe Y, Kakizaki Y, Yamamura S, Nishihara M. 2007 Production of red-flowered plants by genetic engineering of multiple flavonoid biosynthetic genes Plant Cell Reports 26 1951–1959

34. Nakatsuka T, Haruta KS, Pitaksutheepong C, Abe Y, Kakizaki Y, Yamamoto K, Shimada N, Yamamura S, Nishihara M. 2008b Identification and characterization of R2R3-MYB and bHLH transcription factors regulating anthocyanin biosynthesis in gentian flowers Plant and Cell Physiology 49 1818–1829

35. Nakatsuka T, Nishihara M, Mishiba K, Yamamura S. 2005 Temporal expression of flavonoid biosynthesis-related genes regulates flower pigmentation in gentian plants Plant Science 168 1309–1318

36. Nakatsuka T, Sato K, Takahashi H, Yamamura S, Nishihara M. 2008a Cloning and characterization of the UDP-glucose:anthocyanin 5-O-glucosyltransferase gene from blue-flowered gentian Journal of Experimental Botany 59 1241–1252

37. Nishihara M, Nakatsuka T, Mizutani-Fukuchi M, Tanaka Y, Yamamura S. 2008 Gentians: from gene cloning to molecular breeding. In: Jaime A., da Silva T. , editors, Floricultural and ornamental biotechnology V. UK Global Science Books 57–67

38. Nishihara M, Nakatsuka T, Yamamura S. 2005 Flavonoid components and flower color change in transgenic tobacco plants by suppression of chalcone isomerase gene FEBS Letters 579 6074–6078

39. Quattrocchio F, Baudry A, Lepiniec L, Grotewold E. 2006 The regulation

of flavonoid biosynthesis. In: Grotewold E. , editor, The science of flavonoids New York Springer 97–122

40. Quattrocchio F, Wing JF, Leppen H, Mol J, Koes RE. 1993 Regulatory genes controlling anthocyanin pigmentation are functionally conserved among plant species and have distinct sets of target genes The Plant Cell 5 1497–1512

41. Rabinowicz PD, Braun EL, Wolfe AD, Bowen B, Grotewold E. 1999 Maize R2R3 Myb genes: sequence analysis reveals amplification in the higher plants Genetics 153 427–444

42. Spelt C, Quattrocchio F, Mol JN, Koes R. 2000 Anthocyanin1 of petunia encodes a basic helix-loop-helix protein that directly activates transcription of structural anthocyanin genes The Plant Cell 12 1619–1632

43. Stracke R, Ishihara H, Huep G, Barsch A, Mehrtens F, Niehaus K, Weisshaar B. 2007 Differential regulation of closely related R2R3-MYB transcription factors controls flavonol accumulation in different parts of the Arabidopsis thaliana seedling The Plant Journal 50 660–677

44. Stracke R, Werber M, Weisshaar B. 2001 The R2R3-MYB gene family in Arabidopsis thaliana Current Opinion in Plant Biology 4 447–456

45. Tamura K, Dudley J, Nei M, Kumar S. 2007 MEGA4: Molecular Evolutionary Genetics Analysis (MEGA) software version 4.0. Molecular Biology and Evolution 24 1596–1599

46. Tanaka Y, Yonekura K, Fukuchi-Mizutani M, Fukui Y, Fujiwara H, Ashikari T, Kusumi T. 1996 Molecular and biochemical characterization of three anthocyanin synthetic enzymes from Gentiana triflora Plant and Cell Physiology 37 711–716

47. Tanner GJ, Francki KT, Abrahams S, Watson JM, Larkin PJ, Ashton AR. 2003 Proanthocyanidin biosynthesis in plants. Purification of legume leucoanthocyanidin reductase and molecular cloning of its cDNA Journal of Biological Chemistry 278 31647–31656

48. Urao T, Yamaguchi-Shinozaki K, Urao S, Shinozaki K. 1993 An Arabidopsis myb homolog is induced by dehydration stress and its gene product binds to the conserved MYB recognition sequence The Plant Cell 5 1529–1539

49. Winkel BSJ. 2006 The biosynthesis of flavonoid. In: Grotewold E. , editor, The science of flavonoids New York Springer 71–96

50. Xie D-Y, Sharma SB, Paiva NL, Ferreira D, Dixon RA. 2003 Role of anthocyanidin reductase, encoded by BANYULS in plant flavonoid

biosynthesis Science 299 396–399

51. Yoshida K, Mori M, Kondo T. 2009 Blue flower color development by anthocyanins: from chemical structure to cell physiology Natural Product Reports 26 884–915

52. Yoshida K, Toyama Y, Kameda K, Kondo T. 2000 Contribution of each caffeoyl residue of the pigment molecule of gentiodelphin to blue color development Phytochemistry 54 85–92

53. Zhang P, Wang Y, Zhang J, Maddock S, Snook M, Peterson T. 2003 A maize QTL for silk maysin levels contains duplicated Myb-homologous genes which jointly regulate flavone biosynthesis Plant Molecular Biolog

Chapter 8

MODULATION OF FLOWER COLOR BY RATIONALLY DESIGNED DOMINANT-NEGATIVE CHALCONE SYNTHASE

Mamatha Hanumappa[1], Goh Choi[1], Sunhyo Ryu[1] and Giltsu Choi[2]

[1] Kumho Life and Environmental Science Laboratory, Gwangju 500-712, Korea

[2] Department of Biological Sciences, KAIST, Daejeon 305-701, Korea

ABSTRACT

The intensity of flower colour, mainly determined by the amount of anthocyanin, is an important horticultural trait. To modulate flower colour intensity, post-transcriptional gene silencing (PTGS)-based technology has been widely used. The constraint of PTGS, however, is that it requires a high degree of conservation in the nucleotide sequences of the target and the silencer. Further, it is difficult to restrict PTGS to the desired tissue or organ due to its systemic spread. To overcome these problems, dominant-negative chalcone synthase (CHS) enzymes have been developed by mutating a cysteine that is essential for the catalytic activity and a methionine that protrudes into the adjoining CHS monomer, as shown through crystallography. The dominant-negative action of mutated CHS enzymes from *Mazus japonicus* are demonstrated using transgenic *Arabidopsis*. Also, the modulation of *Petunia* flower colour intensity by the dominant-negative CHS is shown. The data support the crystallography result showing the importance of the protruding methionine for the function of the adjoining CHS monomer. Furthermore, the modulation of anthocyanin production by the mutated *Mazus* CHS in *Arabidopsis* and petunia suggests that the dominant-negative CHS can be used even in distantly related species.

INTRODUCTION

Flower colour is an important horticultural trait and is mainly produced by the flavonoid pigments, anthocyanins. Primarily produced to attract pollinators, flavonoids also protect the plant and its reproductive organs from UV damage, pests, and pathogens (Brouillard and Cheminat, 1988;Gronquist *et al.*, 2001). Classical breeding methods have been extensively used to develop cultivars with flowers varying in both the colour and its intensity. The recent advance of knowledge on flower colouration at the biochemical and molecular level has made it possible to achieve this by genetic engineering (Tanaka *et al.*, 1998).

Three different classes of anthocyanidins are responsible for the primary shade of flower colour in many angiosperms: pelargonidin (orange to brick red), cyanidin (red to pink), and delphinidin (purple to blue). The anthocyanidin biosynthetic pathway is well established and most of the enzymes involved in the synthesis have been identified (Holton and Cornish, 1995; Winkel-Shirley, 2001). It starts with the condensation of 4-coumaroyl-CoA and malonyl-CoA by chalcone synthase (CHS) to synthesize anthocyanidins that are then glycosylated by flavonoid 3-*O*-glucosyl transferase to produce anthocyanins. Further modification by rhamnosylation, methylation, or acylation results in a wide variety of anthocyanins (Kroon *et al.*, 1994; Ronchi *et al.*, 1995; Fujiwara *et al.*, 1997; Yoshida *et al.*, 2000; Yabuya *et al.*, 2001). The spectral difference in flower colour is mainly determined by the ratio of different classes of anthocyanins and other factors such as vacuolar pH, co-pigmentation, metal ion complexation, and molecular stacking (Holton *et al.*, 1993;Markham and Ofman, 1993; Mol *et al.*, 1998; Tanaka *et al.*, 1998; Aida *et al.*, 2000). The final shade may be altered further by various factors including the shape of the epidermal cells or the presence of starch that gives creaminess (Markham and Ofman, 1993; Noda *et al.*, 1994; Mol *et al.*, 1998; van Houwelingen *et al.*, 1998).

Genetic engineering to alter flower colour has been attempted using various genes. Some species lack a particular colour due to the absence of a biosynthetic gene or the substrate specificity of an enzyme in the pathway. For example, carnation lacks blue/purple-coloured flowers due to the absence of flavonoid 3'5'-hydroxylase (F3'5'H), while petunia lacks orange and brick-red flowers due to the inability of its dihydroflavonol 4-reductase (DFR) to reduce dihydrokaempferol (Gerats *et al.*, 1982;Forkmann and Ruhnau, 1987). Spontaneous mutations of the flavonoid 3'-hydroxylase (F3'H) gene confer reddish flowers in blue- and purple-flowered morning glory species (Hoshino *et al.*, 2003). Genetic engineering of blue/purple-coloured carnation was achieved by introducing the petunia *F3 5 H* gene and orange-coloured petunia was developed by introducing *DFR* from other species (Meyer *et al.*, 1987; Brugliera *et al.*, 2000; Johnson *et al.*, 2001). The modulation of colour intensity has been another target for genetic engineering. Expression of biosynthetic genes such as *CHS*, *F3H*, and *DFR* in sense or antisense directions has been the most exploited method (van der Krol *et al.*, 1990; Courtney-Gutterson *et al.*, 1994; Jorgensen *et al.*, 1996; Tanaka *et al.*, 1998). The resulting sense suppression or antisense inhibition is collectively called post-transcriptional gene silencing (PTGS). Alternatively, transcription factors that can either activate or repress the transcription of anthocyanin biosynthetic genes have been shown to be useful in regulating colour intensity in model plants such as *Arabidopsis*, tobacco, and petunia (Lloyd *et al.*, 1992; Mol *et al.*, 1998; Borevitz *et al.*, 2000; Aharoni *et al.*, 2001). The overexpression of transcription factors, however, generally alters the expression of many genes, thus the commercial viability of such transgenic flowers is yet to be determined (Lloyd *et al.*, 1994; Bruce *et al.*, 2000).

The biochemical and structural characterization of CHS suggests the possibility of designing a dominant-negative CHS that can be used to regulate flower colour intensity. CHS is the first enzyme in the synthesis of various flavonoids including anthocyanins. It functions as a homodimer and carries out a series of decarboxylation, condensation, cyclization, and aromatization reactions at a single active site (Tropf *et al.*, 1995). The enzyme condenses a molecule of 4-coumaroyl-CoA with three of malonyl-CoA and folds the tetraketide intermediate into an aromatic ring structure to yield chalcone (Ferrer *et al.*, 1999; Schroder, 1999). Site-directed mutagenesis and inhibitor studies have identified the conserved cysteine and histidine residues that are important for the catalytic function of CHS (Lanz *et al.*, 1991; Suh *et al.*, 2000). The mutation of this cysteine to either serine or alanine has been shown to inactivate the CHS (Lanz *et al.*, 1991; Tropf *et al.*, 1995; Jez *et al.*, 2000). The crystal structure of alfalfa CHS confirms that the conserved Cys164, Phe215, His303, and Asn336 form the catalytic active site (Ferrer *et al.*, 1999). The structure also revealed that CHS functions as a homodimer, forming a symmetric dimer with each monomer where the N-terminal helices entwine with each other. The crystal structure also shows that Met137 from the adjoining monomer extends into the cyclization pocket of CHS. Using substrate and product analogues, Ferrer *et al.* (1999) confirm that each monomer consists of two structural domains and two functionally independent active sites as previously reported (Tropf *et al.*, 1995). Point mutations confirm the role of Cys164 as an active site nucleophile, elucidate the importance of His303 and Asn336 in the malonyl-CoA decarboxylation reaction, and suggest that Phe215 may help orient substrates at the active site during elongation of the polyketide intermediate (Jez *et al.*, 2000). Three interconnected cavities, a CoA binding tunnel, a coumaroyl binding pocket, and a cyclization pocket, intersect with these four residues to form the active site architecture of CHS. Each active site consists of residues from a single monomer, with the only exception being Met137 which comes from the adjoining monomer. This suggests that a CHS monomer requires the methionine from the adjoining monomer for its activity. Based on this structural information by Ferrer *et al.* (1999), CHS has been generated that has alanine instead of cysteine at the active site and either glycine or lysine instead of the methionine. The mutation of cysteine to alanine will result in the inactive form of CHS, while the mutation of methionine to glycine or lysine is expected to inactivate the function of an adjoining CHS if the methionine is really important, as suggested by the crystal structure. Using transgenic *Arabidopsis*, it is demonstrated that the mutated CHS is indeed dominant-negative. The present results confirm the importance of the methionine residue and demonstrate the utility of the dominant-negative CHS in modulating flower colour intensity even in a distantly related species.

MATERIALS AND METHODS

Cloning and Characterization of *CHS* from *Mazus japonicus*

A fragment of a putative *CHS* gene was cloned from *Mazus japonicus*, a common garden plant belonging to the Scrophulariaceae family, using the degenerate primers (5'-TAY CAR CAR GCN TGY TTY GCN GG-3', 5'-NAG DAT NGC NGG NCC NCC-3'). The full-length cDNA was cloned using the Marathon RACE kit with specific primers (5'-GTT GTC TGC TCC GAG ATC ACT-3' for 3' RACE; 5'-AGT GAT CTC GGA GCA GAC AAC-3' for 5' RACE) and the AP2 primer provided by the manufacturer (Clontech, Palo Alto, CA, USA).

To determine if the putative *CHS* gene encodes a functional CHS, the cDNA was amplified with primers (5'-GAG ATC TAG AAA AAT GAC GCC GAC CGT CGA GGA G-3' and 5'-GAG ATC TAG ATC AAT TCA TGA AGG GCA CAC T-3'), and the amplified coding sequence cut with *Xba*I was cloned into the *Xba*I site of GUS-deleted pBI121 vector, driven by the CaMV 35S promoter. The gene cloned into the vector was introduced into *Agrobacterium* strain GV3101 and transformed into *Arabidopsis thaliana* wild-type Landsberg*erecta* (L*er*) or the *tt4* mutant. Of the several independent homozygous lines established, three lines were chosen randomly for further analysis.

To determine the ability of the putative *CHS* to complement the *tt4* mutation, the transgenic *tt4* lines were grown for 5 d on water agar plates containing 3% sucrose (0.05% MES, 0.8% phytoagar, 3% sucrose, pH 5.7).

Construction and characterization of dominant-negative *CHS*

The coding sequence of *MjCHS* was also cloned into pTOPOII vector (Invitrogen, Carlsbad, CA, USA). To mutate the cysteine at the 165th residue to alanine (Cys165Ala), the *MjCHS* was amplified in pTOPOII with Pfu polymerase (Stratagene, La Jolla, CA, USA) using the appropriate primer set (5'-TTC GCC CGC GGG ACG GTC CTC-3', 5'-AGC ACC CTG CTG GTA CAT CAT-3'). The amplified product was phosphorylated by polynucleotide kinase and ligated by T4 DNA ligase. The ligated product was transformed into *Escherichia coli*. The mutated clone was confirmed by sequencing. To mutate Met138 to either lysine (mCHSK) or glycine (mCHSG) together with the Cys165Ala mutation, the Cys165Ala-mutated CHS clone was amplified in pTOPOII with either the K-primer set (5'-CCC GGT GCC GAC TAC CAG CTC-3', 5'-CTT GTC GAC CCC GCT GGT GGT-3') or the G-primer set (5'-CCC GGT GCC GAC TAC CAG CTC-3', 5'-GCC GTC GAC CCC GCT GGT GGT-3'). The amplified products were phosphorylated by polynucleotide kinase and ligated by T4 DNA ligase. The mutated genes were sequenced to confirm the mutations. The mutated full-length *MjCHS*genes, driven by the CaMV 35S promoter, were cloned into GUS-deleted pBI121 vector, and introduced into *Arabidopsis*. The homozygous lines were renumbered after quantitating anthocyanin.

Quantitation of anthocyanin

To determine anthocyanin levels in the transgenic plants, cold-imbibed seeds were sown on MS-agar plates containing 2% sucrose (1× MS salts, 0.05% MES, 0.8% phytoagar, 2% sucrose, pH 5.7) and grown for 5 d under continuous white light (2 mW cm^{-2}). The quantitation was done as described before (Shirley *et al.*, 1995). Briefly, 50 seedlings were picked and soaked in 0.5 ml of the extraction solution (100% methanol+0.5% HCl) overnight at 4 °C. The next morning, samples were centrifuged briefly and the supernatant was used for spectrophotometric assay. Anthocyanin was quantitated by absorbance at OD_{530}. The OD_{530} values of *tt4* were subtracted by the OD_{530} value of samples as an indicator of anthocyanin content. Each experiment was run in triplicate.

Northern analysis

Northern analysis was done as described before (Shin *et al.*, 2002). Briefly, total RNA was extracted from 5-d-old seedlings grown on MS-agar plates containing 2% sucrose under continuous white light as for anthocyanin extraction. Fifteen micrograms of total RNA was loaded into each lane and transferred to a nylon membrane. The membrane was probed with ^{32}P-labelled coding sequence of *Arabidopsis CHS* or *Mazus CHS*.

Petunia transformation

Petunia (*Petunia×hybrida* cv. Blue) was transformed with vector containing the *mCHSK* gene as described by Johnson *et al.* (1999). Transformants were grown in regular potting medium until flowering. As a control, transformants having the vector alone were generated and grown side by side.

RESULTS

Cloning and functional characterization of *Mazus japonicus CHS*

A putative *CHS* was cloned from *Mazus japonicus*, a common garden plant belonging to the Scrophulariaceae (Genbank accession no. AY131328). *Mazus* bears bilaterally symmetrical white flowers with a lavender-shaded corolla tube. Phylogenetic analysis indicated that the putative CHS from*Mazus* is very similar to other known CHS enzymes, especially to snapdragon (*Antirrhinum majus*) and torenia (*Torenia hybrida*), which also belong to the Scrophulariaceae (Fig. 1).

Though it is likely that the gene encodes a real CHS, further experimental proof was required as stilbene synthase (STS) has been reported phylogenetically to group with CHS from related plants, suggesting that it has evolved from CHS (Tropf *et al.*, 1994). Known STS enzymes have high sequence similarity to CHS and use the same precursor molecules and reaction mechanism to form a common tetraketide intermediate. However, while CHS catalyses a Claisen condensation to form chalcone, STS modulates an aldol condensation to yield resveratrol (Schroder *et al.*, 1988; Lanz *et al.*, 1991; Ferrer *et al.*, 1999).

To determine if the putative *CHS* isolated from *Mazus* encodes a functional CHS, the gene was expressed both in the wild-type *Arabidopsis* ecotype Landsberg *erecta* (L*er*) and the *chs* mutant (*tt4*; Koornneef, 1990; Shirley *et al.*, 1995; Saslowski *et al.*, 2000) backgrounds, driven by the CaMV 35S promoter. Several independent homozygous lines were established and three lines of each were selected randomly for further analysis. To determine if the putative *CHS* can complement the *tt4* mutation, the seedlings were grown on water agar plates containing 3% sucrose. *tt4* has yellow cotyledons in contrast to the purple cotyledons of the wild-type plants (Koornneef, 1990; Shirley *et al.*, 1995). As seen in Fig. 2, wild-type L*er* plants and all three transgenic lines in *tt4* background expressing the putative *CHS* (tc1, 2, and 3) showed purple cotyledons while *tt4* showed yellow cotyledons. Consistently, both wild-type and transgenic seeds showed a brown coat colour unlike the yellow seed coat colour of the *tt4*mutant. The recovery of anthocyanin production in the transgenic lines indicated that the putative *CHS* from *Mazus* encodes a functional CHS. Henceforth, this gene is referred to as *MjCHS*.

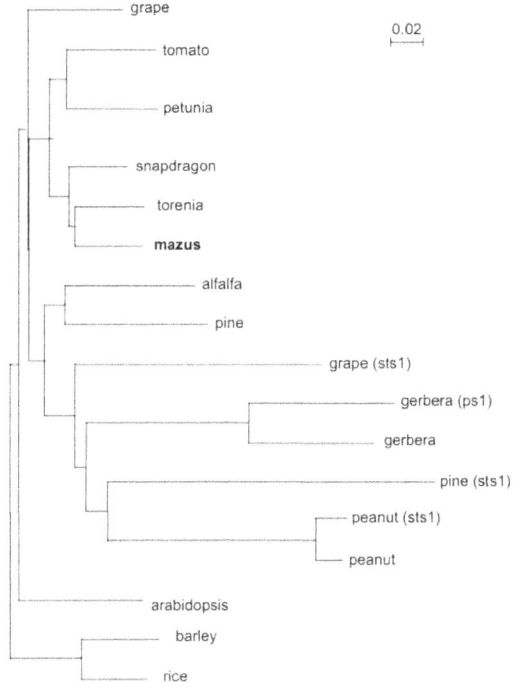

Figure1. Phylogenetic tree showing the homology between CHS from *Mazus* and other species: (ps1), pyrone synthase 1; (sts1), stilbene synthase 1.

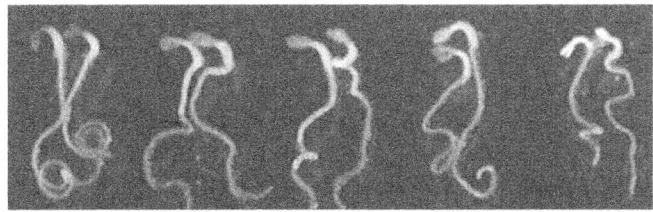

Figure 2. Functional analysis of *MjCHS*in *Arabidopsis*. *MjCHS* can rescue the yellow cotyledon colour of the CHS mutant*tt4*. tC1, 2, and 3 are independent homozygous lines of *tt4* expressing *CHS*from *Mazus*. The purple cotyledon of L*er* is shown on the left.

To test if the overexpression of *MjCHS* can increase anthocyanin in*Arabidopsis*, both wild-type and transgenic *Arabidopsis* plants were grown on

MS-agar plates containing 2% sucrose and anthocyanin content was quantitated using a spectrophotometer. Though it was difficult to distinguish visually, two out of the three randomly chosen homozygous lines expressing *MjCHS* in L*er* background (LC1, 2, and 3) accumulated more anthocyanin compared with the non-transgenic wild type (Fig. 3A). To investigate if this increase reflects the expression of *MjCHS*, northern analysis was carried out. As seen in Fig. 3B, the two lines that showed a higher amount of anthocyanin expressed *MjCHS*, while the third line that showed an amount similar to that of the wild type did not express the transgene. The results indicate that the overexpression of *MjCHS* can increase the anthocayanin levels in *Arabidopsis*. No significant phenotypic difference was observed in adult plants overexpressing the MjCHS enzyme either in *tt4* background or in L*er* background.

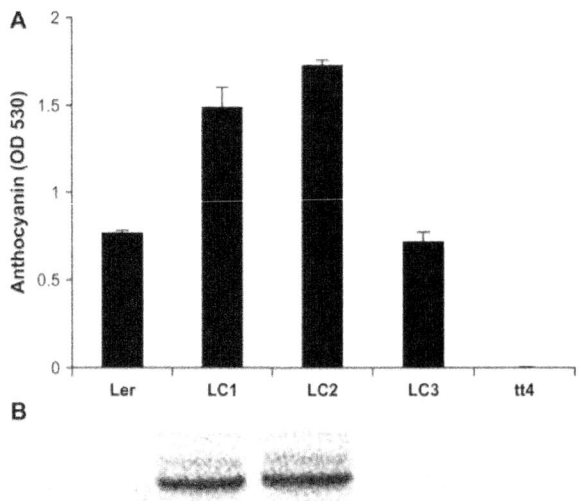

Figure 3. (A) Anthocyanin level (mean ±SD) in L*er* seedlings expressing *CHS* from *Mazus*. LC1, 2, and 3 represent independent homozygous transgenic lines. Wild-type L*er* and CHS mutant *tt4* are shown on the extremes. (B)*MjCHS* expression analysis in the overexpressor lines. Lack of expression in lane 1 indicates that *MjCHS* does not cross-hybridize with *AtCHS*; lack of expression in LC3 and*tt4* corresponds to anthocyanin levels in (A); 15 μg of total RNA extracted from wild-type L*er*, transgenic L*er* expressing *MjCHS*, and mutant *tt4* plants was loaded in each corresponding lane. The full-length coding sequence of *Mazus CHS* was used as a probe.

Development of two dominant-negative *MjCHS*

The crystal structure of alfalfa CHS suggests that a conserved methionine from a monomer is required for the function of an adjoining monomer (Ferrer *et al.*,

1999). It was hypothesized that if the protruding methionine is functionally important, its alteration to other amino acids would inhibit the function of the adjoining CHS, thus the mutation will be dominant-negative. To test this, two mutated *MjCHS* genes were generated by site-directed mutagenesis (Fig. 4A). One mutated MjCHS (mCHSG) has glycine instead of methionine at the 138th residue. Since glycine has a shorter side chain compared with that of methionine, the substrate-binding pocket of the adjoining monomer will be altered. The other mutated MjCHS (mCHSK) has lysine instead of methionine at the 138th residue. Since lysine is positively charged, the electrochemical property of the substrate-binding pocket of the adjoining monomer will be altered. To eliminate the catalytic activity of the mutated MjCHS, the catalytically important 165th cysteine was also changed to alanine.

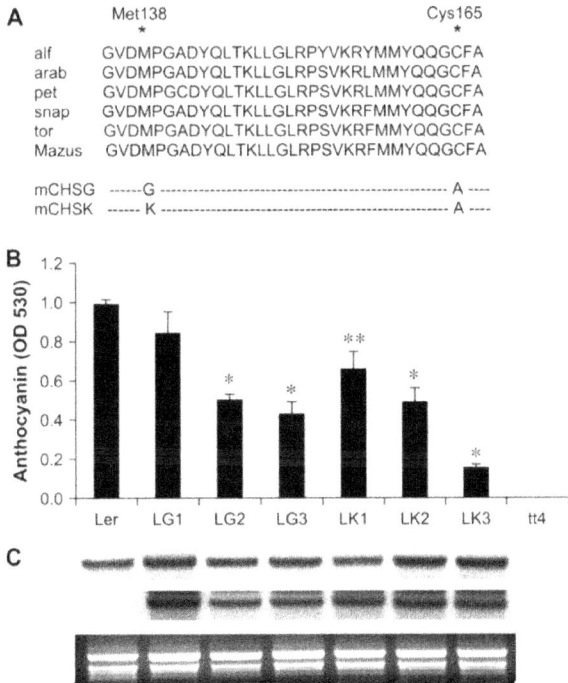

Figure 4. Analysis of the dominant-negative lines. (A) Amino acid alignment of CHS showing residues Met138, which is predicted to be important for the function of its adjoining monomer, and Cys165, shown to be the catalytic cysteine. Numbering is based on the *Mazus* sequence. mCHSG and mCHSK represent the mutations induced. alf, Alfalfa; arab, *Arabidopsis thaliana*; pet, petunia; snap, snapdragon; tor, torenia. (B) Anthocyanin accumulation (mean ±SE) in seedlings of L*er* expressing the dominant-negative *MjCHS*. *, significantly different from L*er* at 1% level; **, significantly

different from L*er* at 5% level. (C) Endogenous *CHS*(*AtCHS*; upper panel) and transgene (*MjCHS*; middle panel) expression in L*er* and the dominant-negative lines. The full-length coding sequence of *AtCHS* and *MjCHS* was used as a probe. Lower panel shows the ribosomal RNA.

To determine if the mutated MjCHS enzymes behave in a dominant-negative way, several transgenic *Arabidopsis* expressing the mutated*mCHSG* and *mCHSK* were generated, and three homozygous lines were chosen randomly for further analysis. For the analysis, both wild-type and the transgenic lines were grown on MS-agar plates containing 2% sucrose. Anthocyanin content was quantitated by a spectrophotometer (Fig. 4B). Both *mCHSG* and *mCHSK* transgenic lines showed significant reduction in anthocyanin levels. This reduction was not due to the lower expression of endogenous *Arabidopsis CHS* in the transgenic lines (Fig. 4C). No phenotypic differences were observed in plants overexpressing the mutated enzyme compared with L*er*. The degree of dominant-negativity depends on the amount of dominant-negative protein. Therefore, anthocyanin content in the transgenic plants is expected to be inversely correlated with the expression level of the mutated MjCHS enzymes. To test this hypothesis, a northern analysis of*MjCHS* mRNA was carried out. However, as shown in Fig. 4C, a direct correlation between anthocyanin content and the level of *MjCHS* mRNA was not detected. The lack of correlation could be due to the amount of the mutant proteins actually translated.

Modulation of flower colour by the dominant-negative MjCHS

To test further if the dominant-negative MjCHS can be used to modulate the flower colour intensity in a heterologous system, *Petunia*×*hybrida* cv. Blue was transformed with *mCHSK*. Several transformants were obtained and all of them showed a similar phenotype. As shown in Fig. 5, petunia transformants expressing *mCHSK* showed reduced flower colour intensity compared with wild type, indicating that the mutated MjCHS can also inhibit petunia CHS. The empty vector control was similar to wild type. Unlike the various colour patterns observed in the sense suppression lines or antisense inhibition lines, all transgenic petunia lines expressing the dominant-negative CHS showed an even decrease in flower colour intensity. The dominant-negative action of the mutated MjCHS enzymes depends on their ability to heterodimerize with the intrinsic CHS enzyme. The inhibition of anthocyanin production indicated that MjCHS can heterodimerize with both *Arabidopsis* and petunia CHS. The ability to regulate anthocyanin production in two different species suggests further utility of the dominant-negative MjCHS to modulate flower colour intensity in distantly related horticultural species.

Figure 5. Flower colour intensity in two independent transgenic petunia lines expressing a dominant-negative *MjCHS*(mCHSK). Wild type is shown on the left.

DISCUSSION

The rational design of dominant-negative CHS enzymes is reported and their utility in modulating flower colour intensity demonstrated. A *CHS*gene was cloned from *Mazus japonicus* and was functionally characterized by complementing the *Arabidopsis chs* mutant (*tt4*). To develop the dominant-negative CHS, both the cysteine, that has been shown to be important for catalysis, and the methionine, that is hypothesized to participate in the function of the adjoining monomer by extending into its cyclization pocket, were mutated. The dominant-negative action of the mutated CHS enzymes was confirmed by the decrease in anthocyanin production in transgenic *Arabidopsis* lines expressing the mutated *CHS*genes. Further, it was shown that the colour intensity of petunia flowers can be modulated by the dominant-negative CHS.

The dominant-negative CHS provides an alternative tool to modulate flower colour intensity by genetic engineering. Currently, the most widely used method is to suppress anthocyanin biosynthetic genes either by antisense overexpression or by sense suppression, which is based on the PTGS phenomenon (Tanaka *et al.*, 1998). Though the method has been very successful in manipulating flower colour intensity and inducing striking patterns, it has some limitations. First, it is necessary to clone the gene of interest from the same or closely related species. However, due to the relatively high functional conservation of CHS enzymes from different species, the dominant-negative CHS is a useful tool in a wider range of species. Therefore, the source does not necessarily have to be closely related. Secondly, it is difficult to down-regulate a gene in a tissue-specific manner using PTGS. Several studies have shown that the small-sized RNA molecules generated by PTGS can travel into different tissues and result in systemic gene silencing (Palauqui *et al.*, 1997; Voinnet and Baulcombe, 1997; Vaucheret *et al.*, 2001). Therefore, the resulting patterns and intensities cannot be rationally designed. Since a dominant-negative CHS requires simple overexpression, tissue-specific promoters can be used to achieve a rationally designed pattern. Thirdly, because the establishment of stable PTGS over generations is more difficult than simple overexpression, a dominant-negative strategy has an advantage in species that have low transformation efficiency. Additionally, though overexpression of maize *CHS* in *Arabidopsis* did not increase anthocyanin production (Dong *et al.*, 2001), the present results show that the overexpression of *MjCHS* can increase the anthocyanin levels in*Arabidopsis*. Regardless of the ability of introduced CHS to increase anthocyanin production, a dominant-negative enzyme can reduce anthocyanin content simply by dimerizing with the native enzyme.

The dominant-negative function of the mutated CHS enzymes that have either glycine or lysine instead of Met138, indicates that the methionine is functionally important. The crystal structure indicates that the methionine protrudes and is positioned at the substrate binding pocket of the adjoining monomer (Ferrer *et al.*, 1999). Each monomer in a CHS dimer can function independently and a mutation in the cysteine residue abolishes enzyme activity of the monomer (Lanz *et al.*, 1991) while not affecting the activity of the adjoining monomer (Tropf *et al.*, 1995).

Therefore, the reduced anthocyanin level in the mCHSG and mCHSK transgenic lines indicates that the mutated MjCHS enzymes behave dominant-negatively. As a corollary, the methionine which extends into the cyclization pocket of the dimerizing partner is functionally important for the activity of the adjoining monomer. The dominant-negative action of both mCHSG and mCHSK further suggests that altering the substrate binding pocket either by changing the side chain length or altering the electrochemical property can inhibit CHS activity.

The rapid accumulation of biochemical and structural information on enzymes provides an opportunity to engineer the functions of various enzymes through a rational approach (Shao and Arnold, 1996; Regan, 1999; Cedrone *et al.*, 2000). Specific amino acids can be substituted, added, or deleted by site-directed mutagenesis to confer the desired functional properties. The engineering of flavonoid biosynthetic enzymes by rational design has also been reported. CHS enzymes engineered to have different substrate specificity or to generate different condensation products have been developed by mutating amino acid residues in the substrate-binding pocket (Jez *et al.*, 2000, 2001, 2002; Lukacin *et al.*, 2001). The conversion of acridone synthase to CHS has been achieved by site-directed mutagenesis (Lukacin *et al.*, 2001). Further downstream of CHS, the substrate preference of gerbera DFR has been changed by mutating an amino acid in the putative substrate-binding region (Johnson*et al.*, 2001). The current development of dominant-negative CHS by rational design based on the available structural information and its use in modulating flower colour demonstrate the utility of this approach for the*in planta* metabolic engineering of flavonoid biosynthesis.

Acknowledgments

We thank Dr Pill-Soon Song, former laboratory members, Sean Blake, Dr In-Jeong Cho, and Dr Fakruddin Bashasab for helpful discussions. The work was partially supported by Plant Metabolism Research Center (Kyung Hee University, Korea).

REFERENCES

1.   Aharoni A, De Vos CH, Wein M, Sun Z, Greco R, Kroon A, Mol JN, O'Connell AP. The strawberry FaMYB1 transcription factor suppresses anthocyanin and flavonol accumulation in transgenic tobacco. The Plant Journal 2001;28:319-332.

2.   Aida R, Yoshida K, Kondo T, Kishimoto S, Shibata M. Copigmentation gives bluer flowers on transgenic torenia plants with the antisense dihydroflavonol-4-reductase gene. Plant Science 2000;160:49-56.

3.   Borevitz JO, Xia Y, Blount J, Dixon RA, Lamb C. Activation tagging identifies a conserved MYB regulator of phenylpropanoid biosynthesis. The Plant Cell 2000;12:2383-2394.

4.   Brouillard R, Cheminat A. Flavonoids and plant color. Progress in Clinical and Biological Research 1988;280:93-106.

5.   Bruce W, Folkerts O, Garnaat C, Crasta O, Roth B, Bowen B. Expression profiling of the maize flavonoid pathway genes controlled by estradiol-inducible transcription factors CRC and P. The Plant Cell 2000;12:65-80.

6.   Brugliera F, Tull D, Holton TA, Karan M, Treloar N, Simpson K, Skurczynska J, Mason JG. Proceedings of the Sixth International Congress of Plant Molecular Biology. 2000. Introduction of cytochrome b5 enhances the activity of flavonoid 3'5' hydroxylase in transgenic carnation. University of Laval, Quebec, S6–S8.

7.   Cedrone F, Menez A, Quemeneur E. Tailoring new enzyme functions by rational redesign. Current Opinion in Structural Biology 2000;10:405-410.

8.   Courtney-Gutterson N, Napoli C, Lemieux C, Morgan A, Firoozabady E, Robinson KE. Modification of flower color in florist's chrysanthemum: production of a white-flowering variety through molecular genetics. Biotechnology (NY) 1994;12:268-271.

9.   Dong X, Braun EL, Grotewold E. Functional conservation of plant

secondary metabolic enzymes revealed by complementation of Arabidopsis flavonoid mutants with maize genes. Plant Physiology 2001;127:46-57.

10. Ferrer JL, Jez JM, Bowman ME, Dixon RA, Noel JP. Structure of chalcone synthase and the molecular basis of plant polyketide biosynthesis. Nature Structural Biology 1999;6:775-784.

11. Forkmann G, Ruhnau B. Distinct substrate specificity of dihydroflavonol 4-reductase from flowers of Petunia hybrida. Zeitschrift für Naturforschung 1987;42:1146-1148.

12. Fujiwara H, Tanaka Y, Fukui Y, Nakao M, Ashikari T, Kusumi T. Anthocyanin 5-aromatic acyltransferase from Gentiana triflora: purification, characterization and its role in anthocyanin biosynthesis. European Journal of Biochemistry 1997;249:45-51.

13. Gerats AG, Vlaming P, Doodeman M, Al B, Schram AW. Genetic control of the conversion of dihydroflavonols and anthocyanins in flowers of Petunia hybrida. Planta 1982;155:364-368.

14. Gronquist M, Bezzerides A, Attygalle A, Meinwald J, Eisner M, Eisner T. Attractive and defensive functions of the ultraviolet pigments of a flower (Hypericum calycinum). Proceedings of the National Academy of Sciences, USA 2001;98:13745-13750.

15. Holton TA, Brugliera F, Tanaka Y. Cloning and expression of flavonol synthase from Petunia hybrida. The Plant Journal 1993;4:1003-1010.

16. Holton TA, Cornish EC. Genetics and biochemistry of anthocyanin biosynthesis. The Plant Cell 1995;7:1071-1083.

17. Hoshino A, Morita Y, Choi JD, Saito N, Toki K, Tanaka Y, Iida S. Spontaneous mutations of the flavonoid 3′-hydoxylase gene conferring reddish flowers in the three morning glory species. Plant and Cell Physiology 2003;44:990-1001.

18. Jez JM, Austin MB, Ferrer J, Bowman ME, Schroder J, Noel JP. Structural control of polyketide formation in plant-specific polyketide synthases. Chemistry and Biology 2000;7:919-930.

19. Jez JM, Bowman ME, Noel JP. Structure-guided programming of polyketide chain-length determination in chalcone synthase. Biochemistry 2001;40:14829-14838.

20. Jez JM, Bowman ME, Noel JP. Expanding the biosynthetic repertoire of plant type III polyketide synthases by altering starter molecule specificity. Proceedings of the National Academy of Sciences, USA 2002;99:5319-5324.

21. Johnson ET, Ryu S, Yi H, Shin B, Cheong H, Choi G. Alteration of a single amino acid changes the substrate specificity of dihydroflavonol 4-reductase. The Plant Journal 2001;25:325-333.

22. Johnson ET, Yi H, Shin B, Oh BJ, Cheong H, Choi G. Cymbidium hybrida dihydroflavonol 4-reductase does not efficiently reduce dihydrokaempferol to produce orange pelargonidin-type anthocyanins. The Plant Journal 1999;19:81-85.

23. Jorgensen RA, Cluster PD, English J, Que Q, Napoli CA. Chalcone synthase cosuppression phenotypes in petunia flowers: comparison of sense vs. antisense constructs and single-copy vs. complex T-DNA sequences. Plant Molecular Biology 1996;31:957-973.

24. Koornneef M. Mutations affecting the testa color in. Arabidopsis. Arabidopsis Information Service 1990;28:1-4.

25. Kroon J, Souer E, de Graaff A, Xue Y, Mol J, Koes R. Cloning and structural analysis of the anthocyanin pigmentation locus Rt of Petunia hybrida: characterization of insertion sequences in two mutant alleles. The Plant Journal 1994;5:69-80.

26. Lanz T, Tropf S, Marner FJ, Schroder J, Schroder G. The role of cysteines in polyketide synthases: site-directed mutagenesis of resveratrol and chalcone synthases, two key enzymes in different plant-specific pathways. Journal of Biological Chemistry 1991;266:9971-9976.

27. Lloyd AM, Schena M, Walbot V, Davis RW. Epidermal cell fate determination in Arabidopsis: patterns defined by a steroid-inducible regulator. Science 1994;266:436-439.

28. Lloyd AM, Walbot V, Davis RW. Arabidopsis and Nicotiana anthocyanin production activated by maize regulators R and C1. Science 1992;258:1773-1775.

29. Lukacin R, Schreiner S, Matern U. Transformation of acridone synthase to chalcone synthase. FEBS Letters 2001;508:413-417.

30. Markham KR, Ofman DJ. Lisianthus flavonoid pigments and factors influencing their expression in flower colour. Phytochemistry 1993;34:679-685.

31. Meyer P, Heidmann I, Forkmann G, Saedler H. A new petunia flower colour generated by transformation of a mutant with a maize gene. Nature 1987;330:677-678.

32. Mol J, Grotewold E, Koes R. How genes paint flowers and seeds. Trends in Plant Science 1998;3:212-217.

33. Noda K, Glover BJ, Linstead P, Martin C. Flower colour intensity depends

on specialized cell shape controlled by a Myb-related transcription factor. Nature 1994;369:661-664.

34. Palauqui JC, Elmayan T, Pollien JM, Vaucheret H. Systemic acquired silencing: transgene-specific post-transcriptional silencing is transmitted by grafting from silenced stocks to non-silenced scions. EMBO Journal 1997;16:4738-4745.

35. Regan L. Protein redesign. Current Opinion in Structural Biology 1999;9:494-499.

36. Ronchi A, Petroni K, Tonelli C. The reduced expression of endogenous duplications (REED) in the maize R gene family is mediated by DNA methylation. EMBO Journal 1995;14:5318-5328.

37. Saslowsky DE, Dana CD, Winkle-Shirley B. An allelic series for the chalcone synthase locus in Arabidopsis. Gene 2000;255:127-138.

38. Schroder J. Probing plant polyketide biosynthesis. Nature Structural Biology 1999;6:714-716.

39. Schroder G, Brown JWS, Schroder J. Molecular analysis of resveratrol synthase: cDNA, genomic clones and relationship with chalcone synthase. European Journal of Biochemistry 1988;172:161-169.

40. Shao Z, Arnold FH. Engineering new functions and altering existing functions. Current Opinion in Structural Biology 1996;6:513-518.

41. Shin B, Choi G, Yi H, Yang S, Cho I, Kim J, Lee S, Paek NC, Kim JH, Song PS. AtMYB21, a gene encoding a flower-specific transcription factor, is regulated by COP1. The Plant Journal 2002;30:23-32.

42. Shirley BW, Kubasek WL, Storz G, Bruggemann E, Koornneef M, Ausubel FM, Goodman HM. Analysis of Arabidopsis mutants deficient in flavonoid biosynthesis. The Plant Journal 1995;8:659-671.

43. Suh DY, Kagami J, Fukuma K, Sankawa U. Evidence for catalytic cysteine-histidine dyad in chalcone synthase. Biochemical and Biophysical Research Communications 2000;275:725-730.

44. Tanaka Y, Tsuda S, Kusumi T. Metabolic engineering to modify flower color. Plant and Cell Physiology 1998;39:1119-1126.

45. Tropf S, Karcher B, Schroder G, Schroder J. Reaction mechanisms of homodimeric plant polyketide synthase (stilbenes and chalcone synthase): a single active site for the condensing reaction is sufficient for synthesis of stilbenes, chalcones, and 6'-deoxychalcones. Journal of Biological Chemistry 1995;270:7922-7928.

46. Tropf S, Lanz T, Rensing SA, Schroder J, Schroder G. Evidence that stilbene synthases have developed from chalcone synthases several times

in the course of evolution. Journal of Molecular Evolution 1994;38:610-618.

47. van der Krol AR, Mur LA, Beld M, Mol JN, Stuitje AR. Flavonoid genes in petunia: addition of a limited number of gene copies may lead to a suppression of gene expression. The Plant Cell 1990;2:291-299.

48. van Houwelingen A, Souer E, Spelt K, Kloos D, Mol J, Koes R. Analysis of flower pigmentation mutants generated by random transposon mutagenesis in Petunia hybrida. The Plant Journal 1998;13:39-50.

49. Vaucheret H, Beclin C, Fagard M. Post-transcriptional gene silencing in plants. Journal of Cell Science 2001;114:3083-3091.

50. Voinnet O, Baulcombe DC. Systemic signalling in gene silencing. Nature 1997;389:553.

51. Winkel-Shirley B. Flavonoid biosynthesis: a colorful model for genetics, biochemistry, cell biology, and biotechnology. Plant Physiology 2001;126:485-493.

52. Yabuya T, Yamaguchi M, Fukui Y, Katoh K, Imayama T, Ino II. Characterization of anthocyanin p-coumaroyltransferase in flowers of Iris ensata. Plant Science 2001;160:499-503.

53. Yoshida K, Toyama Y, Kameda K, Kondo T. Contribution of each caffeoyl residue of the pigment molecule of gentiodelphin to blue color development. Phytochemistry 2000;54:85-92.

Chapter 9

TRANSCRIPTOME SEQUENCING AND METABOLITE ANALYSIS REVEALS THE ROLE OF DELPHINIDIN METABOLISM IN FLOWER COLOR IN GRAPE HYACINTH

Qian Lou[1,2,3], Yali Liu[2,3,4], Yinyan Qi[1,2,3], Shuzhen Jiao[2,3,4], Feifei Tian[2,3,4], Ling Jiang[2,3,4] and Yuejin Wang[1,2,3]

[1] College of Horticulture, Northwest A & F University, Yangling 712100, Shaanxi, PR China

[2] Key Laboratory of Biology and Genetic Improvement of Horticultural Crops (Northwest Region), Ministry of Agriculture, Yangling, Shaanxi 712100, PR China

[3] State Key Laboratory of Crop Stress Biology in Arid Areas, Northwest A&F University, Yangling 712100, Shaanxi, PR China

[4] College of Forestry, Northwest A & F University, Yangling 712100, Shaanxi, PR China

ABSTRACT

Grape hyacinth (*Muscari*) is an important ornamental bulbous plant with an extraordinary blue colour. *Muscari armeniacum*, whose flowers can be naturally white, provides an opportunity to unravel the complex metabolic networks underlying certain biochemical traits, especially colour. A blue flower cDNA library of *M. armeniacum* and a white flower library of *M. armeniacum* f. album were used for transcriptome sequencing. A total of 89 926 uni-transcripts were isolated, 143 of which could be identified as putative homologues of colour-related genes in other species. Based on a comprehensive analysis relating colour compounds to gene expression profiles, the mechanism of colour biosynthesis was studied in *M. armeniacum*. Furthermore, a new hypothesis explaining the lack of colour phenotype of the grape hyacinth flower is proposed. Alteration of the substrate competition between flavonol synthase (FLS) and dihydroflavonol 4-reductase (DFR) may lead to elimination of blue pigmentation while the multishunt from the limited flux in the cyanidin (Cy) synthesis pathway seems to be the most likely reason for the colour change in the white flowers of *M. armeniacum*. Moreover, mass sequence data obtained by the deep sequencing of *M. armeniacum* and its white variant provided a platform for future function and molecular biological research on *M. armeniacum*.

INTRODUCTION

Grape hyacinth (*Muscari*) is an important ornamental bulbous plant with a unique flower shape, extraordinary blue colour, and sweet fragrance (Qi *et al.*, 2013). These quality traits are largely determined by the metabolic composition of the flower. For example, anthocyanins are the principal flower pigments in *Muscari* flowers (Mori *et al.*, 2002). It is reported that the varying shades found in the blue flowers are attributable to delphinidin (Del), while the reddish hues are attributable to cyanidin (Cy) (Qi *et al.*, 2013).

Anthocyanins are among the most studied and best understood compounds in plant science, and their metabolic pathway has been extensively described (Grotewold, 2006; Tanaka *et al.*, 2008). Nevertheless, the mechanisms that control anthocyanin catabolism in different plant species are far from conclusive. It is reasonable to expect that such loss-of-colour adaptations are relatively unconstrained because they can be achieved in many ways (Clark and Verwoerd, 2011). The numerous diverse metabolic pathways by which plant compounds can be produced makes it more difficult to clarify this matter.

The increased ease and efficiency of RNA sequencing (RNA-Seq) tools will facilitate the study of the mechanisms underlying metabolite variation. However, it is still hard to imagine a direct correlation between the transcript abundance and the level of respective metabolite. After all, there are always too many variable factors to reach a clear conclusion. On the basis of metabolite analysis, a stringent logical filter for high-throughput approaches could be set up and used to identify the relevant factors and to circumvent the ambiguities resulting from the transcriptome comparison between different varieties. By choosing an integrative approach, where not only are transcript levels investigated, but also the metabolic products are compared, it is possible to gain an insight into metabolic flows, which would not be possible from transcript analysis alone. Thus the natural variation in blue *M. armeniacum* flower (the white form of *M. armeniacum*) provides opportunities for insight into complex metabolic networks and certain biochemical traits, especially colour.

In the present study, the first RNA-Seq project for *M. armeniacum* and its white variant was performed using the Illumina sequencing technique. Through a combination of chemical analysis with bioinformatics, the major metabolic pathways related to *Muscari* flower pigmentation were deduced and the candidate genes targeting the loss of pigmentation in the plants were examined.

MATERIALS AND METHODS

Plant Material

The little florets just before blooming of *M. armeniacum* and its white form, *M. armeniacum* f. album were collected at 08:00h on 10 April 2012 at Xi'an Botanical Garden, Shaanxi, PR China (Fig. 1A–D). All samples were immediately frozen in liquid nitrogen and stored at –80 °C for RNA extraction and flavonoid analysis.

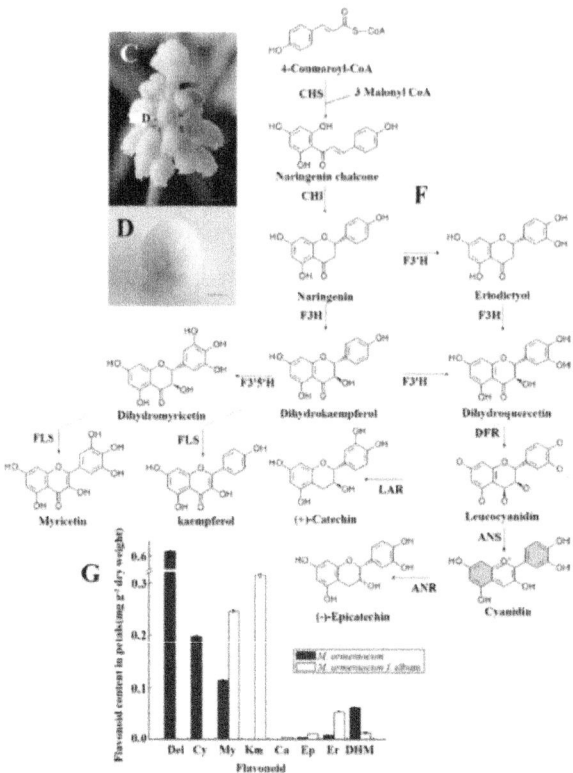

Figure 1. A diagram of the putative anthocyanin metabolic process in blue or white *M. armeniacum* flowers. (A) Mature inflorescence of *M. armeniacum*. Arrows represent small flower buds just before bloom. (B) Flower bud of *M. armeniacum* just before bloom used in deep sequencing. (C) Mature inflorescence of *M. armeniacum* f. album. (D) Flower bud of *M. armeniacum* f. album just before bloom used in deep sequencing. The scale bar=2mm in A and C, and 1mm in B and D. (E) The putative anthocyanin metabolic process in blue *M. armeniacum* flowers. (F) The putative anthocyanin metabolic process in white *M. armeniacum* flowers. (G) Flavonoid composition obtained by HPLC from blue and white flowers of *M. armeniacum*. ANR, anthocyanidin reductase; ANS, anthocyanidin synthase; Ca, catechin; CHI, chalcone isomerase; CHS, chalcone synthase; Cy, cyanidin; Del, delphinidin; DFR, dihydroflavonol 4-reductase; DHM, dihydromyricetin; Ep, epicatechin; Er, eriodictyol; F3H, flavanone 3-hydroxylase; F3′H, flavonoid 3′-hydroxylase; F3′5′H, flavonoid 3′5′-hydroxylase; FLS, flavonol synthase; Km, kaempferol; LAR, leucoanthocyanidin reductase; UFGT, anthocyanidin 3-*O*-glucosyltransferase. (This figure is available in colour at *JXB* online.)

Measurement of flower flavonoids

The anthocyanins were determined using high-performance liquid chromatography (HPLC) as previously described (Qi *et al.*, 2013). For extraction of other flavonoids, freeze-dried flowers were finely ground and 50mg was extracted in 500 μl of MeOH for 48h at 4 °C in darkness. After samples were centrifuged, the supernatants were transferred to fresh tubes and the pellet was resuspended and incubated in 500 μl of 1% MeOH at 4 °C for 24h, and then the supernatant was combined for further HPLC analysis. HPLC was performed as previously described (Qi *et al.*, 2013). Cyanidin, cyanidin-galactoside, dihydroquercetin, dihydrokaempferol, (+)-catechin, (−)-epicatechin, luteolin, naringenin, and quercetin were obtained from Sigma-Aldrich China (Shanghai). Standards of afzelechin, (−)-epiafzelechin, (+)-gallocatechin, and (−)-epigallocatechin were purchased from BioBioPha (Yunnan, China). The delphinidin chloride (ChromaDex, Santa Ana, CA, USA), petunidin chloride (ChromaDex), and other flavonoids such as dihydromyricetin (YiFang S&T, Tianjin, China) equivalents were used as standards for quantification. Mean values and SDs were obtained from three biological replicates.

RNA extraction, library construction, and RNA-Seq

Total RNA of each sample was isolated using a Quick RNA isolation kit (Bioteke Corporation, Beijing, China) and then characterized on a 1% agarose gel and examined with a NanoDrop ND1000 spectrophotometer (NanoDrop Technologies, Wilmington, DE, USA). The RIN (RNA integrity number) values (>8.0) of these samples were assessed using an Agilent 2100 Bioanalyzer (Santa Clara, CA, USA). The construction of the libraries and the RNA-Seq were performed by the Biomarker Biotechnology Corporation (Beijing, China). mRNA was enriched and purified with oligo(dT)-rich magnetic beads and then broken into short fragments. Taking these cleaved mRNA fragments as templates, first- and second-strand cDNA were synthesized. The resulting cDNAs were then subjected to end-repair and phosphorylation using T4 DNA polymerase and Klenow DNA polymerase. After that, an 'A' base was inserted as an overhang at the 3′ ends of the repaired cDNA fragments and Illumina paired-end solexa adaptors were subsequently ligated to these cDNA fragments to distinguish the different sequencing samples. To select a size range of templates for downstream enrichment, the products of the ligation reaction were purified and selected on a 2% agarose gel. Next, PCR amplification was performed to enrich the purified cDNA template. Finally, the four libraries were sequenced using an Illumina HiSeq™ 2000.

De novo transcriptome assembly and annotation

After removing those reads with only adaptor and unknown nucleotides >5%, or those that were of low quality, the clean reads were filtered from the raw reads. The clean reads were then assembled *de novo* using the Trinity platform (http://trinityrnaseq.sourceforge.net/) with the parameters of 'K-mer=25, group pairs distance=300' (Grabherr *et al.*, 2011). For each library, short reads were first assembled into longer contigs based on their overlap regions. Then different contigs from another transcript and their distance were further recognized by mapping clean reads back to the corresponding contigs based on their paired-end information, and thus the sequence of the transcripts was produced. Finally, the potential transcript sequences were clustered using the TGI Clustering tool to obtain uni-transcripts (Pertea *et al.*, 2003). Uni-transcripts were aligned to a series of protein databases using BLASTx (E-value $\leq 10^{-5}$), including the NCBI non-redundant (Nr), the Swiss-Prot, the Trembl, the Kyoto Encyclopedia of Genes and Genomes (KEGG) (http://www.genome.jp/kegg/kegg2.html), and gene ontology (http://wego.genomics.org.cn/cgi-bin/wego/index.pl) databases. To determine the gene coverage, the reference sequences for all three colour-related pathways were downloaded from the public databases (Supplementary Fig. S1, Supplementary Table S1 available at *JXB* online). All isoforms of all colour-related genes present in the databases examined were aligned against corresponding reference sequences using BLASTx. The deduced amino acid sequences of uni-transcripts were required to be longer than 70% of the corresponding sequences. If a uni-transcript met the criteria, it was assumed to contain a near full-length contig. If not, targeted assembly was performed to obtain even greater coverage of the respective genes. All reads in the databases examined were mapped to the reference sequences and the mapped reads were then assembled using clustering and CAP3 assembly (http://compbio.dfci.harvard.edu/tgi/software/).

Expression Annotation

To evaluate the depth of coverage, all usable reads were realigned to each uni-transcript using SOAPaligner (http://soap.genomics.org.cn/soapaligner.html), then normalized into RPKM values (reads per kb per million reads; Mortazavi *et al.*, 2008). After that, uni-transcript abundance differences between the samples were calculated based on the ratio of the RPKM values, and the false discovery rate (FDR) control method was used to identify the threshold of the *P*-value in multiple tests in order to compute the significance of the differences in transcript abundance (Benjamini and Yekutieli, 2003). Here, only uni-transcripts with an absolute value of log2 ratio ≥ 2 and an FDR significance score <0.001 were used for subsequent analysis.

Gene Validation and Expression Analysis

All the colour-related uni-transcripts were subjected to real-time quantitative PCR (q-PCR) with specific primers identified by Primer Premier software (Supplementary Table S1 at *JXB* online). cDNA synthesis and q-PCR were performed as described previously (Qi *et al.*, 2013). SYBR Green was used for detection of PCR products on a MyiQ Single-Color Real-Time Detection System (Bio-Rad). The actin gene was used as the internal control for normalization of gene expression. At least two independent biological replicates and three technical replicates of each biological replicate for each sample were analysed by q-PCR to ensure reproducibility and reliability. The correlation between expression profiles of colour-related genes measured by q-PCR and RNA-Seq was determined using the R package.

RESULTS AND DISCUSSION

Major classes of colour compounds in *M. armeniacum* flowers

To examine the biochemical basis of the lack of colour phenotype of grape hyacinth, the metabolomic profiles of petals was compared, with a focus on the compounds related to colour pigmentation. As expected, blue *M. armeniacum* flowers contain two anthocyanin compounds responsible for colour pigmentation: Del and Cy. In contrast, no colour anthocyanins and no derivatives were detected in the white flowers of *M. armeniacum* f. album (Fig. 1G). Furthermore, to determine why some steps in the ABP (anthocyanin biosynthetic pathway) are blocked in white flowers, the intermediate products involved in the metabolic process and its main branches were compared. Figure 1 shows a diagram of the anthocyanin metabolic process with its core metabolites and enzymes in blue or white*M. armeniacum* flowers. Although anthocyanins were absent, petal extracts of white flowers contained all the other core metabolites involved in the process that had been detected in blue extracts (Fig. 1G). The presence of myricetin and catechin in white petals indicated that the ABP must be blocked fairly far downstream, in one of the late-acting genes such as dihydroflavonol 4-reductase (*DFR*), anthocyanidin synthase (*ANS*), or anthocyanidin 3-*O*-glucosyltransferase (*UFGT*). It is worth noting that epicatechin was detected in white flowers at a concentration three times higher than that in blue flowers (Fig. 1G). This suggests that the colour pigment Cy might be present in the white flowers, but at a level so low as to be barely detectable. Another possible explanation is that Cy exists in white grape hyacinth but only for a very short time. As soon as it formed, the unstable Cy would be converted to colourless epicatechin, which would permanently prevent Cy from changing to stable colour pigments by later glycosylation and other reactions. The compositions of common co-pigment flavonoids, such as flavones, flavonols, flavanones, caffeoyl quinic acid, and coumalic acid, were also examined to obtain a general overview of colour metabolism (Supplementary Table S2 at *JXB* online). White flowers have more typres and higher levels of the flavonoid compounds than do blue flowers, with some exceptions. Not surprisingly, the upstream flux must flow into other branches of the flavonoid metabolic route when the ABP is restrained in white flowers.

RNA-Seq and assembly

To understand the molecular basis of flower colour polymorphism in grape hyacinth, blue flowers of *M. armeniacum* and white flowers of *M. armeniacum* f. album were used to build two libraries for high-throughput sequencing (Fig. 1B, D). The two libraries (Ma1 and Ma2) produced 2031 Mbyte and 2772 Mbyte of raw data (NCBI accessions: SRR998575 and SRR998853), respectively, from paired-end reads with a single read length of ~101bp and Q20 percentages (percentage of sequences with sequencing error rates <1%) and GC percentages of 99.45% and 50.03%, 99% and 53.12%, respectively. These data showed that the throughput and sequencing quality were high enough to warrant further analysis.

Short reads from the two libraries were assembled into 1 634 539 and 1 416 136 contigs with mean lengths of 81bp and 82bp, respectively. These were assembled into scaffolds and uni-transcripts, taking the distance of paired-end reads into account (Supplementary Table S3 at *JXB* online). All sequences were assembled to give 89 926 non-redundant uni-transcripts with a mean length of 633bp.

Genes related to blue colour development

Genes involved in three secondary metabolic pathways (flavonoid biosynthesis, anthocyanin biosynthesis, and flavone and flavonol biosynthesis pathways) that are related to flower pigmentation were analysed using *M. armeniacum* uni-transcripts. They were searched based on standard gene names and synonyms in the combined functional annotations (Table 1). By mapping to the KEGG reference pathways, a total of 143 uni-transcripts were assigned to the three pathways (Supplementary Table S1 at *JXB* online). The data set includes annotated sequences for >88% of genes in the flavonoid biosynthesis pathway (Supplementary Fig. S1). However, only a small percentage of genes in the other two pathways was found (Supplementary Fig. S1). Possible reasons for this might be the metabolite diversification in different species. In support of this, no sequences for methoxylation genes involved in anthocyanin modification were assembled, which was consistent with the absence of methylated anthocyanin in the flowers of *M. armeniacum* (such as petunidin and malvidin; Supplementary Table S2). Therefore, it is reasonable to conclude that the ABP in grape hyacinth is unlike the pathways used in many other blue flowers in that it relies mainly on glycosylation and hydroxylation rather than methoxylation to maintain the stability of its blue pigments (Yoshida *et al.*, 2009). Moreover, an average of 71% of the full-length sequences for each of the ABP genes were obtained (Supplementary Table S1). These genes were thus the focus of further study.

View this table:

Table 1. Candidate genes related to flower pigmentation of *M. armeniacum*

Function	Gene	Enzyme	KO id (EC no.)	No. All[a]	No. Up[b]	No. Down[c]
Anthocyanin biosynthesis	CHS	Chalcone synthase	K00660 (2.3.1.74)	17	3	4
	CHI	Chalcone isomerase	K01859 (5.5.1.6)	3	0	1
	F3H	Flavanone 3-hydroxylase	K00475 (1.14.11.9)	3	0	0
	F3'H	Flavonoid 3'-hydroxylase	K05280 (1.14.13.21)	7	0	1
	F3'5'H	Flavonoid 3',5'-hydroxylase	K13083 (1.14.13.88)	4	1	1
	DFR	Dihydroflavonol 4-reductase	K13082 (1.1.1.219)	18	3	1
	ANS	Anthocyanidin synthase	K05277 (1.14.11.19)	4	0	1
	UFGT	Anthocyanidin 3-O-glucosyltransferase	K12930 (2.4.1.115)	25	0	9
Anthocyanin modification	UGT75C1	Anthocyanin 5-O-glucosyltransferase	K12338 (2.4.1.298)	1	0	0
	5AT	Anthocyanin 5-aromatic acyltransferase	K12936 (2.3.1.153)	3	0	1
	GT1	Anthocyanin 5,3-O-glucosyltransferase	K12938 (2.4.1.-)	12	1	0
	3'GT	UDP-glucose anthocyanin 3'-O-beta-glucosyltransferase	K12939 (2.4.1.238)	4	0	0
	5MaT1	Anthocyanin 5-O-glucoside-6'''-O-malonyltransferase	K12934 (2.3.1.172)	2	0	0
Flavone and flavonol biosynthesis	FNS	Flavone synthase	K13077 (1.14.11.22)	5	2	0
	FLS	Flavonol synthase	K05278 (1.14.11.23)	10	0	2
	C12RT1	Flavanone 7-O-glucoside 2''-O-beta-L-rhamnosyltransferase	K13080 (2.4.1.236)	5	0	0
	FOMT	Flavonol 3-O-methyltransferase	K05279 (2.1.1.76)	6	0	0
	CROMT2	Myricetin O-methyltransferase	K13272 (2.1.1.149)	3	0	1
	LjOMT	Luteolin O-methyltransferase	¥[d] (2.1.1.75)	1	0	0
	F4ST	Flavonol 4'-sulphotransferase	K13271 (2.8.2.27)	2	0	0
	GUSB	beta-Glucuronidase	K01195 (3.2.1.31)	2	0	0
	UF3GT	Flavonol 3-O-glucosyltransferase	K10757 (2.4.1.91)	5	0	0
Flavanone biosynthesis	ANR	Anthocyanidin reductase	K08695 (1.3.1.77)	1	0	0

[a] No. All, the total number of uni-transcripts analysed.
[b] No. Up, the number of uni-transcripts with expression significantly up-regulated in blue flowers of *M. armeniacum* compared with in white flowers.
[c] No. Down, the number of uni-transcripts with expression significantly down-regulated in blue flowers of *M. armeniacum* compared with in white flowers.
[d] ¥, omission of numbers for the KO id.

Comparison of transcriptional profiles of genes involved in anthocyanin metabolism between *M. armeniacum* and *M. armeniacum* f. album

Previous research has demonstrated that the colour difference between white and blue flowers of *M. armeniacum* is due to the loss of flower anthocyanins (Del and Cy). The shift from blue to white requires a complete blockage of the ABP, which probably occurs in some reaction before Del and Cy are formed. Therefore, the abundance of the ABP candidate genes was compared in *M. armeniacum* and *M. armeniacum* f. album transcriptomes to find the key transcripts of blue colour metabolism. Core genes in the pathway were studied in detail, and the results demonstrated that most of the uni-transcripts with significant changes in expression level, regardless of whether they were early [chalcone isomerase (*CHI*), etc] or late genes (*ANS*, *UFGT*, etc.), showed higher transcript abundance in white flowers than in blue flowers (Fig. 2A,B). Interestingly, this result is in sharp contrast to the results of some other studies. In many cases, changes in anthocyanin accumulation have corresponded to changes in expression of genes encoding pathway enzymes (Castellarin and Gaspero, 2007; Wang *et al.*, 2010; Feng *et al.*, 2012; Yuan *et al.*, 2013). To elucidate this matter, the metabolomic profiles of blue petals were compared with those of white petals. A large quantity of flavonoid compounds was detected in white petal extracts, many of them sharing the same intermediates or enzymes with anthocyanin. For example, the contents of myricetin and kaempferol are two and three times greater, respectively, in white petals (Fig. 1G). Common enzymatic steps shared by the biosynthesis of these compounds and anthocyanins are catalysed by chalcone synthase (CHS), flavanone 3′-hydroxylase (F3′H), flavonoid 3′5′-hydroxylase (F3′5′H), etc. (Fig. 2A). This could be the reason why anthocyanin content was not correlated with the expression of anthocyanin biosynthetic genes in grape hyacinth.

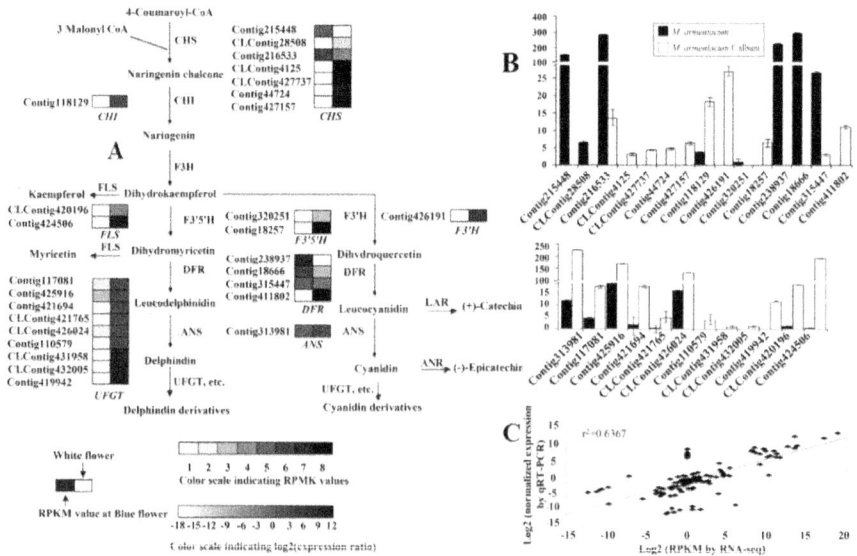

Figure 2. Schematic of physiological and metabolic data related to flower colour development of *M. armeniacum*. (A) A detailed part of the Del and Cy metabolic subnetwork showing the subset of nodes or metabolites that constitute the process. Enzyme

names and expression patterns are indicated at the side of each step. The expression pattern of each uni-transcript is shown on two grids, with the left one representing the RPKM value of blue flowers, and the right one representing the relative log2 (expression ratio) of white flowers. The grids with eight different grey scale levels show the absolute expression magnitude of blue flowers, with the RPKM values 0–10, 10–20, 20–40, 40–80, 80–160, 160–320, 320–640, and 640–1280 represented by grey scale levels 1–8, respectively. (B) Transcript accumulation measurements of colour-related genes involved in the anthocyanin metabolic process. (C) Correlation of gene expression results obtained from q-PCR analysis and RNA-Seq for colour-related genes in blue and white flowers. ANR, anthocyanidin reductase; ANS, anthocyanidin synthase; CHI, chalcone isomerase; CHS, chalcone synthase; DFR, dihydroflavonol 4-reductase; F3H, flavanone 3-hydroxylase; F3′H, flavonoid 3′-hydroxylase; F3′5′H, flavonoid 3′5′-hydroxylase; FLS, flavonol synthase; LAR, leucoanthocyanidin reductase; UFGT, anthocyanidin 3-*O*-glucosyltransferase. (This figure is available in colour at *JXB* online.)

Candidates which are responsible for the loss of blue color in grape hyacinth with white flowers

Even though most Del- and Cy-related reactions may share the same enzymes, not enough is known about how and when they catalyse the corresponding reactions. Accordingly, each event was treated independently. Of all uni-transcripts involved in the Del biosynthesis process, only three *CHS*, three *DFR*, and one *F3′5′H* homologous sequences showed significantly up-regulated expression in blue flowers; these are thought to be the flux-limiting genes leading to Del elimination in white grape hyacinth. It is generally known that CHS catalyses the first reaction for anthocyanin biosynthesis and helps to form the intermediate chalcone, the primary precursor for all classes of flavonoids (Koes *et al.*, 1989). So if CHS reactions are strongly constrained, not only anthocyanin production but also that of nearly all other flavonoids is effectively eliminated (Clark *et al.*, 2011). On the other hand, F3′5′H plays critical roles in the flavonoid biosynthetic pathway, and catalyses the hydroxylation of the B-ring of flavonoids and is necessary to biosynthesize Del (violet to blue)-based anthocyanins (Tanaka and Brugliera, 2013). It was expected that, in the event that the minimal Del path was cut off from F3′5′H, myricetin-related flavonols would be removed along with Del. In fact, however, a great deal of myricetin was found in white flowers, more than twice as much as in blue flowers (Fig. 1G). Yet this is not a satisfactory explanation for the lack of Del in white grape hyacinth. Hence *DFR*, a crucial later gene for anthocyanin formation, was considered. As shown in Fig. 1, DFR reduces dihydroflavonols to colouress leucoanthocyanidins, which are catalysed by ANS to coloured anthocyanidins. No products of the Del synthesis route that occur after dihydromyricetin (the substrate for the DFR enzyme) were detected in white flowers (Fig. 3), suggesting that *DFR* was the most likely target for Del suppression in *M. armeniacum* f. album. It is noteworthy that the transcripts of three *DFR*-like sequences showed significantly higher levels of gene transcripts in blue flowers than in white flowers, in some cases >1000 times higher (Fig. 2A). Although this was unexpected, it is a reasonable explanation for the fact that the Del synthesis reactions are constrained to zero. Additionally, the dihydroflavonols represent a branch point in flavonoid biosynthesis, being the intermediates in the production of both the coloured anthocyanins, through DFR, and the colourless flavonols, through flavonol synthase (FLS) (Davies *et al.*, 2003). As a result of the competition for substrate (dihydroflavonols), the up-regulation of FLS and flavonols might be closely accompanied by a decrease in DFR and anthocyanin accumulation. In support of this, inhibition of FLS production through the introduction of an FLS antisense RNA construct led to anthocyanin production and gave the white-flowered petunia a novel pink hue (Davies *et al.*, 2003). In the present study, the abundance of myricetin (a downstream flavonol product of dihydromyricetin) and that of two FLS-like sequences were far greater in white grape hyacinth than in the blue-flowered strain, confirming the hypothesis by another approach. Combining the information with data from HPLC, it could be inferred that*DFR* might be the target gene for the loss of blue pigmentation (Del) in white grape hyacinth. In addition, strong competition between FLS and DFR for common dihydromyricetin substrates might partially block the synthesis of Del and cause the production of other flavonoid compounds such as myricetin, thereby furthering the process of elimination of blue pigmentation and shifting the flavonol:anthocyanin ratio in *M. armeniacum*.

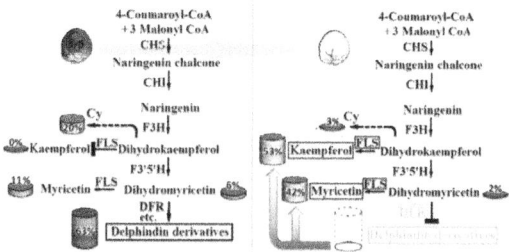

Figure 3. A model for the process of Del elimination in the white flowers of *M. armeniacum*. When *DFR* is suppressed, the substrates used for Del synthesis are then available for synthesis of myricetin and kaempferol. Moreover, an increase of flavonol production occurs through the up-regulation of *FLS*, furthering the process of blue pigmentation elimination in the white flowers of *M. armeniacum*. The global output from the minimal anthocyanin subnetwork in flowers of*M. armeniacum* was considered to

be 100% and was used to define the relative level of each product. The black boxes indicate the genes or the compounds which had a higher relative abundance in white flowers of *M. armeniacum* than that in blue flowers. The grey boxes indicate the genes or the compounds which had a lower abundance in white flowers than that in blue flowers. CHI, chalcone isomerase; CHS, chalcone synthase; Cy, the global output from the minimal cyanidin subnetwork; DFR, dihydroflavonol 4-reductase; F3H, flavanone 3-hydroxylase; F3'5'H, flavonoid 3'5'-hydroxylase; FLS, flavonol synthase. (This figure is available in colour at *JXB* online.)

Reasons for loss of red Cy Accumulation in white-Flowered Grape Hyacinth

To select the target genes for Cy suppression in grape hyacinth, the expression and metabolomic profiles of blue and white petals were compared in whole Cy metabolic reactions. The presence of catechin and epicatechin in white petals indicated that the red Cy must be present in the white flowers, even if only for a very short time or in a very small quantity, hinting at a complex metabolic mechanism underlying the loss of Cy pigmentation. There may be multiple reasons for this phenomenon. First, *DFR* and *FLS* were good candidates for the limitation of Cy accumulation, as discussed earlier. When *FLS* is up-regulated, the substrates used for Cy synthesis are then available for synthesis of kaempferol in white flowers (Fig. 4). The down-regulation of *DFR* could decreased Cy production, but obviously it cannot produce a complete blockage of the process on its own. Secondly, the metabolism of Del plays a particularly important role in the flower coloration system of *M. armeniacum*, whereas the metabolism of Cy is less significant (Fig. 4). In blue flowers, the total content of Del (blue) was three times higher than that of Cy (red), which might also explain why blue is the predominant colour hue in *M. armeniacum* flowers. Even in white flowers, the 44% yield from the Del metabolic pathway was much higher than the 3% yield from the Cy metabolic pathway (Fig. 4). The low level of productive forces might limit the flux through Cy metabolism in grape hyacinth and explain the small amounts of Cy that accumulate in the white flowers. Thirdly, as is known, the last product before Cy formation is leucyanidin, which can generate two different products, colourless catechin and red Cy, in reactions catalysed by leucoanthocyanidin reductase (LAR) and ANS, respectively. In *M. armeniacum*, catechin was detected only in white flowers and not in blue flowers (Fig. 4). Therefore, it could be concluded that the alteration in competition from LAR for the substrate might redirect Cy biosynthesis towards catechin and further restrict the flux through its subsequent biosynthesis process. Fourthly, the next step after Cy formation should convert unstable anthocyanin to stable coloured compounds, but the white flowers contain increased concentrations of epicatechin and undetectable levels of Cy (Fig. 4). It is suggested that the low amounts of Cy might be reduced to colourless epicatechin by anthocyanidin reductase (ANR) and thus redirect anthocyanin biosynthesis away from the production of stable Cy-based pigments. Above all, the limitation of flux in upstream reactions and the multishunt process in downstream reactions led to the process of elimination of red pigmentation in the white flowers of *M. armeniacum*.

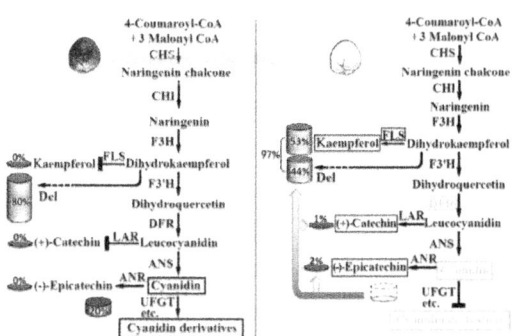

Figure 4. A model for Cy elimination in white flowers of *M. armeniacum*. The fluxes through Cy metabolism were limited. The multishunt process in downstream reactions further promoted Cy turnover and degradation in white flowered grape hyacinth. The global output from the minimal anthocyanin subnetwork in flowers of *M. armeniacum* was considered as 100% and was used to define the relative level of each product. The black boxes indicate the genes or the compounds which had a higher relative abundance in white flowers of *M. armeniacum* than that in blue flowers. The grey boxes indicate the genes or the compounds which had a lower abundance in white flowers than that in blue flowers. ANR, anthocyanidin reductase; ANS, anthocyanidin synthase; CHI, chalcone isomerase; CHS, chalcone synthase; Del, the global output from the minimal delphinidin subnetwork; DFR, dihydroflavonol 4-reductase; F3H, flavanone 3-hydroxylase; F3′H, flavonoid 3′-hydroxylase; FLS, flavonol synthase; LAR, leucoanthocyanidin reductase; UFGT, anthocyanidin 3-*O*-glucosyltransferase. (This figure is available in colour at *JXB*online.)

Recently, Clark *et al.* (2011) considered the advantages of targeting *DFR* in order to eliminate floral pigmentation: the production of only a few compounds is affected; it does not operate too late in the ABP pathway; it is more essential for anthocyanin production than are other earlier genes, etc. It seems to be a very attractive means, for both plants and breeders, by which to change flower colour from blue to white by the down-regulation of a single *DFR*. Nevertheless, it seems that such loss-of-colour adaptations are relatively unconstrained in different species because they can be achieved in many ways. For example, the mutation of a single *CHS*enzyme is often observed. It leads to white flower lines in the petunia (Saito *et al.*, 2006; Spitzer *et al.*,

2007), violet (Hemleben *et al.*, 2004), and arctic mustard flower (Dick *et al.*, 2011). Blocking an early-acting gene such as *CHS* could be more efficient. Perhaps this is why *CHS* mutation is the most common means of producing loss of colour in the literature (Clark and Verwoerd, 2011). Another common reason for pigmentation loss is the absence of more than one enzyme in the ABP, such as ANS and DFR (Ma *et al.*, 2004; Bogs *et al.*, 2007; Clark and Verwoerd, 2011). Recent research has described many new ways to determine the lack of colour phenotype by regulating the branching point of anthocyanin biosynthesis. For instance, inhibition of ANR and consequent LAR production by the transient suppression of the *FcMYB1* gene in white strawberry fruit leads to increased concentrations of anthocyanins and undetectable levels of flavan-3-ols (Salvatierra *et al.*, 2013). Similarly, introduction of apple *ANR* genes into tobacco inhibits expression of both *CHI* and *DFR* genes in flowers, finally leading to loss of anthocyanin (Han *et al.*, 2012). Here, a new hypothesis is proposed explaining a lack of colour phenotype of grape hyacinth flowers. The truth of the matter is probably more complex than what has been described here, the elucidation of which could be an interesting and challenging subject.

SUPPLEMENTARY DATA

Supplementary data are available at *JXB* online.

Figure S1. KEGG reference mappings for flavonoid synthesis, anthocyanin biosynthesis, and flavone and flavonol biosynthesis pathways.

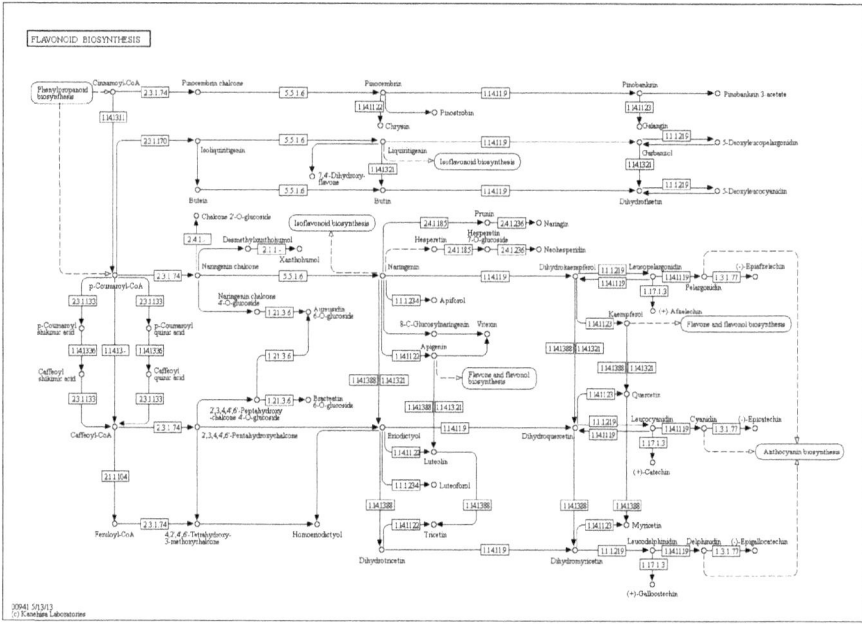

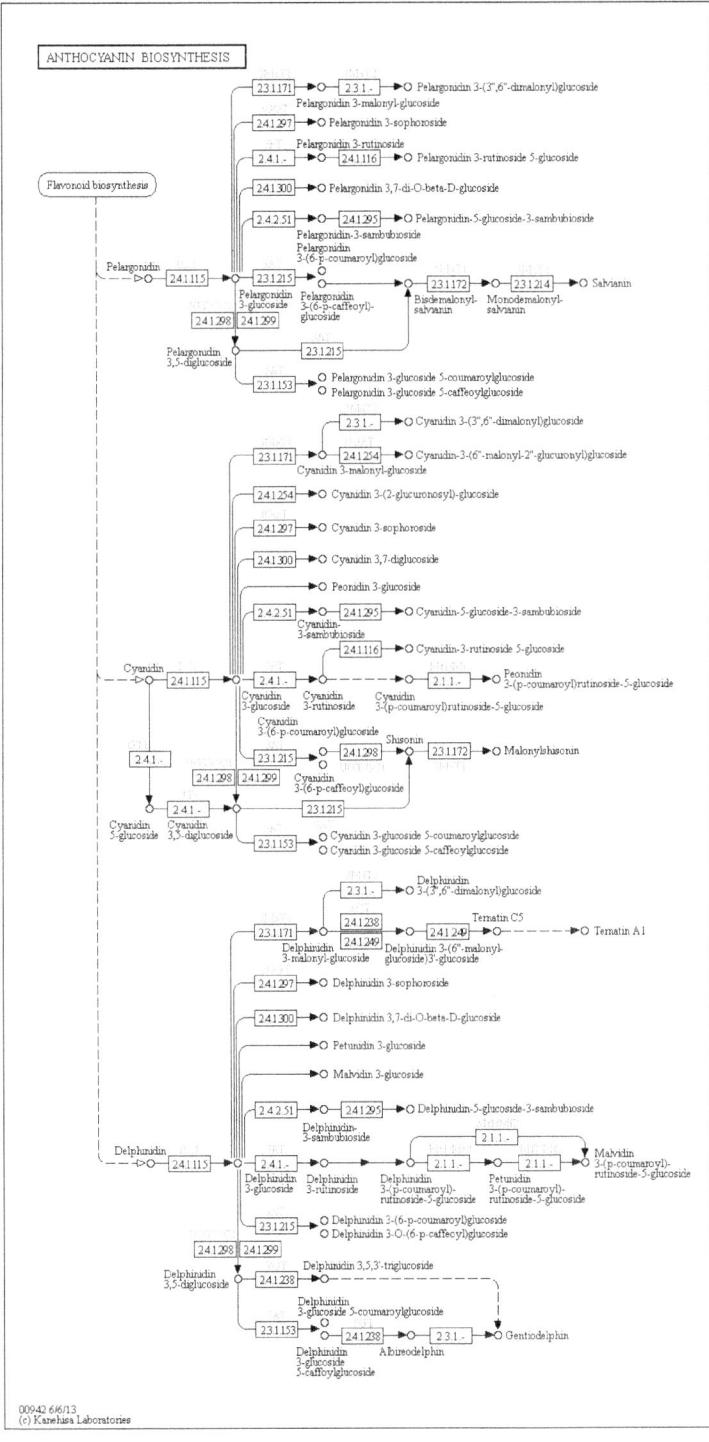

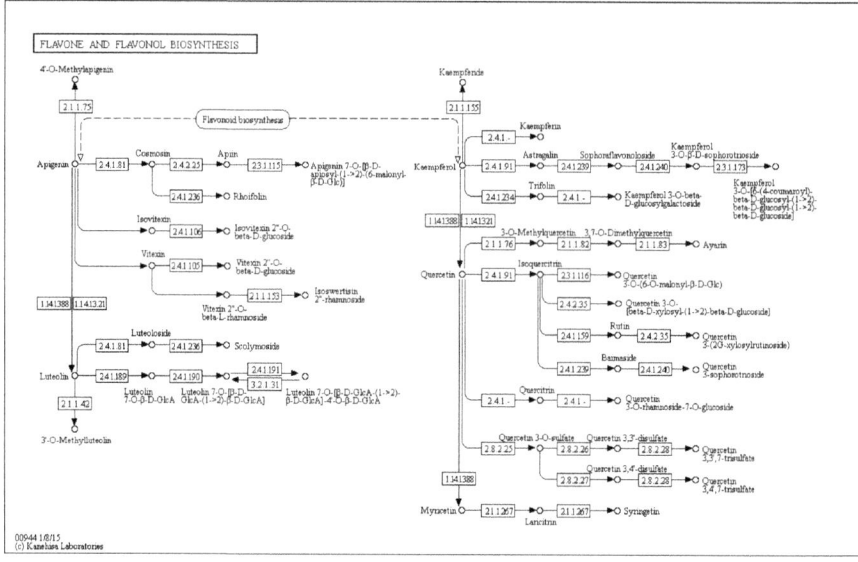

Table S1. List of relative uni-transcripts in the three secondary metabolic pathways in the *M. armeniacum* transcriptome.

Uni-tran-script id	Length	Sequence integrity	Reference annotation	Reference sequence id	Forward primer sequence (5'-3')	Reverse primer sequence (5'-3')
Con-tig47274	1067	90%	Chalcone synthase 2 OS=Daucus carota GN=CHS2 PE=2 SV=1	sp|Q9ZS40|CHS2_DAUCA	ACGAGGGAGGTGAT-GTGGGACGG	ATTCAGGGGCAGCG-GAGGACGAT
Con-tig46819	504	35%	putative UDP-rhamnose:rhamnosyltransferase [Fragaria x ananassa]	gi|51705431|gb|AAU09445.1|	CACCAGTTCGATTTTC-CAGCATG	CGAG-GAGGGAGGGGAGA-GTTACC
Con-tig44724	602	50%	Chalcone synthase C2 OS=Zea mays GN=C2 PE=3 SV=1	sp|P24825|CHS2_MAIZE	GTCACCGCCGT-CACCTTCCG	GCCCGCTC-CACCCCTCCTC
Con-tig440305	494	46%	Putative UDP-glucose:flavonoid 3-O-glucosyl transferase (Fragment) OS=Punica granatum PE=2 SV=1	tr|D3K523|D3K523_PUNGR	CACAAACAATATCCAG-CAGGCTC	AGGGGGAG-GAGGGGAAGACGA
Con-tig439801	464	34%	Anthocyanidin 3-O-glucosyl-transferase OS=Iris hollandica GN=h3GT PE=2 SV=1	tr|Q5KTF3|Q5KTF3_IRIHO	ATGATATGCTGGCCAT-GTTCC	CTCACCGTTCAT-CAGTTCTTTTATC
Con-tig43854	606	55%	Caffeic acid O-methyltransferase OS=Iris hollandica GN=IhCOMT PE=2 SV=1	tr|Q5KSL8|Q5KSL8_IRIHO	CAGGAAGCACCCTCA-CATCAAG	GGCGTCTCCACTAG-GAACACT
Con-tig43831	251	17%	UDP-glucosyl transferase 73B2 OS=Arabidopsis thaliana GN=UGT73B2 PE=1 SV=1	sp|Q94C57|U73B2_ARATH	GGGGAGAGTT-GAAGAGAAGGG	GGAAGATGCAG-GAAAGCTGGG
Con-tig437845	226	19%	Chalcone synthase-like protein OS=Arabidopsis thaliana GN=T11I11.90 PE=2 SV=1	tr|Q9SW49|Q9SW49_ARATH	CTGATCGAGGACAAC-GTGGA	CAACCCTGTTCAG-TATCGCC
Con-tig437584	279	30%	Flavonol synthase/flavanone 3-hydroxylase, putative [Ricinus communis]	gi|255549726|ref|XP_002515914.1|	GAGGTGGCGGGTCT-CACGTTCAC	TCCTTCACCCCACG-CACCCTAAC

Contig	Length	%	Description	Accession	Sequence 1	Sequence 2
Con-tig43697	1753	100%	Flavonoid 3'-hydroxylase OS=Vitis vinifera GN=VvF3'h2 PE=3 SV=1	tr|Q3C213|Q3C213_VITVI	GGCGAGCACATC-GCCTACAACTACCA	CCCTTGCGAACGTG-CATGAAGCGTAT
Con-tig435601	201	17%	Chalcone synthase, putative OS=Ricinus communis GN=RCOM_0661520 PE=3 SV=1	tr|B9STD8|B9STD8_RICCO	GTGACCCAACGGCTG-GAGAT	CTGGAGGAGACG-TAGACGAG
Con-tig434608	227	16%	UDP-glucosyl transferase 73B2 OS=Arabidopsis thaliana GN=UGT73B2 PE=1 SV=1	sp|Q94C57|U73B2_ARATH	AACTCGACGGTGGAG-GCGGCG	GGCATCCCCCTTCAC-GGGCAC
Con-tig433031	320	32%	Flavonol synthase/flavanone 3-hydroxylase OS=Citrus unshiu GN=FLS PE=1 SV=1	sp|Q9ZWQ9|FLS_CITUN	GAAGCATG-CAAGGACTGGGGTTT	GTGTGTAAGTTTT-GATGGAGGAA
Con-tig432219	226	76%	Chalcone synthase RJ5 (Fragment) OS=Fragaria ananassa PE=2 SV=1	sp|P51076|CHSY_FRAAN	--	--
Con-tig432025	252	18%	Anthocyanidin 5,3-O-glucosyl-transferase OS=Rosa hybrid cultivar GN=RhGT1 PE=2 SV=1	sp|Q4R1I9|ANGLT_ROSHC	--	--
Con-tig430117	404	28%	Anthocyanin 3'-O-beta-glucosyl-transferase OS=Gentiana triflora PE=1 SV=1	sp|Q8H0F2|ANGT_GENTR	CCGACTCACCCTCAC-CGTATTTAT	TGAAGCTATTCGAC-CAGGCTCTCA
Con-tig429637	310	23%	UDP-glucose: anthocyanidin 3-O-glucosyltransferase OS=Freesia hybrid cultivar PE=2 SV=1	tr|E2IQH7|E2IQH7_9ASPA	AAGAACTTCAC-GAACTTGCTGACG	AATACTCCATTCTCA-CACCACCCC
Con-tig427157	676	100%	chalcone synthase [Robinia pseudoacacia]	gi|194740626|gb|ACF94727.1|	TCGTCTCCGA-CATCTTTTACCCG	ATCTCGATATGGT-GAGGCATCCC

Contig	Length	%	Description	Accession	Sequence 1	Sequence 2
Con-tig426774	1302	100%	Dihydroflavonol-4-reductase OS=Zea mays GN=A1 PE=3 SV=1	sp|P51108|DFRA_MAIZE	CTCCTCCCCTCGTTC-TACGGCAAG	AGCTCACAAACAC-GACCGACCTGC
Con-tig426191	1922	100%	--	vvi:100263810	CGATGTCACAAATAGC-GAGCACAAGA	CGTAGGCCAACCAG-GAAGGGG
Con-tig425916	1668	100%	putative UDP-rhamnose:rhamnosyltransferase [Fragaria x ananassa]	gi|5170543|gb|AAU09445.1|	TTGAAATTGAGTGAGA-GCAGAGTT	GGGCGTTGTGGCTGA-CAGAGAGAA
Con-tig42485	351	49%	Anthocyanidin 3-O-glucosyltransferase 4 (Fragment) OS=Manihot esculenta GN=GT4 PE=2 SV=1	sp|Q40286|UFOG4_MANES	TCATAAGCGGGCTG-GTCGTGGGG	GCGGCGAAGGC-GATTTCAAGGTG
Con-tig424506	1257	100%	Flavonol synthase/flavanone 3-hydroxylase, putative [Ricinus communis]	gi|255549726|ref|XP_002515914.1|	GAACCTGT-CAGACTTCTTCCCG	GCCCATTTTCATCCTCT-CATCA
Con-tig423314	1481	100%	Putative dihydroflavonol reductase OS=Oryza sativa PE=2 SV=1	tr|Q9XHC8|Q9XHC8_ORYSA	GCATTGCACAC-GTTTCCTTGT	CGGTTCCG-TACCCTTTTTCTT
Con-tig422905	1787	100%	Anthocyanidin 5,3-O-glucosyl-transferase OS=Rosa hybrid cultivar GN=RhGT1 PE=2 SV=1	sp|Q4R1I9|ANGLT_ROSHC	CCCCACCCTGAAT-GAAACAACG	TCTGGACTGGAG-CAATCCGACC
Con-tig422688	1264	100%	Leucoanthocyanidin dioxygenase OS=Malus domestica GN=ANS PE=2 SV=1	sp|P51091|LDOX_MALDO	TGGTGACGTGGCCG-CACTTTG	CCCCCCTCACAT-GCTCCTTGC
Con-tig422641	2288	39%	Flavonol synthase/flavanone 3-hydroxylase OS=Medicago truncatula GN=MTR_3g117490 PE=4 SV=1	tr|G7J7T1|G7J7T1_MEDTR	CCTCACCTCCTTCTC-CGCCTTCC	CCGCTTCCGGCTAT-CATCATCA

Contig						
Con-tig422133	373	38%	Putative dihydroflavonol reductase OS=Oryza sativa PE=2 SV=1	tr|Q9XHC8|Q9XHC8_ORYSA	CAAGCAGGGAGA-CTGGGTCGAAT	GGCGTCAGGAAA-CAGCGCGTAGGA
Con-tig421694	1509	100%	UDP-glucose: anthocyanidin 3-O-glucosyltransferase OS=Freesia hybrid cultivar PE=2 SV=1	tr|E2IQH7|E2IQH7_9ASPA	TGGAGATGAAGAA-GATGGTGGTG	AGAGGAGGAGGAG-GTGTGGAAGA
Con-tig421605	1900	100%	--	vvi:100266013	AAATAGTTGGATAC-CAAGGGTAAGGG	AAATTGCAGGAGGAT-GTTGAGAGTG
Con-tig419989	719	51%	Anthocyanidin 5,3-O-glucosyl-transferase OS=Rosa hybrid culti-var GN=RhGT1 PE=2 SV=1	sp|Q4R1I9|ANGLT_ROSHC	CAACCTG-GAAGAACCCCCACCCT	ATACATCAAACCGC-CGTCCGAAC
Con-tig419942	371	26%	Flavonoid glucosyl-transferase OS=Allium cepa GN=UGT73G1 PE=2 SV=1	tr|Q7XJ50|Q7XJ50_ALLCE	GTTGCAGTTGAG-GAAGTGGGAT	TGTCGGGAGTGTTG-TAGGATGA
Con-tig419941	1071	75%	Flavonoid glucosyl-transferase OS=Allium cepa GN=UGT73G1 PE=2 SV=1	tr|Q7XJ50|Q7XJ50_ALLCE	ATTGGACGGAGC-GAAGGAGAGG	CAAGAC-GGGAAAAGCAGT-GTGG
Con-tig419570	910	63%	putative UDP-rhamnose:rhamnosyltransferase [Fragaria x ananassa]	gi|5170543||gb|AAU09445.1|	GCGTGCTTGGCGAT-GTTTTC	GCCATCCCATGATC-GACGAC
Con-tig419569	676	47%	putative UDP-rhamnose:rhamnosyltransferase [Fragaria x ananassa]	gi|5170543||gb|AAU09445.1|	CCAAGGCGAAGATACT-CAAAGCA	CGCACACTAAACA-CACTCACAAG
Con-tig419474	1000	70%	Anthocyanidin 5,3-O-glucosyl-transferase OS=Rosa hybrid culti-var GN=RhGT1 PE=2 SV=1	sp|Q4R1I9|ANGLT_ROSHC	CCGCTT-GAAGCATCTCTTTCCC	AATCGTCGTCGCC-GTGGTCTGT
Con-tig419441	1295	94%	--	RCOM_1469920	TGGGCATCTCCCCT-CACT	ACCATCCCCC-GCTTCTTC

Contig			Description	Accession	Sequence 1	Sequence 2
Con-tig418986	921	92%	Flavonol synthase OS=Allium cepa PE=3 SV=1	tr|Q6DTI1|Q6DTI1_ALLCE	CACCTCCTTGATCT-GCTCCCC	CCC-GAACTTCCCACCT-GTCTT
Con-tig417695	651	57%	Dihydroflavonol-4-reductase OS=Arabidopsis thaliana GN=DFRA PE=1 SV=2	sp|P51102|DFRA_ARATH	CCGAGCAGGTCCA-CAACGAGAG	CGCCGCCCATCAGC-CGCC
Con-tig415063	1761	100%	Chalcone synthase OS=Antirrhinum majus GN=CHS PE=3 SV=1	sp|P06515|CHSY_ANTMA	GACTCT-GATTTTCTTGGGCTTCACACC	TTTCTTCGCACAAAC-GTAACCCTCTTC
Con-tig415062	329	95%	Putatative chalcone synthase (Fragment) OS=Musa balbisiana GN=chs PE=4 SV=1	tr|Q1RP58|Q1RP58_MUSBA	CCCGCTTGTCATGCT-GCCTCTG	TCCTTCATCTCCCCT-GCTTTCT
Con-tig414839	1073	100%	Myricetin O-methyltransferase OS=Catharanthus roseus PE=1 SV=1	sp|Q8GSN1|MOMT_CATRO	GCGGAAGGAGAG-GTTGGATGGGA	CGGCTTCATCGT-GTTCTGGGGGC
Con-tig4148	713	49%	Anthocyanin 3'-O-beta-glucosyl-transferase OS=Gentiana triflora PE=1 SV=1	sp|Q8H0F2|ANGT_GENTR	GGGACACCAG-GACCTTGCCC	TACGCCCGACCCTC-CACTC
Con-tig414102	675	47%	putative UDP-rhamnose:rhamnosyltransferase [Fragaria x ananassa]	gi|5170543|gb|AAU09445.1|	ACCGTGATCACAGTG-GAATGGG	AGGTATGGGAG-GTTGGGGAAGT
Con-tig41204	546	39%	Malonyl-l-coenzyme:anthocyanin 5-O-glucoside-6'''-O-malonyl-transferase OS=Salvia splendens GN=5MAT1 PE=1 SV=1	sp|Q8W1W9|5MAT1_SALSN	GTTTGGGGGAAGGGATCAG-GTGG	AGGGTCG-GAGTGGGCGGATAG-GC
Con-tig411802	784	79%	Putative dihydroflavonol reductase OS=Oryza sativa PE=2 SV=1	tr|Q9XHC8|Q9XHC8_ORYSA	TGGAAGGCAAAGGC-TAATGGATG	CCTGTGAAAGGGGA-CAAGAGAAC
Con-tig410456	890	85%	Myricetin O-methyltransferase OS=Catharanthus roseus PE=1 SV=1	sp|Q8GSN1|MOMT_CATRO	AGCAGGACGCGCTA-ACGTAACTAAA	AAAGGATGGGAAGAAT-GACAAATGG

Contig			Description	Accession		
Con-tig410317	302	21%	UDP-glycosyltransferase 73B3 OS=Arabidopsis thaliana GN=UGT73B3 PE=2 SV=1	sp|Q8W49|U73B3_ARATH	AGGGAGGTGCTGC-GGGACTACG	AGGACGACGGTCTC-GACGGTGA
Con-tig410211	985	68%	UDP-glucosyl transferase 73B2 OS=Arabidopsis thaliana GN=UGT73B2 PE=1 SV=1	sp|Q94C57|U73B2_ARATH	CCACAACAACAAC-CACCACCAC	GCAGAATCGCTC-CATTCCTCGC
Con-tig39576	1504	100%	Flavonol 4'-sulfotransferase OS=Flaveria chlorifolia PE=1 SV=1	sp|P52837|F4ST_FLACH	TACGGCACCAAC-GCTTTCAGCT	CGGCTTCCTT-GTCCCTCATC
Con-tig38727	1235	100%	Flavonol synthase/flavanone 3-hydroxylase, putative [Ricinus communis]	gi|255549726|ref|XP_002515914.1|	TACTTCTCC-GTCTTCCTCCAG	GGCTTCGTCGTCAT-GCTCTCC
Con-tig37618	1850	100%	Anthocyanin 5-aromatic acyltransferase OS=Gentiana triflora PE=1 SV=1	sp|Q9ZWR8|ANTA_GENTR	GCACCCTCTGGTTCC-GCTCCCC	CAAGCCTCTTCTG-CACGTCGTC
Con-tig37306	718	48%	Flavone synthase II OS=Iris hollandica GN=IhFNSII PE=2 SV=1	tr|A9ZMH5|A9ZMH5_IRIHO	CTCTCC-GACTTCCTCCCCCT	TGCATGTCAAGCTC-CGCCTC
Con-tig35557	1339	100%	Os01g0735500 protein OS=Oryza sativa subsp. japonica GN=B1060H01.31 PE=4 SV=1	tr|Q5JL50|Q5JL50_ORYSJ	TAACCCCTCTGC-GCCTCGCCCTC	CAGCAAT-CAACCCCTATTCTTCG
Con-tig35556	672	45%	Flavonoid glucosyltransferase OS=Crocus sativus GN=GT45 PE=3 SV=1	tr|B9UYP6|B9UYP6_CROSA	CAGTTCCAGGGCGT-CAAGTC	ATCAGCATATCCG-CAGGGGG
Con-tig35393	1234	100%	Leucoanthocyanidin dioxygenase OS=Arabidopsis thaliana GN=LDOX PE=1 SV=1	sp|Q96323|LDOX_ARATH	--	--
Con-tig347559	514	32%	Anthocyanidin 3-O-glucosyltransferase OS=Medicago truncatula GN=MTR_7g067330 PE=4 SV=1	tr|G7L6J2|G7L6J2_MEDTR	GTGATGCTCTCTCCTC-GTCCTCC	GCACAGTTCTGC-CAATCCTTTCT
Con-tig346604	1263	89%	Anthocyanidin 5,3-O-glucosyl-transferase OS=Rosa hybrid culti-var GN=RhGT1 PE=2 SV=1	sp|Q4R1I9|ANGLT_ROSHC	AGCGAATGCGGAGA-CTGCTGTGAAC	TCTTGCG-GAAGAAAGCGCCG-GACCT

Contig	Length	%	Description	Accession	Forward Primer	Reverse Primer
Contig346253	509	36%	Anthocyanin 5-aromatic acyltransferase OS=Gentiana triflora PE=1 SV=1	sp|Q9ZWR8|ANTA_GENTR	TGGCGGAGATGATGCAGAGGAT	TGTTGCGGTCGGTGTGGGGGCT
Contig337519	441	60%	Anthocyanidin synthase (Fragment) OS=Litchi chinensis GN=ANS PE=2 SV=1	tr|E3W5V9|E3W5V9_LITCN	GGTCACCTTCACCGCCATCACCG	TCACCTACCTCCCTCCTCCGCC
Contig335540	245	--	Cellular Component: nucleus (GO:0005634);; Cellular Component: cytosol (GO:0005829);; Cellular Component: plasma membrane (GO:0005886);; Biological Process: polyamine catabolic process (GO:0006598);; Cellular Component: plasmodesma (GO:0009506);; Biological Process: response to wounding (GO:0009611);; Biological Process: coumarin biosynthetic process (GO:0009805);; Biological Process: lignin biosynthetic process (GO:0009809);; Biological Process: positive regulation of flavonoid biosynthetic process (GO:0009963);; Biological Process: sterol biosynthetic process (GO:0016126);; Molecular Function: luteolin O-methyltransferase activity (GO:0030744);; Molecular Function: quercetin 3-O-methyltransferase activity (GO:0030755);; Molecular Function: myricetin 3'-O-methyltransferase activity (GO:0033799);; Biological Process: cellular modified amino acid biosynthetic process (GO:0042398);; Molecular Function: caffeate O-methyltransferase activity (GO:0047763);; Biological Process: flavonol biosynthetic process (GO:0051555);;	--	GGTGGTGGGGCTGGTGGTGGGGA	AACATCGAGCATGGACTGGCGGG
Contig332939	707	46%	Flavonoid 3'-hydroxylase OS=Vitis vinifera GN=VvF3'h2 PE=3 SV=1	tr|Q3C213|Q3C213_VITVI	CCTCTCCGCCTTCGTCTCCAT	ACTGCTCGTACGCCACGCTCT

Contig	Length	%	Description	Accession	Sequence 1	Sequence 2
Con-tig328139	523	53%	Anthocyanidin reductase OS=Arabidopsis thaliana GN=BAN PE=1 SV=2	sp|Q9SEV0|BAN_ARATH	TACTTCCACTTCTTC-CACGGCTAC	CTGAATAGGGTCAG-GATCTTTGAG
Con-tig32509	461	32%	putative UDP-rhamnose:rhamnosyltransferase [Fragaria x ananassa]	gi|51705431|gb|AAU09445.1|	TATTATTCACCTT-GCCCCCAGCC	CCAAAACCCAC-TCCCCATCCTTC-
Con-tig32178	793	55%	Anthocyanin 3'-O-beta-glucosyl-transferase OS=Gentiana triflora PE=1 SV=1	sp|Q8H0F2|ANGT_GENTR	GCTCCTCCATCAGCC-GGGAGT	CGGGTCG-GTGGGGGGTTTCTT
Con-tig32138	571	56%	Dihydroflavonol-4-reductase OS=Vitis vinifera GN=DFR PE=1 SV=1	sp|P51110|DFRA_VITVI	CGGCCAAAGAAT-GCCTCCAGT	GAGAGCGAGC-CAAAGCTACCA
Con-tig320251	363	26%	--	RCOM_1469920	CGACTTCCGATGGC-CAAGCAA	CCGCCGGTCCCGGA-CAATTACT
Con-tig319051	421	29%	Anthocyanidin 3-O-glucosyltrans-ferase 5 OS=Manihot esculenta GN=GT5 PE=1 SV=1	sp|Q40287|UFOG5_MANES	AATCAACTAC-TACCCAAAATGCCC	ATCTCTATGGTGTCTC-CGATGTGC
Con-tig315447	2412	100%	Putative dihydroflavonol reductase OS=Oryza sativa PE=2 SV=1	tr|Q9XHC8|Q9XHC8_ORYSA	ATCCAGGCGGGGCTT-GTCTTG	CTTGCTCAGTGATC-GAGGGCG
Con-tig315020	2002	--	Molecular Function: beta-gluc-uronidase activity (GO:0004566); Biological Process: carbohydrate metabolic process (GO:0005975);; Cellular Component: plant-type cell wall (GO:0009505); Biological Process: unidimensional cell growth (GO:0009826);; Cellular Component: membrane (GO:0016020);; Cellular Component: cytoplasmic membrane-bounded vesicle (GO:0016023);; Molecular Function: cation binding (GO:0043169);;	--	GATGATGGTATCGTC-GCTAAGGT	TTTGCCAAGCATGGT-GCCAGGAG

Contig	Length	%	Description	Accession	Sequence 1	Sequence 2
Con-tig313981	1332	100%	Anthocyanidin synthase OS=Iris hollandica GN=IhANS PE=2 SV=1	tr|A5HUP4|A5HUP4_IRIHO	TTCCGCAGTGT-CATCTTCA	TTCTTCCTTCTTGGGTC-GT
Con-tig313675	1485	100%	Putative dihydroflavonol reductase OS=Oryza sativa PE=2 SV=1	tr|Q9XHC8|Q9XHC8_ORYSA	TGACGGATCTCCC-GAACCCTT	CACCGGCTTGTC-GAAAACCTC
Con-tig313084	1309	100%	Dihydroflavonol-4-reductase OS=Zea mays GN=A1 PE=3 SV=1	sp|P51108|DFRA_MAIZE	GAACACGATCCTCG-GTATGCCGA	GAAC-GTCTCCCTCCCTCCCCTC
Con-tig312774	1989	100%	Flavone synthase II OS=Iris hollandica GN=IhFNSII PE=2 SV=1	tr|A9ZMH5|A9ZMH5_IRIHO	TATCTGCCTGAT-GCTTCACGCTTC	CTTTG-GTTTTGGGTCCTCTTCTCC
Con-tig311644	1616	100%	--	RCOM_1469920	CAAGAAGAGTTA-AAGAATGTAGTTGGC	ATGTG-GAATTAGAAGTGGTG-TAGTAGG
Con-tig310938	1344	100%	Putative dihydroflavonol reductase OS=Oryza sativa PE=2 SV=1	tr|Q9XHC8|Q9XHC8_ORYSA	CAGT-GAGTGGGGGGTTG-GATTG	ACTTCGTCGCATTTTC-CGCAGC
Con-tig310302	1407	100%	Dihydroflavonol 4-reductase OS=Agapanthus praecox GN=ApDFR PE=4 SV=1	tr|Q2L6T0|Q2L6T0_AGAPR	CAGCCACGAAGCAAC-GAACCCAC	GCCAAGTA-CAAGCAGCCACCGCA
Con-tig310081	1195	100%	Dihydroflavonol-4-reductase OS=Zea mays GN=A1 PE=3 SV=1	sp|P51108|DFRA_MAIZE	TTTT-TATTCTTTCCCCCCCTTTT-GTCAG	TTTTCAGGATGCCCTC-GTTCTATTTGC
Con-tig28256	498	69%	Anthocyanidin 3-O-glucosyltransferase 4 (Fragment) OS=Manihot esculenta GN=GT4 PE=2 SV=1	sp|Q40286|UFOG4_MANES	GAGCTGCTGATGCA-GATGAAGA	CGTTGTTGGGAGGAT-GAAGGAGA
Con-tig239073	466	40%	Os01g0735500 protein OS=Oryza sativa subsp. japonica GN=B1060H01.31 PE=4 SV=1	tr|Q5JL50|Q5JL50_ORYSJ	AGAGAGGAGATGGC-GAAGGGAGGA	TTGGACGGTGGCGTG-GACGGAGTA
Con-tig238937	1444	100%	Putative dihydroflavonol reductase OS=Oryza sativa PE=2 SV=1	tr|Q9XHC8|Q9XHC8_ORYSA	CGTCCTCACCTC-CATTTCCACCA	TCCGACCCTTTCGAC-TACCACAG

Contig	Length	%	Description	Accession	Sequence 1	Sequence 2
Con-tig238767	324	95%	Putative chalcone synthase (Fragment) OS=Musa paradisiaca GN=chs PE=4 SV=1	tr|Q1RP55|Q1RP55_MUSPR	TGATTCATTGCTTC-GAGTGGGAG	GCCGTTTTATTTG-TACTGTTTGT
Con-tig236731	234	16%	Flavonoid glucosyl-transferase OS=Allium cepa GN=UGT73G1 PE=2 SV=1	tr|Q7XJ50|Q7XJ50_ALLCE	CAGAGGTGGCGAGCG-GAAACAG	TCTCCCAGAGGACG-GCAATCAT
Con-tig234778	252	18%	--	RCOM_1469920	TA-AATCCCACCCCATCT-CACAAACC	GGACAGCTCGGCGTC-TACAGTC
Con-tig231258	1826	100%	Flavonol synthase/flavanone 3-hydroxylase, putative OS=Ricinus communis GN=RCOM_1037900 PE=3 SV=1	tr|B9RJR8|B9RJR8_RICCO	CGCCTCATCAAA-CATCTCCTCCATT	GCAAGTATCCTCAG-TATCACATCCCACA
Con-tig230553	1495	100%	Caffeic acid O-methyltransferase OS=Iris hollandica GN=IhCOMT PE=2 SV=1	tr|Q5KSL8|Q5KSL8_IRIHO	GTTTCTCTGTTGTT-GCCGTTCTA	AGCCTCCCTCTCAGC-CAATGTCTT
Con-tig229947	1628	100%	flavanone 3-hydroxylase [Allium cepa]	gi|29123528|gb|AAO63022.1|	TCGTACA-GATTCTCTTCGGGGAC	ATCCTTGTATTAT-GTTCTGGGCC
Con-tig227708	1890	100%	--	vvi:100263810	GGATGTGGTCCATGAG-CAGGC	GGGGAGGGGAT-CAGGGAGAGG
Con-tig225832	1797	100%	Flavonol synthase/flavanone 3-hydroxylase, putative OS=Ricinus communis GN=RCOM_1278050 PE=3 SV=1	tr|B9SNS2|B9SNS2_RICCO	AAAGGCTAAA-CATGGGCAAAAAATC-TAGG	AGCGTCGGCAG-CAAAGAGAAGA
Con-tig225830	1257	100%	--	vvi:100266013	TTGTCTCGGAAGGCTG-GATGTAGGT	TTGACTAGCCAGGAT-GCGAAGAAGC
Con-tig225716	1736	100%	Flavone synthase II OS=Iris hollandica GN=IhFNSII PE=2 SV=1	tr|A9ZMH5|A9ZMH5_IRIHO	TCCCCATCCTTCAT-TAGCCTGCC	AACAAACAAACAT-CAAATTCCAT

Contig	Length	%	Description	Accession	Sequence 1	Sequence 2
Con-tig222850	1681	100%	--	POPTR_828087	TCCTTCGTCCCCGAG-CAGATG	CTCCGGCAAAAAC-GCTTCCCC
Con-tig221459	670	61%	Caffeic acid O-methyltransferase OS=Iris hollandica GN=IhCOMT PE=2 SV=1	tr|Q5KSL8|Q5KSL8_IRIHO	CCTTGATGTCCTT-GTTGGTGAGC	TCAACGATTCTAC-GACAGGCTC
Con-tig220284	1561	100%	Anthocyanidin 5,3-O-glucosyl-transferase OS=Rosa hybrid culti-var GN=RhGT1 PE=2 SV=1	sp|Q4R1I9|ANGLT_ROSHC	CTCTCAGATC-CATTCCCCCTTCC	CGCACAACTTCTT-TTCTTCCCGC
Con-tig219088	869	88%	Putative dihydroflavonol reductase OS=Oryza sativa PE=2 SV=1	tr|Q9XHC8|Q9XHC8_ORYSA	CGTAGGCGCAGGG-TAGACGGGAAG	CCGGGGTTGAGCAT-GTTGGAGGAG
Con-tig218007	480	33%	Anthocyanidin 3-O-glucosyltrans-ferase 5 OS=Manihot esculenta GN=GT5 PE=1 SV=1	sp|Q40287|UFOG5_MANES	CTCTTGGGGG-TCCTTTCG	TGAGTTCGGGTTGC-GGG
Con-tig216533	1485	100%	Chalcone synthase OS=Medicago truncatula GN=MTR_7g016820 PE=3 SV=1	tr|G7KR33|G7KR33_MEDTR	CCCCTATAAATACAC-CATCATGCCCACAAA-CA	AGAGGTAGAA-CACCCAAGGCTG-CAAATCC
Con-tig215448	1771	100%	Chalcone synthase OS=Perilla frutescens GN=CHS PE=2 SV=1	sp|O04111|CHSY_PERFR	CATTCTGATGCGGTT-GCCT	TGGATTTATAAACT-GCCCTGTGC
Con-tig214099	763	50%	--	vvi:100232999	ATTCTTCTCCGTA-ACCTCCCGCC	CAAAC-GTTCTTCAAGCCCCCAA
Con-tig211565	374	34%	Caffeic acid O-methyltransferase OS=Iris hollandica GN=IhCOMT PE=2 SV=1	tr|Q5KSL8|Q5KSL8_IRIHO	CCACCACCACCTTCT-TATTATGT	CTCTCCACCGCCAG-GCTCCAGTC
Con-tig19721	470	44%	Dihydroflavonol-4-reductase OS=Zea mays GN=A1 PE=3 SV=1	sp|P51108|DFRA_MAIZE	ATGGGAATACATAG-CACCAGTAAAAGG	TGGCTCCA-CAAAATACG-CAATAGAATA

Contig	Length	%	Description	Accession	Sequence 1	Sequence 2
Con-tig18666	1339	100%	Putative dihydroflavonol reductase OS=Oryza sativa PE=2 SV=1	tr|Q9XHC8|Q9XHC8_ORYSA	GGATAAGAAATGGGT-GTGGGGGTA	TGTGGGTGAT-GATGGGATGCTGTG
Con-tig18257	337	21%	Flavonoid 3',5'-hydroxy-lase OS=Campanula medium GN=CYP75A6 PE=2 SV=1	sp|O04773|C75A6_CAMME	AGGTCACTCC-GCCCCCTTACAA	AAATCGAGGAC-GAGCCCACACA
Con-tig17084	1000	100%	Chalcone--flavonone isom-erase 1B-1 OS=Glycine max GN=CHI1B1 PE=2 SV=1	sp|Q53B75|CF1B1_SOYBN	GAAGAAGCGAT-CAAAGAGTCCGAGGC	TCAAACGGGAAGA-CATTCAACAAAGC
Con-tig17019	1133	80%	Anthocyanidin 5,3-O-glucosyl-transferase OS=Rosa hybrid culti-var GN=RhGT1 PE=2 SV=1	sp|Q4R1I9|ANGLT_ROSHC	TATCGTTTTGCGGT-GATTCTTTG	AGGGTGCTGGCT-GATTCTGTTG
Con-tig135502	859	85%	Flavonol synthase OS=Allium cepa PE=3 SV=1	tr|Q6DTI1|Q6DTI1_ALLCE	AGCATGAGGCCC-GACTCGCGTAGCT	CCCTGCAAGGACCA-CATCCAGAACC
Con-tig13188	352	49%	Anthocyanidin 3-O-glucosyltrans-ferase 4 (Fragment) OS=Manihot esculenta GN=GT4 PE=2 SV=1	sp|Q40286|UFOG4_MANES	AGGTCATCAACAATG-GAGAGCG	CCGAAAGTAGAAGTC-GGCGAAA
Con-tig12738	430	37%	Os01g0735500 protein OS=Oryza sativa subsp. japonica GN=B1060H01.31 PE=4 SV=1	tr|Q5JL50|Q5JL50_ORYSJ	GGTCGAGTC-CATCTTCACCGTCT	CCTTGTGATGTTGTT-GAATGCCA
Con-tig12320	418	29%	Anthocyanidin 5,3-O-glucosyl-transferase OS=Rosa hybrid culti-var GN=RhGT1 PE=2 SV=1	sp|Q4R1I9|ANGLT_ROSHC	CACGACAGTCTGCAC-GATAGTTGTAG	CCAAGTGGAGACGC-TAITTTGATGAA
Con-tig123109	463	32%	putative UDP-rhamnose:rhamnosyltransferase [Fragaria x ananassa]	gi|51705431|gb|AAU09445.1|	CCCTCAGCCCCCT-GCCCTCCTCC	TTAATCACTCG-CAATCCCCCACT
Con-tig122770	463	40%	Chalcone synthase OS=Iris ger-manica GN=IgCHS1 PE=2 SV=1	tr|Q2WFX2|Q2WFX2_IRIGE	TTTGCGAGG-TATGGGAGGATGGGAT	TTCAGGGGCTTCGTG-TACTGATGAA

Contig						
Con-tig121708	1741	100%	Anthocyanin 5-aromatic acyltransferase OS=Gentiana triflora PE=1 SV=1	sp|Q9ZWR8|ANTA_GENTR	TCGTCCTCAAGAAGGC-TACTCCGCT	CATAAGCAACATCC-GTGTAACCAAC
Con-tig120937	1262	100%	Caffeic acid O-methyltransferase OS=Medicago truncatula GN=MTR_3g092900 PE=1 SV=1	tr|G7J914|G7J914_MEDTR	ATTCACACCCA-CACCCATCACCC	GAAGCACTTAC-CAAAGGAACCAG
Con-tig120431	1662	100%	Flavonoid 3'-hydroxylase OS=Allium cepa PE=2 SV=1	tr|Q6QHJ9|Q6QHJ9_ALLCE	GCAACAAGCG-GAGGGAATGAGA	TGGTGAGGACGGT-GACGGAGAA
Con-tig118886	898	100%	Chalcone isomerase OS=Gossypium hirsutum PE=2 SV=1	tr|D6N3G6|D6N3G6_GOSHI	GGGTGATAGAG-GAGCTGAAACGG	TACTTGAGTCGGGC-CAGACGTTG
Con-tig118597	487	34%	Anthocyanidin 5,3-O-glucosyl-transferase OS=Rosa hybrid cultivar GN=RhGT1 PE=2 SV=1	sp|Q4R1I9|ANGLT_ROSHC	TCTTCCTCTC-CAAACTCTC-TCCCCCT	CCTGCACTCGGAG-CAGTTCATCAATG
Con-tig118129	1048	100%	Chalcone-flavanone isomerase OS=Elaeis oleifera GN=CHI PE=2 SV=1	tr|C3W5J1|C3W5J1_ELAOL	GCCGTCTTCCTC-CACTTCCCCAACCT	AACCCTCGTAG-GCAICGTCGTCCCG
Con-tig117081	503	36%	UDP-glycosyltransferase 78D2 OS=Arabidopsis thaliana GN=UGT78D2 PE=2 SV=1	sp|Q9LFJ8|U78D2_ARATH	CGGCTCTAATGCTCG-GAAIGTG	CATCCTTACC-GCCTCCTCTCCA
Con-tig116541	1381	100%	Chalcone synthase 6 OS=Sorghum bicolor GN=CHS6 PE=3 SV=1	sp|Q9SBL3|CHS6_SORBI	GGAACGCGAGCTTG-GATGGGA	CACGGCGGAGGTG-GTGGAGAG
Con-tig114102	782	74%	Cytosolic sulfotransferase 18 OS=Arabidopsis thaliana GN=SOT18 PE=1 SV=1	sp|Q9C9C9|SOT18_ARATH	GATGAGGAAGAGGTC-GGCGGAGC	TGAGTCGTGACCGAG-GCAGCAAA

Contig113663	1596	100%	Malonyl-coenzyme:anthocyanin 5-O-glucoside-6'''-O-malonyltransferase OS=Salvia splendens GN=5MAT1 PE=1 SV=1	sp|Q8W1W9|5MAT1_SALSN	GACAAGATTTGGATGC-GGAGGG	GTGATGGAGGATCG-GCAAGAGG
Contig11174	786	75%	Myricetin O-methyltransferase OS=Catharanthus roseus PE=1 SV=1	sp|Q8GSN1|MOMT_CATRO	GATATGTTCGACTAC-GTCCCCCCT	GATAACCTTGCCTG-GCCTCTCTTT
Contig110579	757	57%	UDP-glucose: anthocyanidin 3-O-glucosyltransferase OS=Freesia hybrid cultivar PE=2 SV=1	tr|E2IQH7|E2IQH7_9ASPA	CGTTGGTT-GTTCACCCTATTC	GGCCCTTCCATT-TATTCATTT
CLContig434963	434	30%	putative UDP-rhamnose:rhamnosyltransferase [Fragaria x ananassa]	gi|51705431|gb|AAU09445.1|	GATTTCTTCCTT-GCTCTTGTTTCAG	TTCCTCCTCGTCTTC-GTATTTGTCG
CLContig432005	307	23%	Anthocyanin 3-O-galactosyltransferase OS=Aralia cordata GN=acgat PE=2 SV=1	tr|Q76G23|Q76G23_ARACO	GCAGCCAGTTTTTC-CAAGTTAGAATAA	TGAAGAAGGGAG-GAGGGTGAGG
CLContig431958	269	20%	Anthocyanin 3-O-galactosyltransferase OS=Aralia cordata GN=acgat PE=2 SV=1	tr|Q76G23|Q76G23_ARACO	ATCAA-CATAGCCCAGTCCA-CAA	CATAATCCGAGAG-GAGCAAAGG
CLContig427737	499	42%	Chalcone synthase-like protein OS=Pinus strobus PE=2 SV=1	tr|O65873|O65873_PINST	ACAAATTGCCATTAT-GCTTGCGCG	TTTCATCTTCT-TCCCTTTCTCTCC
CLContig426024	1709	100%	UDP-glucose: anthocyanidin 3-O-glucosyltransferase OS=Freesia hybrid cultivar PE=2 SV=1	tr|E2IQH7|E2IQH7_9ASPA	TCAATCATTTCG-GCCTTTA	TAGTCCGGG-TAGTCCTCCTG
CLContig424585	1865	100%	Flavone synthase II OS=Iris hollandica GN=IhFNSII PE=2 SV=1	tr|A9ZMH5|A9ZMH5_1RIHO	GAAGGCTGGCGGC-CAGTCAA	CG-GAAGGGAGGGGTTC-GGAT

CLContig421765	857	60%	putative UDP-rhamnose:rhamnosyltransferase [Fragaria x ananassa]	gi|51705431|gb|AAU09445.1|	GCTCGTTCTGCTGCCC-GTCTTC	TTCCCGTCCTCCTC-GTCCCTCT
CLContig420196	1203	100%	Flavonol synthase/flavanone 3-hydroxylase, putative OS=Ricinus communis GN=RCOM_1278050 PE=3 SV=1	tr|B9SNS2|B9SNS2_RICCO	GGAAGAGCCTCCGAC-CGCCTT	GCCCCTCCCACATC-CACCAA
CLContig41646	332	23%	putative UDP-rhamnose:rhamnosyltransferase [Fragaria x ananassa]	gi|51705431|gb|AAU09445.1|	CGCTAGGC-GAAGGGCTGAGGCT	TCGTAGGTGTGGAA-CAGGGAGA
CLContig41646	1307	100%	Bifunctional dihydroflavonol 4-reductase/flavanone 4-reductase OS=Malus domestica GN=DFR PE=1 SV=1	sp|Q9XES5|DFRA_MALDO	ACGGCGACCTGAG-GAGGGACTG	GTCACCGGGTTCCC-GAACACGC
CLContig4125	469	70%	chalcone synthase [Robinia pseudoacacia]	gi|194740626|gb|ACF94727.1|	TTACACTCCCTGCATT-TAGCCCTG	GGACGAGGAG-GATGGGTCTTTTAC
CLContig318153	507	43%	Chalcone synthase OS=Malus domestica GN=CHS1 PE=2 SV=1	tr|Q4TZJ6|Q4TZJ6_MALDO	GGTTTCCTCCGTTGGT-GCTATTTA	GCCTTACGCTCTGC-TATTGTCTTT
CLContig312358	1105	77%	putative UDP-rhamnose:rhamnosyltransferase [Fragaria x ananassa]	gi|51705431|gb|AAU09445.1|	ATTTTGGACGAAGT-GAGGAGGAAGT	CTGTGAAGCACAA-CAGTCTCGACAG
CLContig28508	387	58%	chalcone synthase [Robinia pseudoacacia]	gi|194740626|gb|ACF94727.1|	GAGGGAGGCGGAGGTC-GCTGGTG	CCGATTGTAGGGAT-GAAGGTGT
CLContig28383	969	89%	Flavone 3'-O-methyltransferase 1 OS=Arabidopsis thaliana GN=OMT1 PE=1 SV=1	sp|Q9FK25|OMT1_ARATH	CAATGGCG-GAAAAGAGAGGAC-TAAG	TGAAGGAAGGATG-GCAATAAAGGAG
CLContig17039	1134	100%	Naringenin,2-oxoglutarate 3-dioxygenase OS=Malus domestica PE=2 SV=1	sp|Q06942|FL3H_MALDO	ACGCCCTCAC-CATCCTCCTCCA	GCCCTGTGCCAAAC-GCTTCTAT

CLCon-tig13516	803	--	Molecular Function: beta-glucuronidase activity (GO:0004566);; Cellular Component: plant-type cell wall (GO:0009505);; Biological Process: unidimensional cell growth (GO:0009826);; Cellular Component: cytoplasmic membrane-bounded vesicle (GO:0016023);;	--	ACATCCTCTTGAGA-ACCCAAATAATCCA	GGGAGAATCTTCT-GCTCTAAATCTGGAC
CLCon-tig123507	449	30%	Flavone synthase II OS=Iris hollandica GN=IhFNSII PE=2 SV=1	tr|A9ZMH5|A9ZMH5_IRIHO	GATAGT-GACCCTCTTCTTCG-CAG	GCTGATGACCGCTTG-CAGGTACG
CLCon-tig117687	1438	100%	putative UDP-rhamnose:rhamnosyltransferase [Fragaria x ananassa]	gi|51705431|gb|AAU09445.1|	CCCAAG-TATCTCCCCCTCCTCC	CCGCATCCCGTC-GCTGTCGTC
CLCon-tig111449	856	60%	putative UDP-rhamnose:rhamnosyltransferase [Fragaria x ananassa]	gi|51705431|gb|AAU09445.1|	GCTCGTTCTGCTGCCC-GTCTTC	TTCCCGTCTCCTC-GTCCCTCT
actin					GCCTGCCATGTATGTT-GCG	CGGAGCTTCCATTC-CGATC

Table S2. The contents of flavonoids in flower petals of *M. armeniacum*.

Standard		Content (mg g-1)	
		M. armeniacum	*M. armeniacum* f. album
Anthocyanin (mg g^{-1})	Delphinidin	0.629±0.004	ND
	Cyanidin	0.197±0.003	ND
	Petunidin	ND	ND
	Pelargonin	ND	ND
	Malvidin	ND	ND
	Cyanidin-galacto-side	ND	ND
	Cyanidin-rutin	ND	ND
Flavanones(mg g-1)	Catechin	ND	0.004±0.000
	Epicatechin	0.004±0.000	0.011±0.000
	Gallocatechin	ND	ND
	Epigallocatechin	ND	ND
	Afzelechin	ND	ND
	Epiafzelechin	ND	ND
	Naringenin	ND	ND
	Eriodictyol	0.007±0.002	0.053±0.001
Flavones(mg g-1)	Apigenin	ND	ND
	Luteolin	ND	0.042±0.006
Flavonols(mg g-1)	Rutin	6.158±0.023	3.593±0.015
	Myricetin	0.114±0.001	0.247±0.002
	Quercetin	ND	ND
	Kaempferol	ND	0.315±0.003
	Dihydromyricetin	0.061±0.001	0.012±0.001
	Dihydroquercetin	ND	ND
	Dihydrokaempferol	ND	ND
Other(mg g-1)	Caffeoyl quinic acid	ND	0.214±0.001
	ρ-Coumaric acid	0.001±0.000	ND
	Coumalic acid	ND	1.139±0.001
ND means not detected this compounds in the flower.			

Table S3. Length and gap distribution of contigs, scaffolds, and uni-transcripts from each library of *M. armeniacum.*

		Length distribution of uni-transcripts						N50	Mean	All uni-transcript	Length of all uni-transcript (nt)
Sheet A: Length distribution of contigs											
Sheet B: Length distribution of scaffolds											
Sheet C: Length distribution of uni-transcripts											
sample	number/ percent	200-300nt	300-500nt	500-1000nt	1000-2000nt	>=2000nt					
Blue-uni-transcript	number	14,986	11,137	8,007	5,616	1,948		991	643	41,694	26,821,522
	percent	35.94%	26.71%	19.20%	13.47%	4.67%					
White-uni-transcript	number	13,014	11,991	8,536	5,422	1,483		887	623	40,446	25,190,346
	percent	32.18%	29.65%	21.10%	13.41%	3.67%					

Acknowledgements

This work was supported by the National Natural Science Foundation of China (grant no. 30871734).

REFERENCES

1. Benjamini Y, Yekutieli D. 2001. The control of the false discovery rate in multiple testing under dependency. Annals of Statistics 29, 1165–1188.

2. Bogs J, Jaffe FW, Takos AM, Walker AR, Robinson SP. 2007. The grapevine transcription factor VvMYBPA1 regulates proanthocyanidin synthesis during fruit development. Plant Physiology 143, 1347–1361.

3. Castellarin SD, Gaspero GD. 2007. Transcriptional control of anthocyanin biosynthetic genes in extreme phenotypes for berry pigmentation of naturally occurring grapevines. BMC Plant Biology 7, 46.

4. Clark ST, Verwoerd WS. 2011. A systems approach to identifying correlated gene targets for the loss of colour pigmentation in plants. BMC Bioinformatics 12, 343.

5. Davies KM, Schwinn KE, Deroles SC, Manson DG, Lewis DH, Bloor SJ, Bradley JM. 2003. Enhancing anthocyanin production by altering competition for substrate between flavonol synthase and dihydroflavonol 4-reductase. Euphytica 131, 259–268.

6. Dick CA, Buenrostro J, Butler T, Carlson ML, Kliebenstein DJ, Whittall JB. 2011. Arctic mustard flower color polymorphism controlled by petal-specific downregulation at the threshold of the anthocyanin biosynthetic pathway. PLoS One 6, e18230.

7. Feng C, Chen M, Xu CJ, Bai L, Yin XR, Li X, Allan AC, Ferguson IB, Chen KS. 2012. Transcriptomic analysis of Chinese bayberry (Myrica rubra) fruit development and ripening using RNA-Seq. BMC Genomics 13, 19.

8. Grabherr MG, Haas BJ, Yassour M, et al. 2011. Full-length transcriptome assembly from RNA-Seq data without a reference genome. Nature Biotechnology 29, 644–652.

9. Grotewold E. 2006. The genetics and biochemistry of floral pigments. Annual Review of Plant Biology 57, 761–780.

10. Han YP, Vimolmangkang S, Soria-Guerra RE, Korban SS. 2012. Introduction of apple ANR genes into tobacco inhibits expression of both CHI and DFR genes in flowers, leading to loss of anthocyanin. Journal of Experimental Botany 63, 2437–2447.

11. Hemleben V, Dressel A, Epping B, Lukačin R, Martens S, Austin M. 2004.

Characterization and structural features of a chalcone synthase mutation in a white-flowering line of Matthiola incana R. Br. (Brassicaceae). Plant Molecular Biology 55, 455–465.

12. Koes RE, Spelt CE, Elzen PJMV, Mol JNM. 1989. Cloning and molecular characterization of the chalcone synthase multigene family of Petunia hybrida . Gene 81, 245–257.

13. Ma H, Zhao X, Yuan Y, Zeng A. 2004. Decomposition of metabolic network into functional modules based on the global connectivity structure of reaction graph. Bioinformatics 20, 1870–1876.

14. Mori S, Asano S, Kobayashi H, Nakano M. 2002. Analyses of anthocyanidins and anthocyanins in flowers of Muscari spp. Bulletin of the Faculty of Agriculture 55, 13–18.

15. Mortazavi A, Williams BA, McCue K, Schaeffer L, Wold B. 2008. Mapping and quantifying mammalian transcriptomes by RNA-Seq. Nature Methods 5, 621–628.

16. Pertea G, Huang XQ, Liang F, et al. 2003. TIGR gene indices clustering tools (TGICL): a software system for fast clustering of large EST datasets. Bioinformatics 19, 651–652.

17. Qi YY, Lou Q, Li HB, Yue J, Liu YL, Wang YJ. 2013. Anatomical and biochemical studies of bicolored flower development in Muscari latifolium . Protoplasma 250, 1273–1281.

18. Saito R, Fukuta N, Ohmiya A, Itoh Y, Ozeki Y, Kuchitsu K, Nakayama M. 2006. Regulation of anthocyanin biosynthesis involved in the formation of marginal picotee petals in Petunia . Plant Science 170, 828–834.

19. Salvatierra A, Pimente P, Moya-León MA, Herrera R. 2013. Increased accumulation of anthocyanins in Fragaria chiloensis fruits by transient suppression of FcMYB1 gene. Phytochemistry Letters 90, 25–36.

20. Spitzer B, Zvi MMB, Ovadis M, et al. 2007. Reverse genetics of floral scent: application of tobacco rattle virus-based gene silencing in Petunia . Plant Physiology 145, 1241–1250.

21. Tanaka Y, Brugliera F. 2013. Flower colour and cytochromes P450. Philosophical Transactions of theRoyal Society B: Biological Sciences 368, 20120432.

22. Tanaka Y, Sasaki N, Ohmiya A. 2008. Biosynthesis of plant pigments: anthocyanins, betalains and carotenoids. The Plant Journal 54, 733–749.

23. Wang KL, Bolitho K, Grafton K, Kortstee A, Karunairetnam S, McGhie TK, Espley RV, Hellens RP, Allan AC. 2010. An R2R3 MYB transcription factor associated with regulation of the anthocyanin biosynthetic pathway

in Rosaceae. BMC Plant Biology 10, 50.

24. Yoshida K, Mori M, Kondo T. 2009. Blue flower color development by anthocyanins: from chemical structure to cell physiology. Natural Product Reports 26, 884–915.

25. Yuan Y, Ma XH, Shi YM, Tang DQ. 2013. Isolation and expression analysis of six putative structural genes involved in anthocyanin biosynthesis in Tulipa fosteriana . Scientia Horticulturae 153, 93–102.

CITATION

CHAPTER 1

Reena Kushwaha, Pankaj Srivastava, and Lal Bahadur, "Natural Pigments from Plants Used as Sensitizers for TiO2 Based Dye-Sensitized Solar Cells," Journal of Energy, vol. 2013, Article ID 654953, 8 pages, 2013. doi:10.1155/2013/654953.

CHAPTER 2

Shi Q, Zhou L, Wang Y, Li K, Zheng B, Miao K (2015) Transcriptomic Analysis of Paeonia delavayi Wild Population Flowers to Identify Differentially Expressed Genes Involved in Purple-Red and Yellow Petal Pigmentation. PLoS ONE 10(8): e0135038. doi:10.1371/journal.pone.0135038.

CHAPTER 3

Zheng J, Hu Z, Guan X, Dou D, Bai G, Wang Y, et al. (2015) Transcriptome Analysis of Syringa oblata Lindl. Inflorescence Identifies Genes Associated with Pigment Biosynthesis and Scent Metabolism. PLoS ONE 10(11): e0142542. doi:10.1371/journal.pone.0142542.

CHAPTER 4

Nakatsuka, T.et al. Genetic engineering of yellow betalain pigments beyond the species barrier. Sci. Rep. 3, 1970; DOI:10.1038/srep01970 (2013).

CHAPTER 5

Chiyomi Uematsu, Hironori Katayama, Izumi Makino, Azusa Inagaki, Osamu Arakawa and Cathie Martin, Peace, a MYB-like transcription factor, regulates petal pigmentation in flowering peach 'Genpei' bearing variegated and fully pigmented flowers, doi:10.1093/jxb/ert456.

CHAPTER 6

Hui Du, Jie Wu, Kui-Xian Ji, Qing-Yin Zeng, Mohammad-Wadud Bhuiya, Shang Su, Qing-Yan Shu, Hong-Xu Ren, Zheng-An Liu and Liang-Sheng Wang, Methylation mediated by an anthocyanin, O-methyltransferase, is involved in purple flower coloration in Paeonia, doi: 10.1093/jxb/erv365.

CHAPTER 7

Takashi Nakatsuka, Misa Saito, Eri Yamada, Kohei Fujita, Yuko Kakizaki and Masahiro Nishihara, Isolation and characterization of GtMYBP3 and GtMYBP4, orthologues of R2R3-MYB transcription factors that regulate early flavonoid biosynthesis, in gentian flowers, doi: 10.1093/jxb/ers306.

CHAPTER 8

Mamatha Hanumappa1, Goh Choi, Sunhyo Ryu and Giltsu Choi, Modulation of flower colour by rationally designed dominant-negative chalcone synthase, doi: 10.1093/jxb/erm104.

CHAPTER 9

Qian Lou, Yali Liu, Yinyan Qi, Shuzhen Jiao, Feifei Tian, Ling Jiang and Yuejin Wang, Transcriptome sequencing and metabolite analysis reveals the role of delphinidin metabolism in flower colour in grape hyacinth, doi: 10.1093/jxb/eru168.

INDEX

.